# 1st International Conference on Metals and Alloys (CMA 2019)

IOP Conference Series: Materials Science and Engineering
Volume 668

Beijing, China
19 - 22 August 2019

ISBN: 978-1-7138-0921-0
ISSN: 1757-8981

**Printed from e-media with permission by:**

Curran Associates, Inc.
57 Morehouse Lane
Red Hook, NY 12571

**Some format issues inherent in the e-media version may also appear in this print version.**



Institute of Physics
Dirac House, Temple Back
Bristol BS1 6BE UK

Phone: 44 1 17 929 7481
Fax: 44 1 17 920 0979

techtracking@iop.org

**Additional copies of this publication are available from:**

Curran Associates, Inc.
57 Morehouse Lane
Red Hook, NY 12571 USA
Phone: 845-758-0400
Fax: 845-758-2633
Email: curran@proceedings.com
Web: www.proceedings.com

# TABLE OF CONTENTS

IOP Conf. Series: Materials Science and Engineering **668** (2019) 011001 doi:10.1088/1757-899X/668/1/011001

# Proceedings of 1st International Conference on Metals and Alloys (CMA), 19-22 August 2019, Beijing, China

## Chairman

**Prof. Magd Abdel Wahab**
Ghent University, Belgium

## Co-chairman

**Dr. Samir Khatir**
Ghent University, Belgium

## International Scientific Committee

Prof. L. Vanegas Useche, Universidad Tecnológica de Pereira, Colombia
Prof. J. Toribio, University of Salamanca, Spain
Prof. T. Rabczuk, Bauhaus University Weimar, Germany
Prof. V. Silberschmidt, Loughborough University, UK
Prof. J. Santos, University of Madeira, Portugal
Prof. N. A. Noda, Kyushu Institute of Technology, Japan
Prof. K. Oda, Oita University, Japan
Prof. C. Zhou, Nanjing University of Aeronautics and Astronautics, China
Prof. S. Abdullah, Universiti Kebangsaan Malaysia, Malaysia
Prof. R. V. Prakash, Indian Institute of Technology, India
Prof. H. N. Xuan, Hutech University, Vietnam
Prof. A. Vladescu, National Institute for Research and Development in Optoelectronics, Romania
Prof. S. Satuluri, Lendi Institute of Engineering & Technology, India
Dr. A. Caggiano, Technische Universität Darmstadt, Germany
Dr. H. T. Ali, University of Bristol, UK
Dr. S. X. Ji, Brunel University London, UK
Dr. S. Khatir, Ghent University, Belgium
Dr. T. Yue, Ghent University, Belgium
Dr. G. Minafo, University of Palermo, Italy
Dr. L. V. Tran, Sejong University, South Korea
Dr. C. Wang, Liaocheng University, China
Dr. Y. L. Zhou, National University of Singapore, Singapore
Dr. I. Hilmy, International Islamic University Malaysia, Malaysia
Dr. L. C. Wei, University of Malaya, Malaysia
Dr. Karol Kyziol, AGH University of Science and Technology, Poland
Dr. Siamak Hoseinzadeh, Department of Mechanical and Aeronautical Engineering, University of Pretoria
Dr Amin Rabiei Baboukani, Department of Mechanical and Materials Engineering, Florida International University, Miami, Florida, USA

IOP Conf. Series: Materials Science and Engineering **668** (2019) 011001 doi:10.1088/1757-899X/668/1/011001

**Editorial**

This volume contains the proceedings of the 1st International Conference on Metals and Alloys (CMA), 19-22 August 2019, Beijing, China.

The overall objective of the conference is to bring together international scientists and engineers in academia and industry in fields related to research in metals and alloys, more specifically related to their mechanical behaviour, material properties, metallurgical behaviour and microstructures etc., and their applications to a wide range of engineering disciplines. All types of metals, e.g. iron, brass, copper, aluminium, steel, etc., and all types of alloys, e.g. stainless steel, alloys of gold and silver, high temperature alloys, magnetic alloys, etc., are considered in CMA conference. The conference covers industrial engineering applications of metals and alloys to Civil Engineering, Aerospace Engineering, Materials Engineering, Mechanical Engineering, and Biomedical Engineering, etc. The presentations of CMA 2019 are divided into two main sessions on metals and alloys. The papers that are presented in CMA 2019 include application to automotive industry, fatigue problems, nanomaterials, steel production, high temperature performance of metals and alloys, microstructural characterization, welded structures, surface treatment and sintering process. The conference gathers international scientists from more than 20 different countries.

The organising committee is grateful to the keynote speaker; Prof. Shouxun Ji, Reader in Materials, BCAST, Institute of Materials and Manufacturing, Brunel University London, UK, for his keynote lecture entitled 'Development of Al-Si-Cu-Mg Based Aluminium Alloys for Automotive Powertrain at Elevated Temperatures'.

Special thanks go to members of the Scientific Committee of CMA 2019 for reviewing the articles published in this volume and for judging their scientific merits. Based on the comments of reviewers and the scientific merits of the submitted manuscripts, the articles were accepted for publication in the conference proceedings and for presentation at the conference venue. The accepted papers are of a very high scientific quality and contribute to advancement of knowledge in all research topics relevant to CMA conference.

Finally, the organising committee would like to thank all authors, who have contributed to this volume and presented their research work at the confrence venue in Beijing, China.

**Professor Magd Abdel Wahab**

Chairman of CMA 2019

IOP Conf. Series: Materials Science and Engineering **668** (2019) 011002 doi:10.1088/1757-899X/668/1/011002

# Peer review statement

All papers published in this volume of *IOP Conference Series: Materials Science and Engineering* have been peer reviewed through processes administered by the proceedings Editors. Reviews were conducted by expert referees to the professional and scientific standards expected of a proceedings journal published by IOP Publishing.

IOP Conf. Series: Materials Science and Engineering **668** (2019) 012001 doi:10.1088/1757-899X/668/1/012001

# Study of the relationship between stacking fault energy and microstructure in different compositions of Fe-Mn steels produced by weld deposit

**G G Ribamar[1,2], T C Andrade[1], H C Miranda[1] and H F G Abreu[1]**

[1]Department of Metallurgical and Materials Engineering, Federal University of Ceará, Ceará, Fortaleza, Brazil

E-mail: giovanigoncalvesr@gmail.com

**Abstract.** The properties of high manganese steels are usually related to the mechanism of austenite deformation, either martensite or twin formation during deformation. This deformation mechanism is often related to the stacking fault energy (SFE) that the material possesses. Studies show that SFE values between 0 and 8 $mJ/m^2$ provide formation of ε martensite during deformation, whereas only twins are formed for energies above 18 $mJ/m^2$. Intermediate values provide the formation of both ε martensite and twins. In this way, it is possible to predict the microstructure and, consequently, mechanical properties with the knowledge of the material's SFE. In this work, a weld consumable with high manganese content was deposited on a SAE 1012 carbon steel plate with the flux-cored arc welding process producing different chemical compositions and consequently different SFE that varied as a function of the dilution. Low dilution conditions led to austenite and ε martensite formation in the molten zone, while in high dilution conditions, α' martensite were found, besides those other two. Manganese was found to be the most influential element in the phase formation change, based on SFE and SFE components analysis.

## 1. Introduction

High manganese steels are alloys with outstanding mechanical properties due to the activation of deformation mechanisms such as twins or martensite formation [1]. Steels with manganese content higher than 15 wt.% have shown to be excellent materials for automotive and cryogenic applications, because of their good ductility and strength combination, besides the stabilization of austenite at room temperature, avoiding the ductile-fragile transition at low temperatures [2]. Nowadays, it is well known the relation between deformation mechanism and the stacking fault energy (SFE) in high manganese steels. For those with low SFE values, ε and/or α' martensite tend to form during their deformation, increasing its strength by hard phase formation, effect called transformation induced plasticity (TRIP) [3]. On the other hand, for those steels with high SFE values, the increase in the strength accounts for twin formation inside austenite phase, effect called twinning induced plasticity (TWIP) [3].

The main reason for these effects is due to the stabilization level of austenite which increase with the manganese and carbon content [4]. Xiong *et al.* (2014) showed the influence of these elements on the SFE, demonstrating that, at SFE values close to 0 $mJ/m^2$, ε martensite has formed even without any deformation, while at higher SFE this martensite was formed just after some deformation level, or even did not formed [4]. Curtze *et al.* (2010) have studied the dependence of tensile deformation behavior on SFE, temperature, and strain, showing that austenite becomes more stable with the increase in temperature [5]. Besides, they showed that the magnetic component is strongly affected by temperature, being the main reason to austenite stabilization [5].

The SFE can be calculated based on model proposed by Olson and Cohen [6] as follows:

$$SFE = 2\rho\Delta G^{\gamma\to\epsilon} + 2\sigma^{\gamma/\epsilon} \tag{1}$$

where $\rho$ is the molar surface density for the $\{111\}_\gamma$ closed packed planes, $\Delta G^{\gamma\to\epsilon}$ is the molar Gibbs energy for $\gamma \rightarrow \varepsilon$ phase transformation and $\sigma^{\gamma/\epsilon}$ is the energy per surface unit of the {111} interface between the $\gamma$ and $\varepsilon$ boundary. The molar Gibbs energy is dependent of two components: chemical and magnetic Gibbs free energy, $\Delta G_{che}^{\gamma\to\epsilon}$ and $\Delta G_{mg}^{\gamma\to\epsilon}$ respectively, and can be calculated as follows:

$$\Delta G^{\gamma\to\epsilon} = \Delta G_{che}^{\gamma\to\epsilon} + \Delta G_{mg}^{\gamma\to\epsilon} \tag{2}$$

Although development has been made in this area, it is difficult to find studies on welding with filler metal for these steels. The main works in this field are limited to autogeneous welding, such as friction stir welding [7], resistance spot welding [8], and high frequency electrical resistance welding [9]. One of the reasons for this absence is due to its composition sensitivity, because a little change in manganese or carbon content can affect the austenite deformation mechanism [10], leading to a very strong decreasing in steel's properties. In order to understand it, the present work aims to study SFE/microstructure relationship using different high manganese steels made by weld deposit. It was used a high manganese filler metal and a SAE 1012 plate steel as base metal through the flux cored arc welding (FCAW) process. The steel compositions were measured by energy dispersive spectroscopy (EDS) and the microstructures were obtained using a scanning electronic microscopy (MEV). The SFE and SFE components were calculated and related to dilution and microstructure obtained.

## 2. Materials and Methods

### *2.1. Welding*

The different steels were made using FCAW process. A high manganese steel electrode as consumable and an SAE 1012 steel as base metal were used. The different steels obtained were taken from the weld zone. The chemical composition of the materials is shown in Table 1. The different steels were obtained varying the weld dilution level, which was calculated using the geometric method [11]. Table 2 shows the welding parameters used.

**Table 1.** Materials Chemical composition. All values are in wt.%.

| | Mn | Ni | Cr | Si | C | Fe |
|---|---|---|---|---|---|---|
| **PT-400HM** | 20.94 | 3.76 | 0.87 | 0.58 | 0.17 | Bal. |
| **SAE 1012** | 0.60 | - | - | - | 0.12 | Bal. |

**Table 2.** Welding Parameters.

| | Power (W) | Current (A) | Tension (V) | Welding Speed (cm/min) | Feed rate (m/min) | Energy (kJ/mm) |
|---|---|---|---|---|---|---|
| **B-11** | 7480 | 220 | 34 | 15 | 10 | 3.0 |
| **B-07** | 7480 | 220 | 34 | 22 | 10 | 2.0 |
| **A-31** | 7480 | 220 | 34 | 30 | 10 | 1.5 |
| **A-38** | 7480 | 220 | 34 | 45 | 10 | 1.0 |

### *2.2. Characterization*

Steels characterization were carried out using standard metallography procedures, which involved grinding with silicon carbide papers, followed by polishing with diamond suspensions, and finally, polishing with colloidal silica for several hours in order to avoid martensite presence due to sample preparation. The samples were etched with a 10% Nital solution (10% Nitric Acid + 90% Alcohol) and the structure was observed with SEM. EDS measurements were carried out in a SEM using 20 kV and working distance of 10 mm.

### *2.3. Stacking Fault Energy*

The SFE were calculated following Equation 1. The equation developments can be found in references [4,6,10]. All functions and constants can be found in [5, 12-22].

## 3. Results and Discussion

Different dilution values were obtained varying the welding speed, forming four different alloys, as can be seen in Table 2. Figure 1 shows scanning electron micrographs of all four steels and the Table 3 shows the chemical composition, SFE values calculated from this composition, dilution level and microhardness values. DuPont (2011) found the similar result for a stainless austenitic steel [23], although the materials are quite distinctives, the arc physics is the same. The microstructure for all steels shows dendritic morphology, with epitaxial grains. For alloys A-38 and A-31, those who achieved high dilution level, α' martensite and austenite/ε martensite were found, while for alloys B-11 and B-07, just austenite phase was present. Besides, twins can be seen in A-31, B-11, and B-07 conditions.

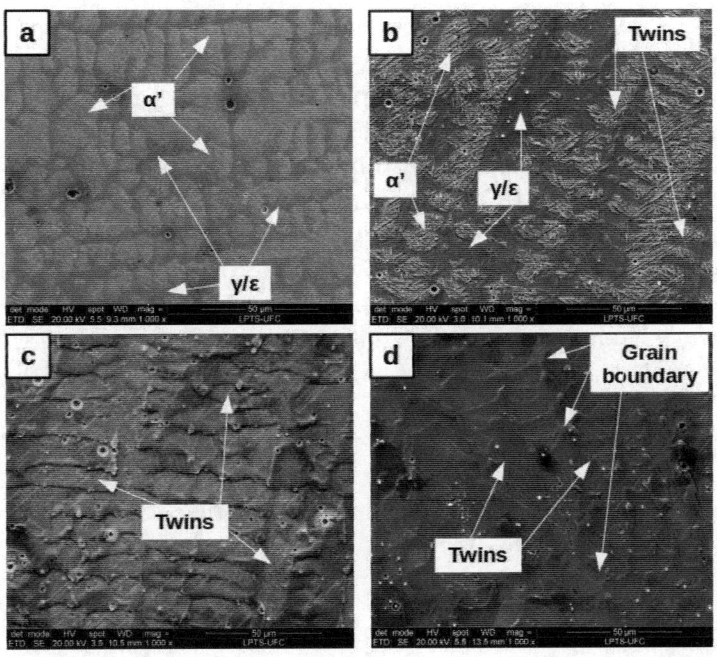


**Figure 1.** Scanning electron micrographs of alloy (a) A-38, (b) A-31, (c) B-11, and (d) B-07.

The dendritic morphology is more apparent in steels with α' martensite, because it is formed in the core of dendrite. During the solidification process, manganese is rejected and the poor manganese regions transform into α' martensite, while the inter dendritic spaces do not, due to their higher manganese content. Comparing A-38 and A-31, the presence of α' martensite decrease. The lower dilution level leads to a higher manganese content in weld zone (Table 3), which provide more regions of austenite stabilized. B-11 and B-07 conditions do not show α' martensite presence and the manganese condition measured is higher than 18 wt.%. For those conditions with higher dilution level, A-31 and A-38, calculated SFE values showed to be negative, while for those with lower dilution, just austenite twinned were observed. Besides extend works about SFE/microstructure relationship [1, 3, 24, 25], no mention has been made about negative SFE and α' martensite presence. The α' martensite absence in low dilution conditions is reflected in microhardness values, which decrease with the decreasing in dilution level, that is, with the α' martensite fraction decrease. The high standard deviation occurs due the manganese microssegregation that increase the hardness by solid solution or even by α' phase transformation, leading to different hardness values.

**Table 3.** Chemical composition, SFE, dilution level and microhardness values obtained from all conditions.

| Condition | Mn | C | Ni | Si | Cr | Fe | SFE ($mJ/m^2$) | Dilution (%) | Hardness (HV) |
|---|---|---|---|---|---|---|---|---|---|
| **A-38** | 12.69 | 0.15 | 2.05 | 0.32 | 0.51 | Bal. | -5.1 | 37.7 | 346 ± 12.5 |
| **A-31** | 14.69 | 0.15 | 2.66 | 0.51 | 0.59 | Bal. | -1.0 | 31.1 | 269 ± 18.5 |
| **B-11** | 18.71 | 0.16 | 3.42 | 0.50 | 0.82 | Bal. | 7.1 | 11.3 | 242 ± 20.0 |
| **B-07** | 18.91 | 0.17 | 3.42 | 0.54 | 0.76 | Bal. | 7.4 | 7.6 | 225 ± 18.5 |

Although the hardness value has been decreased with the decrease in α' martensite fraction, the austenite TRIP or TWIP effect can be present, leading to high strength with higher ductility. Curtze et al. (2010) and Xiong et al. (2015) showed the deformation mechanism influence on the high manganese mechanical properties [5, 26], where martensite formation rose the work hardening rates, while it decreased the ductility. Then, a balance of α' martensite and austenite can be predicted by dilution level for these two materials (base metal and filler metal). Despite the authors do not found a clear evidence of the composition limit to α' martensite formation, it was found in SFE values. In order to understand it, influence of dilution level on the SFE value was calculated, as well as on the chemical and magnetic Gibbs free energy component, as can be seen in Figure 2a.

Figure 2a shows the relation between dilution level and SFE, magnetic, and chemical Gibbs free energy. The markers point out SFE calculated from the steel's composition, showed in Table 3. The values fit very well with the curves obtained for the base and filler metal studied. The SFE value decreased with the increasing in dilution level, as well as the chemical Gibbs component (Figure 2a). However, the magnetic component appears to have little influence on the total Gibbs Free energy, decreasing smoothly with the increase in dilution. The increase in dilution level refers to a change in more than one element [11], and so, it is important to analyze carefully the chemical composition of the two weld materials in order to obtain a good understanding of the dilution influence on the SFE, and consequently, on the microstructure developed.

Figure 2b shows manganese influence on the SFE and its components. The composition calculated was the filler metal composition, varying just manganese content. It can be seen the strong manganese influence on the SFE, passing easily for all fields of phase transformation and twin formation presented by [2, 24], who presented the phase prediction based on the chemical composition of manganese and carbon. Although their diagram has shown a good tool for phase prediction in high manganese steels, additional elements such as nickel, silicon and chromium are not considered. Overcoming this problem, thermodynamic SFE model shows a better tool, enabling additional elements consideration and other ways to use this tool, as the one used in the present work, with dilution level. As thermodynamic SFE's main components are the chemical and magnetic Gibbs free energy, they can show useful information to understand the SFE increasing behavior. For manganese, both chemical and magnetic component increase, leading to a high increasing in SFE value. On the other hand, nickel does not show an increase in magnetic component, despite of SFE increase with nickel addition. Similar results were found by Xiong et al. (2014) [4], however, no mention was made about nickel influence. Both manganese and nickel are austenite stabilizers, with nickel being the most influential (Figures 2b and 2c). However, the high nickel cost justifies the manganese preference.

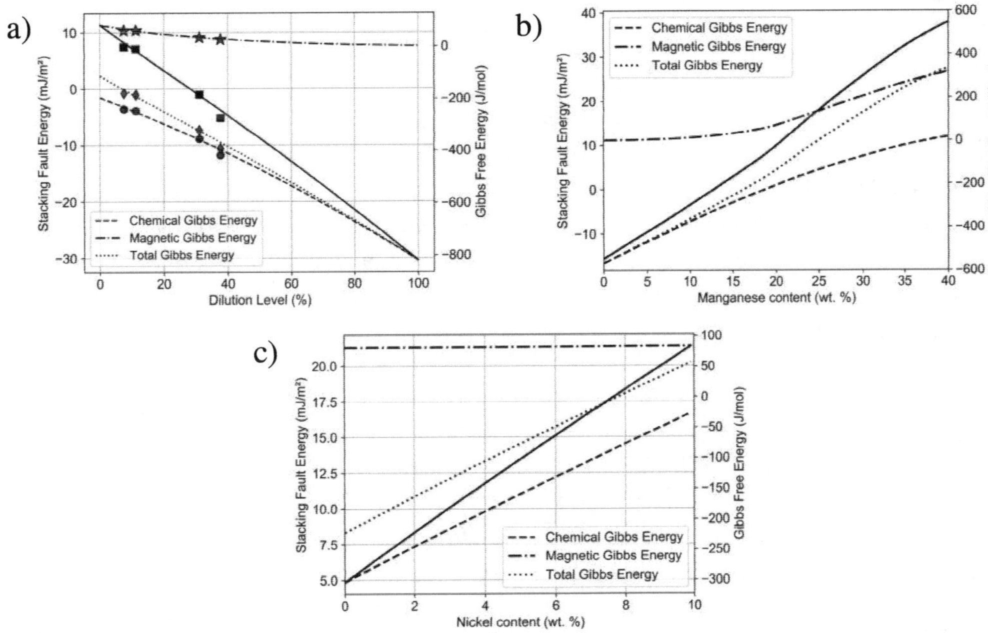


**Figure 2.** SFE, ΔG, dilution relationship, and (b) Manganese, (c) Nickel influence on SFE and ΔG. Squares, circles, stars and diamonds corresponds to SFE, chemical, magnetic and total Gibbs free energy measured for the four alloys, respectively.

The results showed a good relation achieved between dilution, SFE and microstructure. Negative SFE values led to α' martensite formation, while positive SFE values did not. Manganese was the most influential element to increase SFE value, because it was the element with the most composition difference between base and filler metal, and because manganese increases both magnetic and chemical Gibbs free energy component.

## 4. Conclusions

It was obtained four high manganese steels by welding deposit. The chemical composition, microstructure and stacking fault energy was studied and related with dilution level. The following statement can be drowned from the results:

- The increasing in welding speed led to a higher dilution level, and consequently high α' martensite formation and microhardness value;

- The dendritic morphology led to regions with high α' martensite density and others with austenite/ε martensite. The reason for this morphology accounts for the manganese microssegregation, what increased the standard deviation in microhardness;
- SFE calculated in function of dilution level showed a smooth decrease in magnetic Gibbs free energy component, while a rapid decreasing in chemical component were found. It led to a continuous decreasing in SFE value with the increase in dilution level;
- Manganese was the major influential element in the SFE value, due to increase both chemical and magnetic Gibbs free energy component and for the high chemical difference between base and filler metal.

## Acknowledgements

This study was financed in part by the Coordenação de Aperfeiçoamento de Pessoal de Nível Superior - Brasil (CAPES) - Finance Code 001.

## References

[1] De Cooman B C, Kwon O and Chin K G 2012 *Mater. Sci. Technol.* **28** 513-27
[2] Choi J K, Lee S G, Park Y H, Han I W and Morris Jr J W 2012 High manganese austenitic steel for cryogenic applications *22nd International Offshore and Polar Engineering Conference* (South Korea: Technical Research Laboratories) pp 29-35
[3] De Cooman B 2017 *High Mn TWIP steel and medium Mn steel Automotive Steels* (Amsterdam: Elsevier) pp 317-85
[4] Xiong R, Peng H, Si H, Zhang W and Wen Y 2014 *Mat. Sci. Eng. A-Struct.* **598** 376-86
[5] Curtze S and Kuokkala V T 2010 *Acta Mater.* **58** 5129-41
[6] Olson G B and Cohen M 1976 *Metall. Trans. A* **7** 1897-904
[7] Li Y j, Fu R d, Li Y, Peng Y and Liu H j 2018 *J. Mater. Sci. Technol.* **34** 157-62
[8] Paveebunvipak K and Uthaisangsuk V 2018 *Mater. Design* **160** 731-51
[9] Park G, Jeong S, Kang H and Lee C 2018 *Mater. Charact.* **139** 293-302
[10] Curtze S, Kuokkala V T, Oikari A, Talonen J and Hanninen H 2011 *Acta Mater.* **59** 1068-76
[11] DuPont J N and Marder A R 1996 *Metall. Mater. Trans. B* **27** 481-9
[12] Dinsdale A 1991 *Calphad* **15** 317-425
[13] Ishida K 1976 *Phys. Status Solidi A* **36** 717-28
[14] Ishida K and Nishizawa T 1974 *Trans. Japan Institute of Metals* **15** 225-31
[15] Dumay A, Chateau J P, Allain S, Migot S and Bouaziz O 2008 *Mat. Sci. Eng. A-Struct. A* **483-484** 184-7
[16] Yang W S and Wan C M 1990 *J. Mater. Sci.* **25** 1821-3
[17] Chen S, Chung C, Yan C and Hsu (Xu Zuyao) T 1999 *Mat. Sci. Eng. A-Struct. A* **264** 262-8
[18] Adler P H, Olson G B and Owen W S 1986 *Metall. Trans. A* **17** 1725-37
[19] Allain S, Chateau J P, Bouaziz O, Migot S and Guelton N 2004 *Mat. Sci. Eng. A-Struct. A* **387-389** 158-62
[20] Yakubtsov I, Ariapour A and Perovic D 1999 *Acta Mater.* **47** 1271-9
[21] Wu X and Hsu (Xu Zuyao) T 2000 *Mater. Charact.* **45** 137-42
[22] Zhang Y S, Lu X, Tian X and Qin Z 2002 *Mat. Sci. Eng. A-Struct. A* **334** 19-27
[23] Dupont J N 2011 *Dilution in Fusion Welding Welding Fundamentals and Processes* vol 6A (ASM International) pp 115-21
[24] De Cooman B 2012 *Phase transformations in high manganese twinning-induced*

*plasticity (TWIP) steels Phase Transformations in Steels* vol 2 (Amsterdam: Elsevier) pp 295-331
[25] De Cooman B C, Estrin Y and Kim S K 2018 *Acta Mater.* **142** 283-362
[26] Xiong R, Peng H, Wang S, Si H and Wen Y 2015 *Mater. Design* **85** 707-14

IOP Conf. Series: Materials Science and Engineering **668** (2019) 012002 doi:10.1088/1757-899X/668/1/012002

# Engineering design highlights of high-efficiency slab casters relocated at Baosteel Zhanjiang

**X L Ding[1,3], C B Feng[2], Q Liu[1], J W Yan[1], Y D Zhang[1] and G Q Luo[1]**

[1]Continuous Casting Business Section, Iron & Steel Business Division of CISDI Engineering Co., Ltd., Chongqing 401122, China
[2]Steel Making Plant of Baosteel Zhanjiang Iron & Steel Co., Ltd., Zhanjiang 524000, China

E-mail: xiaolin.ding@cisdi.com.cn

**Abstract.** Baosteel Group decided in 2013 to relocate the main segment equipment of the 2 single-strand 2,300 mm slab casters which were shut down at its Luojing Plant to the Zhanjiang Plant and transform them into an upgraded two-strand slab caster. CISDI was contracted to supply the overall design and the main equipment packages. CISDI employed a couple of advanced technologies including online mould width adjustment, top-inserted dummy bar, dual strand tracking and casting at constant speeds to enhance the production capacity of the caster; for sake of Opex saving CISDI also optimized the arrangements of the independent mechanical maintenance shop, utilities of compact design, cooling water valve stations and hydraulic valve stations of compact designs, and open-air cooling water pump station, etc.

## 1. Introduction

The two single-strand 2,300 mm continuous slab casters at Baosteel Luojing which were completed and put into operation in 2007 and 2009 respectively were shut down in 2012. In 2013, Baosteel Group decided to relocate the main segment equipment of the two casters at its Luojing Plant to the Zhanjiang Plant, and transform them into a two-strand slab caster. Main technical data of the slab caster as transformed are shown in table 1. CISDI was contracted to supply the overall design and the main equipment packages. At present, the steel industry of China has been facing a severe overcapacity of common steel products. Meanwhile, the supply of upmarket products is relatively insufficient. In the current scenario when supply-side structural reforms are being continuously implemented nationwide, it will put any late-comers at a disadvantage, if a high-efficiency operation for new slab casters are designed with a differentiated and boutique product portfolio [1-5]. The relocation project was planned adjustment to transform the continuous casters into one caster that could perform high-efficiency, low-cost casting of high-quality products. In May 2013, the project was launched. After three years of engineering design, equipment demolition, transformation, installation and commissioning, the desired slab caster underwent the load test successfully in one go on May 15, 2016.

## 2. Technologies for high-efficiency operation

### *2.1. Online mould width adjustment*

Provided the casting speed is stable, the most effective way to increase the capacity of a caster is to increase the number of continuous casting heats. However, in the traditional continuous casting production, when the slab width size is changed, the casting has to be interrupted in order to replace the mould, or the caster has to be shut down in order to allow mould width adjustment, then perform restranding, resume casting, which result in low production efficiency, high workload, and considerable metal loss.

**Table 1.** Main technical data of the 2,300 mm slab caster at Baosteel Zhanjiang.

| No. | Designation | Unit | Specification/Value |
|---|---|---|---|
| 1 | No. of strands | | 2-2 |
| 2 | Caster design | | Vertical curved, with continuous bending, straightening |
| 3 | Products | | Line pipe steel, pressure vessel steel, structural steel, ship-building steel, steels for special purposes, etc. |
| 4 | Dummy bar insertion | - | Top-inserted |
| 5 | Slab sizes | mm | 200/230/250 × (1,200~2,300) |
| 6 | Production capacity | tpa | 2,750,000 |

After many years of R & D activities, CISDI has successfully developed the on-line mould width adjustment technology [6]. It allows quick adjustments to the mold widths even when casting at full speed, meets the requirements for the production of slabs with different width sizes in an uninterrupted way, thus increases the metal yield and operating rate, and improve the caster productivity.

CISDI has conducted in-depth studies on the mechanism of strain and air gap generation in the strand shell during the mould width adjustment process and established an online mould width adjustment process model with wedge slab envelop as a core, paving the way theoretically for the performance of quick, safe and stable online width adjustments. On this basis CISDI has developed high-precision mould width adjustment device successfully, incorporating the driving device for width adjustment with high control accuracy and self-locking capability, the soft clamping device on the wide face which can ensure the safety of casting operation and the mould body which meets the width adjustment requirements. The width adjustment is performed with electric servo drives. The two narrow faces of the mould are driven by the four sets of worm servo motor transmissions to move in the same or opposite direction at the same speed so as to achieve the width and taper as required by the width adjustment operation [7]. Featuring width adjustment with high precision and good stability, etc., it can effectively suppress the narrow face width and taper drift during the production process.

The 2,300 mm slab caster at Baosteel Zhanjiang equipped with the online mould width adjustment technology developed by CISDI has been running stably since it was put into production. It performs width adjustments more than 2 times a day on average, with width and taper adjustment within ±0.2 mm.

*2.2. Two-strand tracking*

The production capacity of a caster can be effectively increased by reducing the restranding time, provided that the casting speed is stable and the number of continuous casting heats is fixed.

If a bottom-inserted dummy bar is used, it cannot be inserted for the casting of the next heat before the last strand is transported out of the dummy bar storage area, which means the restranding time is as long as 80 minutes or so.

Use of a top-inserted dummy bar is an effective measure to reduce the restranding time. When the last strand comes out of the segment, the dummy bar can be inserted for casting of the next heat. And the restranding time is shortened to about 50 minutes.

A top-inserted dummy bar and two-strand tracking technology, as shown in figure 1 have been applied to the 2,300 mm slab caster at Baosteel Zhanjiang. They have enabled the slab caster to cast the last strand and insert the dummy bar at the same time: Once the last strand leaves the segment 3, the dummy bar can immediately be inserted for casting of the next heat [8], further reducing the restranding time to about 35 minutes. So, the slab caster is more flexible in production and its production capacity can be increased effectively.

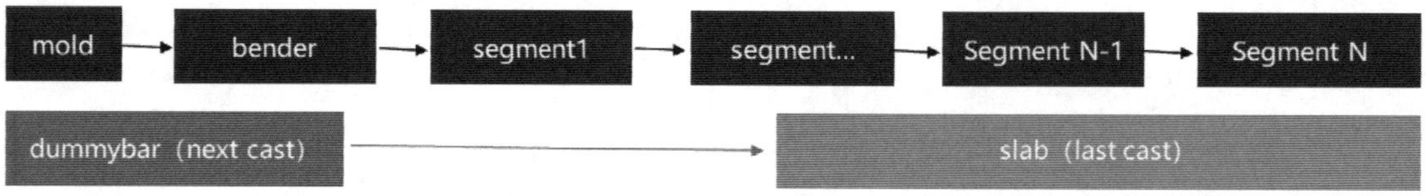


**Figure 1.** Sketch of strand dual-tracking.

### *2.3. Casting at constant speeds*

In the traditional continuous casting production, the continuously cast steel grades, section sizes frequently change due to the restrictions imposed by the composition, temperature of the molten steel and the tempo of the molten steel supply from the upstream (steelmaking) process. It frequently changes also due to the influence of the changes in the slab section sizes and the demand for steel grades by the downstream (rolling mill) process. In order to allow the slab caster to keep up with the tempo of the basic oxygen furnace(s), the casting speed often has to fluctuate greatly, making the casting tempo and the slab quality cannot be stable.

Based on the process of hot rolling with constant width and edge reduction, the continuous slab caster at Baosteel Zhanjiang has been designed to produce slabs of a few typical widths. The hot rolling mill can perform heavy edge reduction to meet the requirements for all the width sizes, so the continuous caster does not need to frequently change the slab widths, allowing for operation with a larger number of continuous casting heats for a single slab width size. Moreover, sufficient secondary metallurgy facilities have been installed upstream to provide qualified molten steel with stable composition and temperature of molten steel at the tempo required by the slab caster, truly conducive to realizing production scheduling with continuous casting as the core. Continuous casting is performed at constant speeds. Stable molten steel flow field distribution in the tundish, stable mould level and stable equipment status [9] have contributed to the high-efficiency operation of the slab caster and high-quality slabs by the slab caster.

## 3. Technologies for low-cost operation

### *3.1. Independent mechanical maintenance shop and utilities of compact design*

In the traditional design, the mechanical maintenance facilities for continuous casters are arranged in the main building of the continuous casting plant, and the utilities such as high-voltage electrical rooms, water treatment plant, and air compressor stations have to be placed in an area far away from the continuous caster itself. However, at Baosteel Zhanjiang, all the mechanical maintenance facilities for the continuous caster have been concentrated, and an independent light-duty Mechanical Maintenance Center has been constructed outside the main building of the Continuous Casting Plant [10]. The capital expenditure (Capex) and operating expenditure (Opex) of the mechanical maintenance facilities are reduced, and on the other hand, as the mechanical maintenance facilities are not in the main building, the spiral-flow settling tank and other water treatment equipment, the secondary cooling zone air compressor station and the high-voltage electrical rooms, etc. can be close to the continuous caster itself. Arrangements like this have not only reduced the length of pipes and cables, but also increased the response speed and sensitivity of the continuous caster itself to the changes in the demand for energy & media, thus helpful to reduce the Capex, Opex and improve the slab quality as well.

### *3.2. Cooling water valve stations and hydraulic valve stations of compact design*

For a slab caster of the traditional design, the cooling water valve stations and the hydraulic valve stations are respectively arranged on multiple-floor platforms including pipe tunnels. In the Continuous Casting Plant at Baosteel Zhanjiang all the cooling water valve stations have been installed on the ground floor on both sides of the slab caster, and all the hydraulic valve stations have been arranged on the second floor on both sides of the slab caster. This has not only greatly reduced the length of the pipes, shortened the distance between the regulating valves to the end-user, but also increased the response speed and sensitivity of the equipment to the changes in the demand for cooling water and hydraulic media, thus helpful to reduce the Capex, Opex and improve the slab quality as well.

### *3.3. Open-air cooling water pump houses*

The cooling water pump station for continuous casters in the traditional design are mainly installed indoors. However, the cooling water pump station for the slab caster at Baosteel Zhanjiang all have been installed open air. The production practice in the past two years has proved the success in the open-air arrangement of the pump station. It has not only reduced the Capex, but also decreased the Opex of the ventilation and lighting facilities for the plant building.

## 4. Conclusions

The 2,300 mm slab caster at Baosteel Zhanjiang, equipped with the technologies of online mould width adjustment, top-inserted dummy bar, two-strand tracking and casting at constant speeds, etc. has achieved the desired high-efficiency casting operation. The production of one year since its start-up has proven that the production capacity of the two-strand slab caster reaches 2.8 million tonnes of slabs per annum, with qualified slabs accounting for over 99%.

The arrangements of independent mechanical maintenance shop, utilities of compact design, cooling water valve stations and hydraulic valve stations of compact design, and open-air cooling pump station have benefited the slab caster by reducing production cost. The production of one year since its start-up has proven that the energy consumption by the continuous casting process at Baosteel Zhanjiang is about 30% lower if compared with a similar continuous caster.

## References

[1] Fan S L, Li G P and Lv J 2006 Baosteel lean cost management application case study *Science Research Management* **27** 89-94
[2] Wang L Y 2009 Lean manufacturing-based MES system application study (Changchun, China: Jilin University)
[3] Xu S L, Shao M Z, Zheng L *et al* 2013 Target-oriented lean management system at Jinan iron & steel Shandong *Metallurgy* **35** 66-75
[4] Li Z H, Shou Y H, Dong W M *et al* 2012 Exploration and practices of implementing lean production mode at Anyang iron & steel *Metallurgical Economics and Management* **2012** 30-39
[5] The Technical Society 2013 The iron and steel institute of Japan. procuction and technology of iron and steel in Japan during 2012 *ISIJ International* **53** 927-35
[6] Chen J, Peng X H *et al* 2013 Online mould width adjustment technology and its applications *Metallurgical Equipment* **2013** 16-8
[7] Gan Y 2010 *Modern Continuous Steel Casting Practice Manual* (Beijing, China: Metallurgical Industry Publishing House)
[8] Xiong Y G 1994 *Continuous Slab Casting* (Beijing, China: Metallurgical Industry Publishing House)
[9] Chen J L 2015 Specialized continuous (Slab) caster design *Steelmaking* **31** 7-14
[10] Zhang N 2017 Technological design of the 2150 mm slab caster at Baosteel Zhanjiang *Continuous Casting* **2017** 71-8

IOP Conf. Series: Materials Science and Engineering **668** (2019) 012003 doi:10.1088/1757-899X/668/1/012003

# Crystallization processes and magnetic properties of FeCoZrMB (M=Nb, V, Cr, Ti) amorphous alloys

**W Q Yu[1,2] and L P Lu[1,3]**

[1]School of Materials Science and Engineering, Changchun University of Science and Technology, Changchun 130022, China
[2]National Demonstration Center for Experimental Physics Education, Jilin Normal University, Siping 136000, China

E-mail: luliping771219@126.com

**Abstract.** A series of $Fe_{40}Co_{40}Zr_8M_2B_{10}$ (M=Nb, V, Cr, Ti, W, Al) alloys were prepared using melt-spinning. The thermal curve, structure and magnetic property of alloys are examined. Because of different negative heat of mixing between elements, only $Fe_{40}Co_{40}Zr_8M_2B_{10}$ (M=Nb, V, Cr, Ti) alloys form amorphous structure. These amorphous alloys are annealed at different temperatures under vacuum conditions. The crystallization processes of four amorphous alloys are similar. In the primary stage of crystallization process, only α-Fe(Co) phase precipitates and Co element mainly distributes in the residual amorphous. For the four alloys after annealing at 550°C, there is a few differences in saturation magnetization ($\boldsymbol{M_s}$) and coercivity ($\boldsymbol{H_c}$) due to their different microstructures.

## 1. Introduction

Since the excellent soft magnetic properties, Fe-based nanocrystalline soft magnetic materials have been attracted lots of attentions for a long time. Among the Fe-based nanocrystalline soft magnetic alloys, Hitperm FeCoMB(Cu) alloys as competitive materials for high temperature soft magnetic applications have been developed and attracted lots of attentions in recent years [1-4]. The magnetism of the materials is closely related to their microstructures composed by α-FeCo nanocrystals embedded in amorphous matrix. Grain size is one of the important factor affecting the properties of nanocrystalline soft magnetic materials. Within the exchange correlation length, coercivity ($\boldsymbol{H}_c$) of nanocrystalline alloys with the two-phase structure follows a $D^6$ power law [5]. Moreover, $\boldsymbol{H}_c$ is also related to the grain-size distribution of α-Fe nanograins [6].

In order to improve the performance of alloys, a wide variety of elements have been added into Fe-based alloy system. Adding transition metal can greatly improve and adjust their microstructure and performance, such as Nb [7-9], V [10,11], Cr [12-14], Ti [15,16], W [17,18] and Al [19-22] element. Nb and W are the most effective elements which can refine the grain size during annealing treatment [17]. The addition of V can improve the amorphous forming ability and the bending ductility of the alloys [10]. Minor Cr addition significantly improves the corrosion resistance [14]. Doping Ti can decrease core loss mainly by reducing its hysteresis loss [16]. Al doping results in the reduction of grain size and coercivity [22].

In this paper, the $Fe_{40}Co_{40}Zr_8M_2B_{10}$ (M=Nb, V, Cr, Ti, W, Al) alloy system is based on the addition of different transition metals to a conventional Hitperm-like amorphous alloy system. The crystallization processes and magnetic properties of $Fe_{40}Co_{40}Zr_8M_2B_{10}$ alloys with different transition metal will be discussed in detail.

## 2. Experimental

*2.1. Preparation*

Ingots of $Fe_{40}Co_{40}Zr_8M_2B_{10}$ (M=Nb, V, Cr, Ti, W, Al) alloys were prepared by arc melting the mixtures of the pure metals above 99.9% and pure B (99.999%), and the ingots were remelted with the magnetic stirring for 4 times to ensure the homogeneity under vacuum conditions. Then alloy ribbons were prepared using a single roller melt-spinning equipment with copper wheel rotating at a surface velocity of 38 m/s. Ribbons were 1 cm in width and about 30 μm in thickness. $Fe_{40}Co_{40}Zr_8M_2B_{10}$ [M=Nb(#1), V(#2), Cr(#3), Ti(#4)] amorphous alloys were post annealed at 500, 550, 600, 650 and 700°C for 1h under vacuum conditions, respectively.

*2.2. Measurements*

Thermal curves were performed by differential scanning calorimetry (DSC, Diamond) instrument. Structures of alloys were identified by transmission electron microscopy (TEM, JEM-2100E) and X-ray diffraction (XRD, D/max 2500/PC, 40 kV, 200 mA, Cu-Kα, λ=1.5406 Å). Thin-film TEM specimens were prepared by rgon ion beam thinning technique. Magnetic hysteresis loops were examined by vibrating sample magnetometer (VSM, Lake Shore M7407).

## 3. Results and discussion

*3.1. Glass forming ability of alloys*

Figure 1 gives the XRD patterns of $Fe_{40}Co_{40}Zr_8M_2B_{10}$ (M=Nb, V, Cr, Ti, W, Al) alloys as-quenched. $Fe_{40}Co_{40}Zr_8M_2B_{10}$ (M=Nb, V, Cr, Ti) alloys as-quenched are all amorphous but not for $Fe_{40}Co_{40}Zr_8M_2B_{10}$ (M=W, Al) alloys.

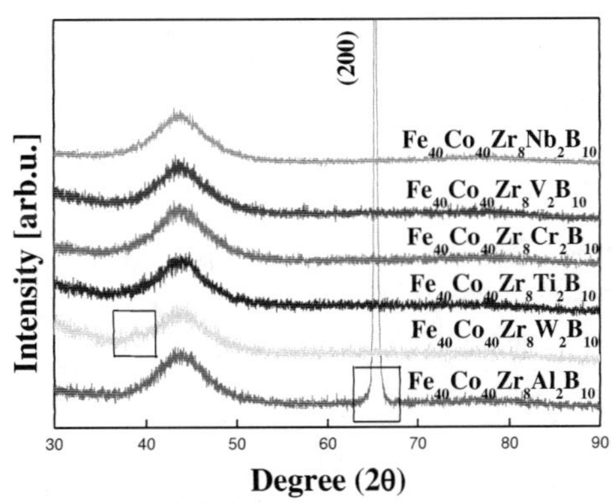


**Figure 1.** XRD patterns of $Fe_{40}Co_{40}Zr_8M_2B_{10}$ (M=Nb, V, Cr, Ti, W, Al) alloys as-quenched.

Table 1 gives the values of the heat of mixing between M element and other elements [23]. For $Fe_{40}Co_{40}Zr_8W_2B_{10}$ alloy, the negative heat of mixing between W and other elements is relatively low. Therefore, $Fe_{40}Co_{40}Zr_8W_2B_{10}$ alloy is not fully amorphous. $Fe_{40}Co_{40}Zr_8Al_2B_{10}$ alloy has large negative heats of mixing between Al element and other elements. However, the (200) diffraction peak of α-Fe(Co) phase can be observed. It should be that Al has a large saturation vapor pressure and evaporates during the smelting process. For $Fe_{40}Co_{40}Zr_8M_2B_{10}$ (M=Nb, V, Cr, Ti) alloys, the negative heat of mixing between M(M=Nb, V, Cr, Ti) and other elements are bigger and form amorphous structure in the as-quenched state.

**Table 1.** The values of heat of mixing between M and other elements (kJ/mol).

| M | Fe | Co | Zr | B |
|---|---|---|---|---|
| Nb | -16 | -25 | 4 | -54 |
| V | -7 | -14 | -4 | -42 |
| Cr | -1 | -4 | -12 | -31 |
| Ti | -17 | -28 | 0 | -58 |
| W | 0 | -1 | -9 | -31 |
| Al | -11 | -19 | -44 | 0 |

For the as-quenched alloys which is not fully amorphous, the existing grains will increase rapidly after heat treatment, resulting in the great difference in grain size. Relatively uniform grains can be obtained by annealing amorphous alloys. Therefore, we mainly study the crystallization processes and magnetic properties of $Fe_{40}Co_{40}Zr_8M_2B_{10}$ (M=Nb, V, Cr, Ti) amorphous alloys. DSC traces of $Fe_{40}Co_{40}Zr_8M_2B_{10}$ (M=Nb, V, Cr, Ti) alloys as-quenched are shown in figure 2. The glass forming ability (GFA) can be calculated by the temperature span of the supercooled liguid region $\Delta T_x$ (difference between $T_g$ and $T_x$) [24]. The glass transition temperature($T_g$), onset crystallization temperature ($T_x$), supercooled liguid region $\Delta T_x$ and crystallization peak temperature ($T_p$) can be obtained by DSC.

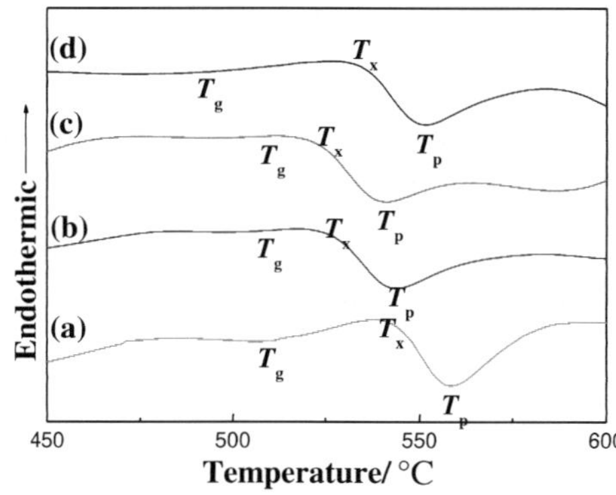


**Figure 2.** DSC traces of four amorphous alloys at a heating rate of 10°C/min.
(a)$Fe_{40}Co_{40}Zr_8Nb_2B_{10}$, (b) $Fe_{40}Co_{40}Zr_8V_2B_{10}$
(c) $Fe_{40}Co_{40}Zr_8Cr_2B_{10}$, (d) $Fe_{40}Co_{40}Zr_8Ti_2B_{10}$

Table 2 gives a list of $T_g$, $T_x$, $\Delta T_x$ and $T_p$ of four kinds of amorphous alloys. The wider the $\Delta T_x$ is, the greater the GFA is. Both $\Delta T_x$ and $T_p$ of the alloys have the following relations: $\Delta T_{x\#4}>\Delta T_{x\#1}>\Delta T_{x\#2}>\Delta T_{x\#3}$ and $T_{p\#1}>T_{p\#4}>T_{p\#2}>T_{p\#3}$. The GFA of four amorphous alloys has the following relations: $GFA_{\#4}> GFA_{\#1}> GFA_{\#2}> GFA_{\#3}$. The negative heat of mixing between M(M=Nb, V, Cr, Ti) and other elements have the following relations: $H_{\#4}>H_{\#1}>H_{\#2}>H_{\#3}$. Therefore, the GFA of four kinds of alloy presents the results above.

**Table 2.** $T_g$, $T_x$, $\Delta T_x$ and $T_p$ of $Fe_{40}Co_{40}Zr_8M_2B_{10}$ (M=Nb, V, Cr, Ti) amorphous alloys.

| | $Fe_{40}Co_{40}Zr_8Nb_2B_{10}$ (#1) | $Fe_{40}Co_{40}Zr_8V_2B_{10}$ (#2) | $Fe_{40}Co_{40}Zr_8Cr_2B_{10}$ (#3) | $Fe_{40}Co_{40}Zr_8Ti_2B_{10}$ (#4) |
|---|---|---|---|---|
| $T_g$/°C | 505.0 | 505.5 | 506.3 | 496.8 |
| $T_x$/°C | 540.1 | 524.7 | 522.2 | 532.6 |
| $\Delta T_x$/°C | 35.1 | 19.2 | 15.9 | 35.8 |
| $T_p$/°C | 558.3 | 543.0 | 539.6 | 551.7 |

### *3.2. Crystallization processes of alloys*

XRD patterns of $Fe_{40}Co_{40}Zr_8M_2B_{10}$ (M=Nb, V, Cr, Ti) alloys after annealing are shown in figure 3. The crystallization processes of these four amorphous alloys are similar. After annealing at 500°C, only α-Fe(Co) phase is formed. The lattice constants of α-Fe(Co) of $Fe_{40}Co_{40}Zr_8M_2B_{10}$ (M=Nb, V, Cr, Ti) alloys after annealing at 550°C are 2.860, 2.859, 2.860 and 2.862, respectively. The lattice constants of pure α-FeCo and α-Fe are 2.8550 and 2.8664 Å [25], respectively. It is deduced that Co element in the alloys mainly distributes into the remained amorphous. It is attributed to the big heat of mixing between Co element and Zr element (-41 kJ/mol). More and more α-Fe(Co) nanocrystals precipitate and the diffraction peaks intensity increases with the increase of annealing temperature. After annealing at 650°C, $ZrCo_3B_2$ phase forms. The precipitation of $ZrCo_3B_2$ phase at high temperature is due to that the Co element in residual amorphous matrix reacts with the Zr and B elements distributed in the residual amorphous matrix [26]. The crystalline phases of alloys precipitated at the final annealing stage are the same and identified to be α-Fe(Co), $ZrCo_3B_2$ and $Fe(Co)_3Zr$ phases. The crystallization processes of these four amorphous alloys are similar: amorphous→remained amorphous+α-Fe(Co)→α-Fe(Co)+$ZrCo_3B_2$+$Fe(Co)_3Zr$.

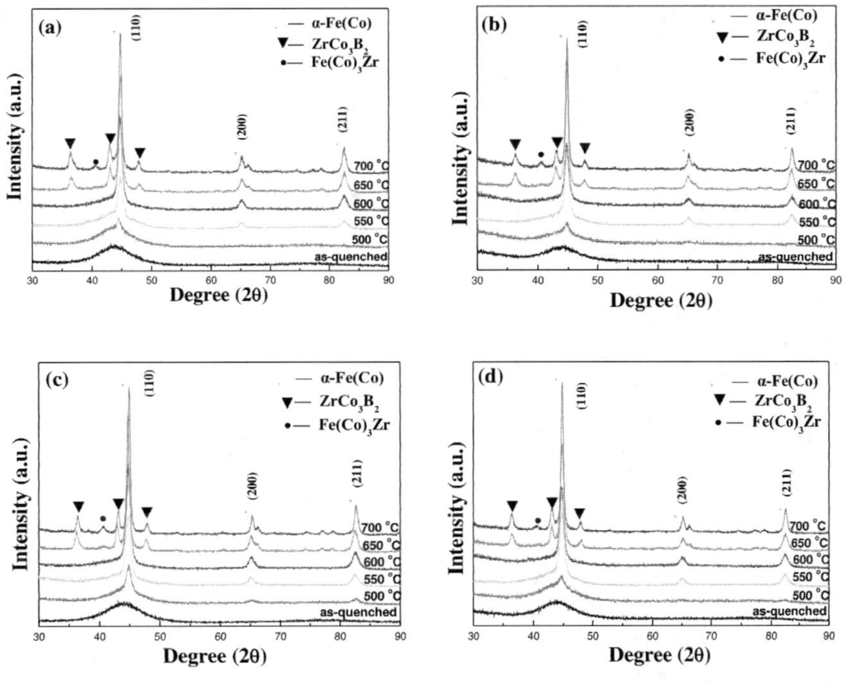


**Figure 3.** XRD patterns of four kinds of alloys after annealing at different temperatures. (a) $Fe_{40}Co_{40}Zr_8Nb_2B_{10}$, (b) $Fe_{40}Co_{40}Zr_8V_2B_{10}$, (c) $Fe_{40}Co_{40}Zr_8Cr_2B_{10}$, (d) $Fe_{40}Co_{40}Zr_8Ti_2B_{10}$.

Figure 4 shows the TEM images, SAED patterns and PSD of the alloys after annealing at 550°C. For the four kinds of alloys after annealing at 550°C, the nanocrystals are surrounded by a residual amorphous matrix. SAED patterns all show the crystallization phase is only α-Fe(Co). They all form a typical two-phase structure characterized by bcc α-Fe(Co) nanocrystalline grains embedding in the amorphous matrix. The volume fraction of $Fe_{40}Co_{40}Zr_8Nb_2B_{10}$ alloy is the lowest and the crystallization volume fraction of $Fe_{40}Co_{40}Zr_8Cr_2B_{10}$ alloy is the biggest. The crystallization volume fraction of $Fe_{40}Co_{40}Zr_8M_2B_{10}$ (M=Nb, V, Cr, Ti) alloys have the following relations:$V_{\#3}>V_{\#2}>V_{\#4}>V_{\#1}$. According to the grain size distribution, the grain size [α-Fe(Co)] of $Fe_{40}Co_{40}Zr_8M_2B_{10}$ (M=Nb, V, Cr, Ti) alloys are 7.8, 7.2, 6.9 and 8.2 nm, respectively. Furthermore, $Fe_{40}Co_{40}Zr_8V_2B_{10}$ and $Fe_{40}Co_{40}Zr_8Ti_2B_{10}$ alloys nanocrystals distribute inhomogeneously. During the annealing process of amorphous alloy, the transition metal will reject from α-Fe particles and accumulate in the residual amorphous matrix, thus reducing the growth of α-Fe particles. The difference in the atomic radius and electronegative between elements leads to the differences in grain refinement.

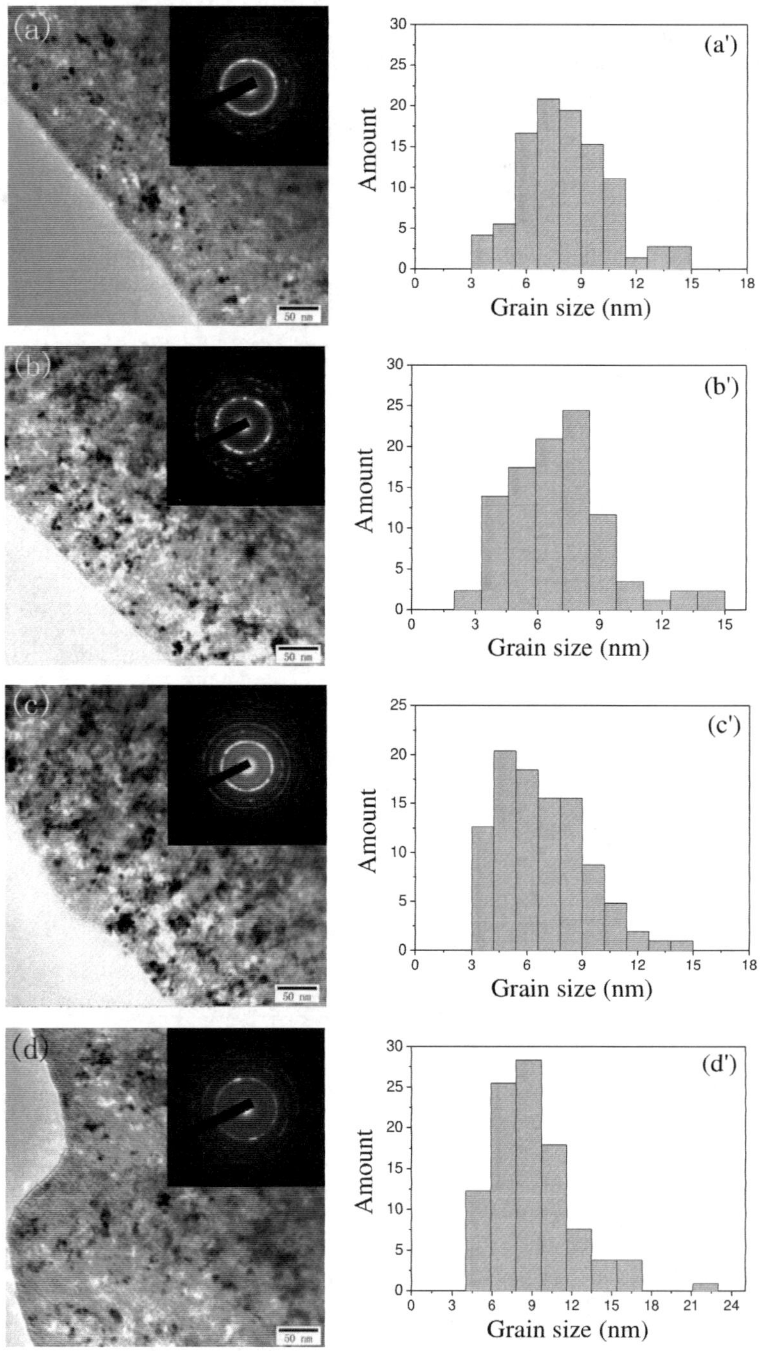


**Figure 4.** Left is the transmission electron microscope (TEM) images and selected-area electron diffraction (SAED) patterns of the alloys after annealing at 550°C; Right is the grain size distribution (PSD) of the alloys after annealing at 550°C. (a, a')$Fe_{40}Co_{40}Zr_8Nb_2B_{10}$, (b, b') $Fe_{40}Co_{40}Zr_8V_2B_{10}$, (c, c') $Fe_{40}Co_{40}Zr_8Cr_2B_{10}$, (d, d') $Fe_{40}Co_{40}Zr_8Ti_2B_{10}$

### *3.3. Magnetic properties of alloys*

The magnetic hysteresis loops of $Fe_{40}Co_{40}Zr_8M_2B_{10}$ (M=Nb, V, Cr, Ti) alloys are shown in figure 5. Four kinds of alloys all show good soft magnetic property from magnetic hysteresis loops. Table 3 gives the saturation magnetization ($M_s$) and coercivity ($H_c$) of $Fe_{40}Co_{40}Zr_8M_2B_{10}$ (M=Nb, V, Cr, Ti) alloys annealed at 550°C.

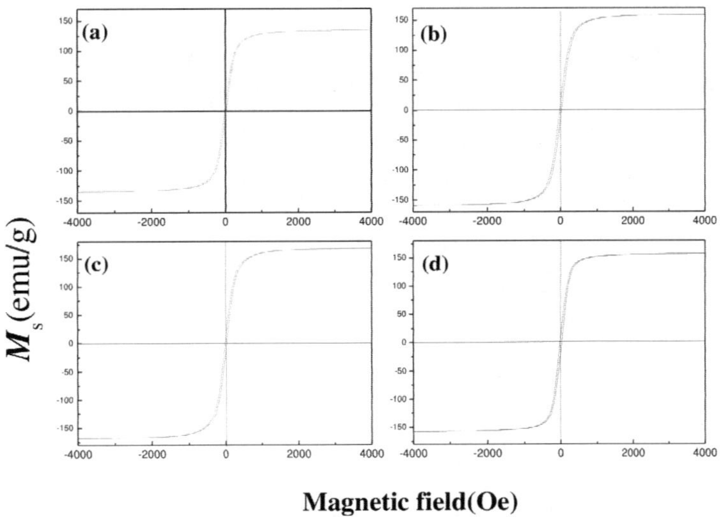


**Figure 5.** Magnetic hysteresis loops of $Fe_{40}Co_{40}Zr_8M_2B_{10}$ (M=Nb, V, Cr, Ti) alloys annealed at 550°C.
(a) $Fe_{40}Co_{40}Zr_8Nb_2B_{10}$, (b) $Fe_{40}Co_{40}Zr_8V_2B_{10}$
(c) $Fe_{40}Co_{40}Zr_8Cr_2B_{10}$, (d) $Fe_{40}Co_{40}Zr_8Ti_2B_{10}$

**Table 3.** $\boldsymbol{M_s}$ **and** $\boldsymbol{H_c}$ of $Fe_{40}Co_{40}Zr_8M_2B_{10}$ (M=Nb, V, Cr, Ti) alloys annealed at 550°C.

| | $M_s$ (emu/g) | $H_c$ (Oe) |
|---|---|---|
| $Fe_{40}Co_{40}Zr_8Nb_2B_{10}$ (#1) | 135.4 | 22.4 |
| $Fe_{40}Co_{40}Zr_8V_2B_{10}$ (#2) | 159.0 | 24.6 |
| $Fe_{40}Co_{40}Zr_8Cr_2B_{10}$ (#3) | 169.0 | 22.5 |
| $Fe_{40}Co_{40}Zr_8Ti_2B_{10}$ (#4) | 158.3 | 24.5 |

There is a little difference in Coercivity ($H_c$) among alloys, which is due to the different grain size and different nanocrystals distribution of α-Fe(Co) phase. The differences in saturation magnetization ($M_s$) among alloys is due to the differences of crystallization volume fraction of α-Fe(Co) phase. $M_s$ of the alloys has the following relations: $M_{s\#3} > M_{s\#2} > M_{s\#4} > M_{s\#1}$.

## 4. Conclusions

$Fe_{40}Co_{40}Zr_8M_2B_{10}$ (M=Nb, V, Cr, Ti, W, Al) alloys were prepared by melt-spinning. Among of them, $Fe_{40}Co_{40}Zr_8M_2B_{10}$ [M=Nb(#1), V(#2), Cr(#3), Ti(#4)] alloys have large negative heat of mixing and form amorphous. The GFA of the four kinds of alloy has the following relations: $GFA_{\#4} > GFA_{\#1} > GFA_{\#2} > GFA_{\#3}$. Only α-Fe(Co) phase precipitates in the primary stage of crystallization process and Co element mainly distributes into the residual amorphous. They all form a typical two-phase structure characterized by bcc α-Fe(Co) nanocrystalline grains embedding in the amorphous matrix. For the four amorphous alloys after annealing at 550°C, there is a few differences in saturation magnetization ($M_s$) and coercivity ($H_c$) due to their different microstructures.

## References

[1] Han B-K, Kim Y K and Choi-Yim H 2014 *Curr. Appl. Phys.* **14** 685-7
[2] Arvindha Babu D, Majumdar B, Sarkar R, Manivel Raja M and Akhtar D 2011 *J. Mater. Res.* **26** 2065-71
[3] Sun Y, Zhong L and Bi X F 2009 *Scripta Mater.* **60** 814-7
[4] Radoń A, Patryk W, Łukasz H, Mariola K G, Piotr G, Ryszard N and Rafał B 2018 *J. Alloy. Compd.* **763** 141-52
[5] Herzer G 1990 *IEEE Trans. Magn.* **26** 1397-402
[6] Bitoh T, Makino1 A, Inoue A and Masumoto T 2003 *Mater. Trans.* **44** 2011-9
[7] Li X L, Liu J B, Qu C R, Song K K and Wang L 2017 *J. Alloy. Compd.* **694** 643-6
[8] Alleg S, Souilah S, Younes A, Bensalem R, Suñol J J and Greneche J M 2012 *J. Alloy. Compd.* **536** S394-7
[9] Silveyra J M and Illeková E 2014 *J. Alloy. Compd.* **610** 180-3
[10] Ji L, Zheng Z G, Qiu Z G, Zeng D C, Yu H Y and Xu B 2018 *J. Alloy. Compd.* **766** 391-7
[11] Liu L, Yang Y R, Qiu M, Pi Y M and Yan G M 2015 *J. Simul.* **3** 22-4
[12] Ma Y, Liu Y, Li J, Zhang H and Yang H 2016 *J. Iron Steel Res. Int.* **23** 625-32
[13] Botta W J, Berger J E, Kiminami C S, Roche V, Nogueira R P and Bolfarini C 2014 *J. Alloy. Compd.* **586** S105-10
[14] Xu D D, Zhou B L, Wang Q Q, Zhou J, Yang W M, Yuan C C, Xue L, Fan X D, Ma L Q and Shen B L 2018 *Corros. Sci.* **138** 20-7
[15] Liu W L, Wang Y G and Chen F G 2014 *J. Mater. Sci-Mater. El.* **25** 5066-70
[16] Wang J, Liu X, Mo J, Mao X H, Fan X A and Luo Z G 2018 *J. Alloy. Compd.* **766** 769-74
[17] Zhang Y, Yan B, Yang Y and Wang Y X 2013 *J. Alloy. Compd.* **574** 556-9
[18] Liang D D, Wei X S, Chang C T, Li J W, Wang X M and Shen J 2017 *J. Alloy. Compd.* **731** 1146-50
[19] Stepanov N D, Shaysultanov D G, Chernichenko R S, Tikhonovsky M A and Zherebtso S V 2019 *J. Alloy. Compd.* **770** 194-203
[20] He J Y, Liu W H, Wang H, Wu Y, Liu X J, Nieh T G and Lu Z P 2014 *Acta Mater.* **62** 105-13
[21] Duan H J, Wang Z and Jia Y R 2019 *Mater. Res. Bull.* **111** 289-93
[22] Murugaiyan P, Abhinav A, Verma R, Panda A K, Mitra A, Baysakh S and Roy R K 2018 *J.Magn. Magn. Mater.* **448** 66-74
[23] Takeuchi A and Inoue A 2005 *Mater. Trans.* **46** 2817-29
[24] Inoue A 2000 *Acta Mater.* **48** 279-306
[25] Yu W Q, Zeng H Q, Sun Y M, Sun Y J and Hua Z 2017 *Phys. Lett. A* **381** 1573-6
[26] Yu W Q, Zeng H Q, Sun Y M and Hua Z 2017 *Vacuum* **137** 175-82

# Effects of alkalis and zinc on the damage of the tuyere composite brick of a blast furnace

**Q Niu[1], S S Cheng[1,3], W X Xu[1], W J Niu[2] and Y G Mei[1]**

[1]School of Metallurgical and Ecological Engineering, University of Science and Technology Beijing, 30 Xueyuan Road, Haidian District, Beijing, 100083 P. R. China
[2]Ironmaking Plant, Anyang Iron and Steel Group Co. Ltd., Yindu District, Anyang, Henan, 455004 P. R. China

E-mail: chengsusen@metall.ustb.edu.cn

**Abstract.** The distribution of foreign elements and its true mineral forms in the tuyere composite brick extracted from a 2800 $m^3$ industrial blast furnace were examined to gain a better understanding of the damage mechanism of the brick. The X-ray diffraction (XRD) and the scanning electron microscopy-energy dispersive spectroscopy (SEM-EDS) were applied to characterize the brick. The results show that the potassium is easier to permeate into the brick than zinc. The potassium is prone to react with Al-Si compounds in the brick to form kalsilite or leucite while the zinc is apt to react with the corundum in the brick to produce zinc-aluminum spinel. In addition, the zinc can interact with the blast furnace gas to form zinc oxide and zinc sulfide filled in the cracks of the brick. The mineral phases of the residual brick change from the original corundum and mullite to corundum, gahnite, zincite, sphalerite, kalsilite and leucite. The main cause of the damage of the brick is found to be potassium and zinc attack. In the case of degradation of the raw materials, it can choose the micro corundum brick instead of the corundum-mullite or mullite brick to prolong the life of the tuyere composite brick.

## 1. Introduction

Silicon-aluminum composite refractory is one of the refractories used in blast furnace. The composite brick arranged around the tuyere is in such harsh environment (high temperature, multiphase coexistence and so on) that its damage is inevitable in a working blast furnace [1-4]. With the deterioration of the raw materials as well as the application of the high blast temperature and pulverized coal injection technology [5,6], the significant damage of the tuyere composite brick is found [4]. Therefore, it is necessary to figure out the damage mechanism to develop and choose proper tuyere composite brick. Previous researches [1-2,7] on tuyere composite brick mostly focused on composition analysis of the residual bricks. However, limited studies have considered inorganic elements in their true mineral forms rather than in oxides or element forms. Besides, previous studies [1,4,7] usually focused on the properties of the tuyere composite brick affected by potassium and zinc separately, while potassium and zinc simultaneously influence the performance of tuyere composite brick in the actual production of the blast furnace. Although researchers have studied the effect of potassium on Al-Si bricks, the studied environment is different from the environment of blast furnace near the tuyere zone [8-12]. Accordingly, in order to prolong the life of the tuyere composite brick, it is necessary to identify its real phases of the residual brick and make clear the actual behavior simultaneously degraded by potassium and zinc vapors.

In this paper, the microstructure of tuyere composite brick and its composition as well as mineral phases were investigated in detail with samples obtained from a 2800 m³ blast furnace during its overhaul. In addition, the formation process of the foreign phases in the tuyere composite brick was investigated thermodynamically. Finally, the damage mechanism of the tuyere composite brick was analyzed.

## 2. Samples and methods

### *2.1. Introduction of blast furnace and the sampling position*

In the present study, the samples were obtained from an industrial blast furnace with an inner volume of 2800 $m^3$, 30 tuyeres and 3 tap holes. The relevant parameters are listed in the previous paper [13]. After the blow-out and cool-down of the furnace, some tuyere composite brick samples collected manually. The tuyere composite brick installed around the tuyeres was the corundum-mullite brick. It was comprised of $Al_2O_3$ (87.72%), $SiO_2$ (11.03%) and $Fe_2O_3$ (0.35%). In this paper, the neighboring part of hot surface, the vicinity part of macroscopic crack and the inside part of tuyere composite brick around No. 6 tuyere at layer 2 were studied, and the sampling positions were shown in the figure 1.

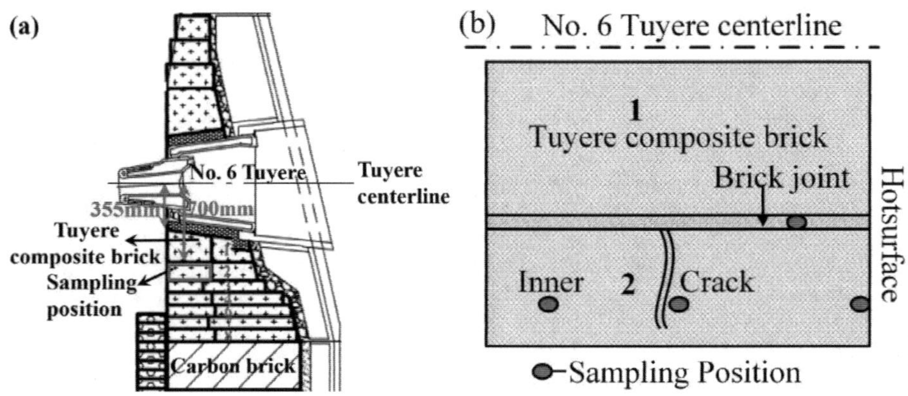


**Figure 1.** The schematic diagram of the sampling positions. (a) Sampling position in the blast furnace; (b) Sampling position in the residual brick.

### *2.2. Analysis methods and sample preparation*

The samples were prepared, ground and polished similar to the previous study [13]. Then the samples were examined with a Zeiss Evo18 Special Edition SEM-EDS. In order to identify the mineral phases of the residual brick, the samples obtained from different positions were crushed to passing less than 74um and then examined using a Rigaku diffractometer (DMAX-RB 12 kW; Rigaku Corporation, Tokyo, Japan) using Cu Ka as radiation. During this analysis, samples were scanned with 2θ in the range of 10 to 90 deg at a scan rate of 5 deg/minute.

## 3. Results and discussion

### *3.1. The composition and microstructure analysis of the hot surface of the tuyere composite brick*

The composition and microstructure of the hot surface of the brick is presented in figure 2. The residual brick is composed of corundum particles (black phases in figure 2(a)), potassium aluminosilicates (grey phases in figure 2(a)), oxide and sulfide of zinc (white phases in figure 2(a)). In the neighboring part of the hot surface of residual brick, various foreign elements (K, Zn and S) penetrated into the brick as shown in figure 2(b). A great deal of zinc was primarily observed in the hot surface and the internal cracks, while the potassium is mainly distributed with Al and Si inside the brick. It is indicated that the permeation ability of potassium into the brick is stronger than that of zinc, which is in agreement with previous studies [7]. A layer of oxide and sulfide of zinc with a thickness of about 2 mm (figure 2(a)) is observed in the hot surface. Thus, the liquid slag and hot metal are prevented from penetrating into the tuyere composite brick. From this perspective, the layer of foreign elements is contribution to the protection of the brick to some extent.

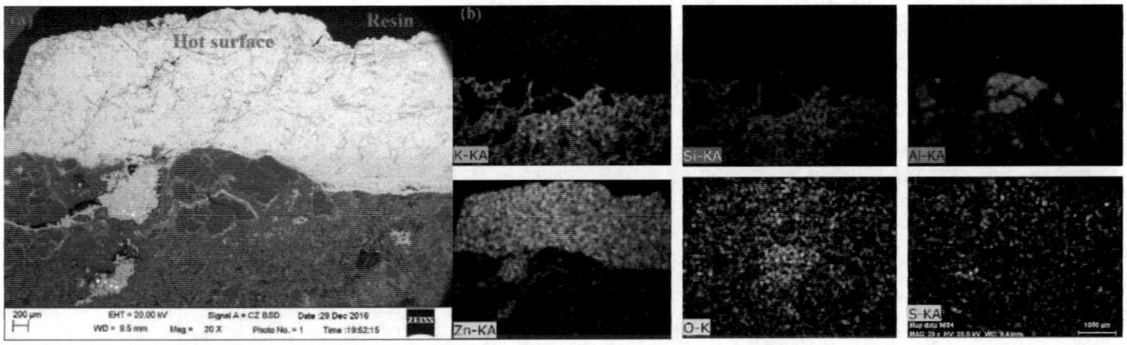


**Figure 2.** Composition and microstructure of the hot surface of the tuyere composite brick. (a) SEM micrograph; (b) EDS maps showing the distribution of Si, Al, K, S, Zn and O in residual brick.

The micro and macro cracks in the vicinity part of hot surface of brick were presented in figure 3. The residual brick was cut along the vertical direction of the hot surface and its profile is shown in figure 3(b). The residual brick cracked with numerous laminations in parallel with the hot surface of the brick. Part of the cracks pass through the brick, as shown in figures 3(a) and 3(b), The micro and macro cracks decrease the strength of the brick and are very detrimental to the life of the brick. The maximum crack width of the residual brick is about 3 mm. Some cracks are completely filled with foreign matters while the others are only partially filled, indicating that the foreign matters in the cracks is gradually accumulated. The cracks provide a special channel for the harmful elements into the brick. The thermal conductivity of the corundum-mullite brick (3.2 W/(m•K) [3]) is about 128 times that of gas (0.025 W/(m•K)). That is to say, the thermal resistance of air gap with a 3 mm thickness is equivalent to that of a brick with a thickness of 384 mm. Therefore, the cracks filled with gas are bad to the heat transfer of the brick. Besides, non-uniform distribution of the cracks cause uneven thermal stress distribution of the bricks. It may leads to the emergence of the new cracks and damage to the brick.

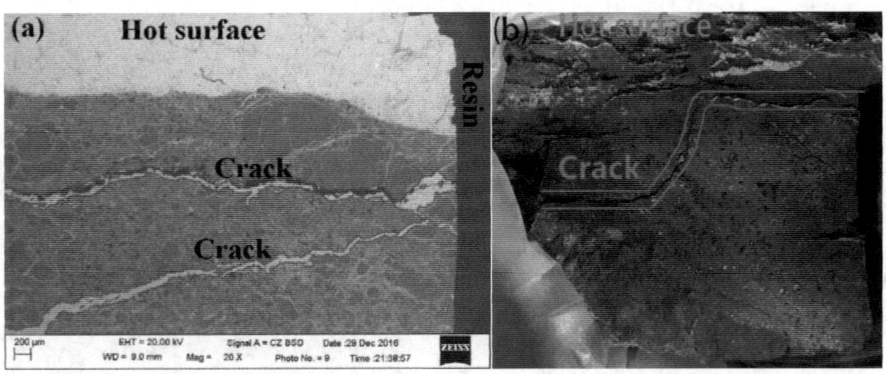


**Figure 3.** The micro cracks and macro cracks in hot surfaces of tuyere composite brick. (a) SEM micrograph; (b) Macro picture.

The XRD spectra of the brick at different positions is shown in figure 4. The original brick is composed of corundum and mullite, which is in accordance with the previous stuydies [14-16]. The absence of mullite peak in the diffraction pattern of the tuyere composite brick was confirmed, while the mineralogical phases in the neighboring part of hot surface of the brick were determined to be zincite (ZnO), sphalerite (ZnS), gahnite ($ZnAl_2O_4$), kalsilite ($KAlSiO_4$) and corundum ($Al_2O_3$), as shown in figure 4. Thus, foreign elements in the brick were proved to be mainly alkalis and zinc. With the exception of the corundum, the other phases are absent from original brick, indicating that mullite in the original brick is easy to interact with potassium to form kalsilite and the corundum can react with zinc to form zinc-aluminum spinel. Also, the blast furnace gas permeating the composite brick can interact with zinc to form ZnO and ZnS concentrated on the brick cracks or pores. Previous studies have shown that the transformation from zinc to zinc oxide, from zinc to zinc sulfide, from zinc to gahnite and from potassium to kalsilite is accompanied by a volume expansion of about 54% [7], 83% [7], 15% [17] and 36% [18], respectively, which may lead to the generation of new cracks and deteriorate the brick.

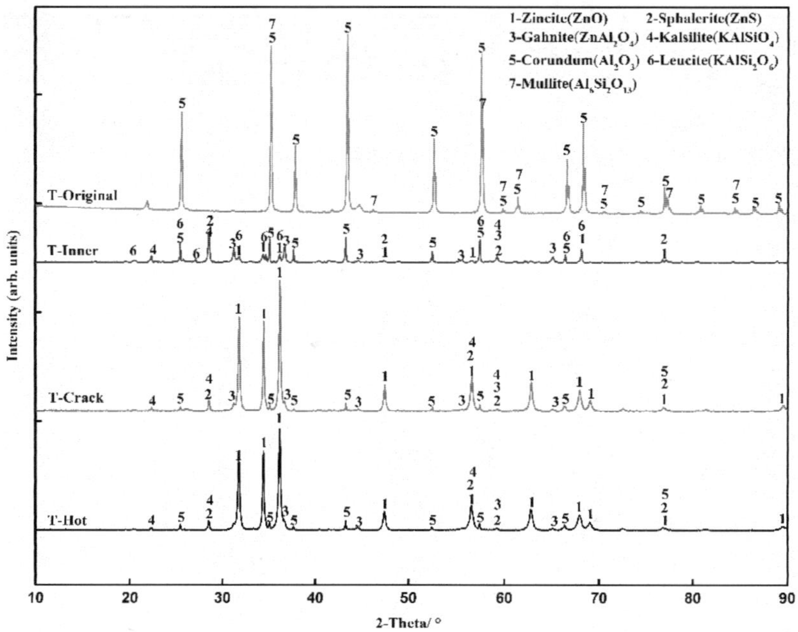


**Figure 4.** XRD spectra of the different positions of the tuyere composite brick. T-Inner: the sample in the internal brick; T-Crack: the sample near the crack; T-Hot: the sample of hot surface of residual brick.

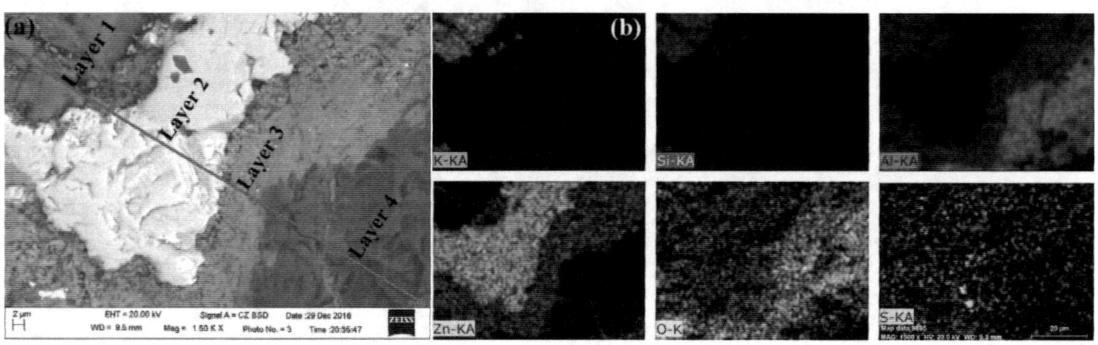


**Figure 5.** The composition and microstructure of hot surface of the tuyere composite brick. (a) SEM micrograph; (b) EDS maps showing the distribution of Si, Al, K, S, Zn and O.

Figure 5 shows the composition and microstructure of the internal cracks in the tuyere composite bricks near the hot surface. The tuyere composite brick can be divided into four different layers along the arrow direction: kalsilite layer (Layer 1 in the figure 5), ZnO-ZnS layer(Layer 2 in the figure 5), gahnite layer (Layer 3 in the figure 5) and corundum layer (Layer 4 in the figure 5). The potassium is not found in corundum layer while the potassium content is high in the place where Al and Si coexist in the original brick. It can be seen that potassium is apt to react with the Al-Si compounds of the original composite brick to form kalsilite or leucite with about 15-36% [18] volume expansion. Volume expansion or uneven thermal stress leads to the new cracks. Subsequently, zinc permeates into the brick along the cracks and reacts with the corundum particles at the edge of the crack to form the zinc-aluminum spinel, which is accompanied by a volume expansion of about 15% [17]. When the Zn-Al spinel accumulates to a certain extent, zinc can not contact with corundum particles, begin to react with the permeated blast furnace gas to form ZnO-ZnS and accumulates in the crack until the entire crack is filled.

### *3.2. The composition and microstructure analysis of the tuyere composite brick near the macro crack*

The composition and microstructure of the residual brick near macro crack presents in figures 6 and 7. Figure 7 is the magnification photo of a partial region of figure 6. The ZnO-ZnS layer with a thickness of about 40 microns was observed at the crack edge. Therefore, zinc permeates the composite brick along the pores and micro cracks, and accumulates continuously in the form of oxides and sulfides. When a certain thickness is reached, the micro cracks are expanded into the macro cracks which is visible to the naked eye. The mineral phases near the macro cracks of the brick were confirmed to be zincite (ZnO), sphalerite (ZnS), gahnite ($ZnAl_2O_4$), kalsilite ($KAlSiO_4$) and corundum ($Al_2O_3$), as shown in figure 4. From the figure 7, we can see that the brick can be divided into five different layers: kalsilite layers, ZnO layers, ZnS layers, gahnite layers and corundum layers. It is similar to the results of the brick near the hot surface (figure 5).

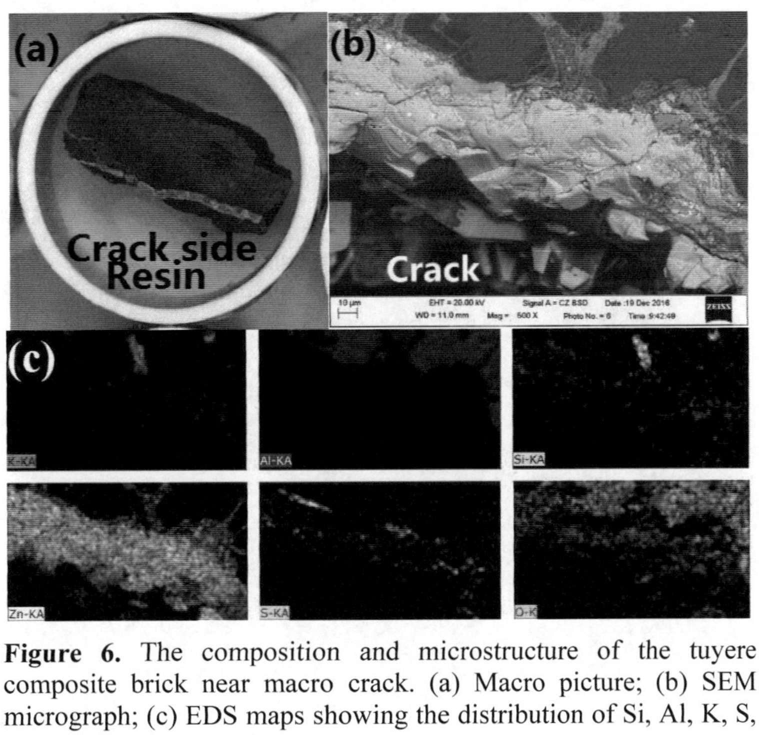


**Figure 6.** The composition and microstructure of the tuyere composite brick near macro crack. (a) Macro picture; (b) SEM micrograph; (c) EDS maps showing the distribution of Si, Al, K, S, Zn and O in residual brick.

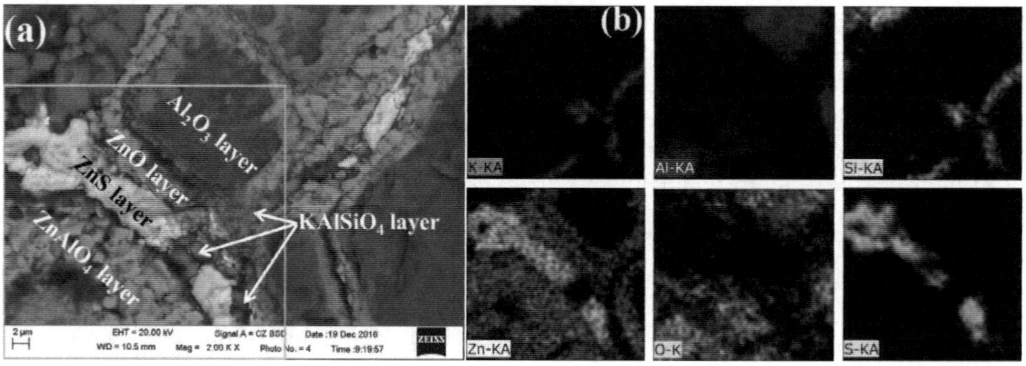


**Figure 7.** The composition and microstructure of tuyere composite brick near macro crack. (a) SEM micrograph; (b) EDS maps showing the distribution of Si, Al, K, S, Zn and O.

### *3.3. The damage mechanism of the tuyere composite brick*

As can be seen from figure 4, the minerals inside the brick were confirmed to be zincite (ZnO), sphalerite (ZnS), gahnite ($ZnAl_2O_4$), kalsilite ($KAlSiO_4$), leucite ($KAlSi_2O_6$) and corundum ($Al_2O_3$). The main reactions between zinc or potassium and the composite brick components are as follows:

$$2K + CO + 2[3Al_2O_3 \cdot 2SiO_2] = K_2O \cdot Al_2O_3 \cdot 4SiO_2 + 5Al_2O_3 + C \quad (1)$$

$$2K + CO + 3Al_2O_3 \cdot 2SiO_2 = K_2O \cdot Al_2O_3 \cdot 2SiO_2 + 2Al_2O_3 + C \quad (2)$$

$$2K + CO + Al_2O_3 = K_2O \cdot Al_2O_3 + C \quad (3)$$

$$Zn + CO + 1/3[3Al_2O_3 \cdot 2SiO_2] = ZnO \cdot Al_2O_3 + 2/3SiO_2 + C \quad (4)$$

$$Zn + CO + Al_2O_3 = ZnO \cdot Al_2O_3 + C \quad (5)$$

$$Zn + CO = ZnO + C \quad (6)$$

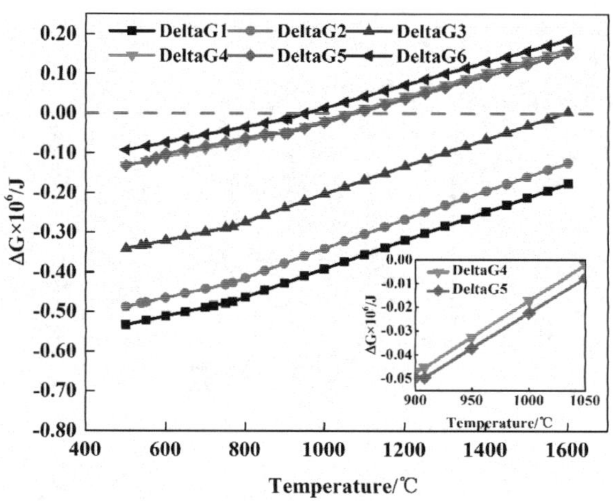


**Figure 8.** The relationship between Gibbs free energy and temperature in the standard state.

The Gibbs free energy of each reaction is calculated using FactSage 7.1 thermodynamic software [19], as shown in figure 8. The Gibbs free energy of the reactions (1) - (3) is always negative at temperatures below 1600°C, and the Gibbs free energy of the generation of kalsilite and leucite is smaller than that of the other reactions at the same temperature. Therefore, potassium is easier to react with mullite in the brick to produce kalsilite or leucite. This is consistent with the previous SEM-EDS analysis. When the temperature is below 1050℃, the free Gibbs energy of reaction (4) and (5) is always negative while the Gibbs free energy of reaction (6) is always negative when the temperature is lower than 950℃. The Gibbs free energy of reaction (5) is less than that of reaction (4) at the temperature range from 900 to 1050℃. As potassium more easily penetrates into the tuyere composite brick than zinc and is apt to react with mullite, the infiltrated zinc primarily reacts with corundum in the brick to generate zinc-aluminum spinel. The Gibbs free energy of the reactions (4) and (5) is always smaller than that of the reaction (6) at the same temperature, indicating that zinc prefers to react with corundum in the tuyere composite brick. When the reaction product (zinc-aluminum spinel) accumulates to a certain extent, zinc begins to react with the permeated blast furnace gas to generate ZnO and ZnS filled in the cracks and the open pores in brick.

Potassium, zinc and blast furnace gas penetrate into the brick as the gaseous phase through the pore and crack, and they react with the brick or each other. Based upon the results of the SEM-EDS and thermodynamics analysis, potassium is apt to react with Al-Si compounds in the composite brick to form kalsilite or leucite while zinc prefers to react with corundum to form zinc alumina spinel. The formation of the new phases results in volume expansion and new cracks. When the thickness of the new phases reaches a certain extent, the contact reaction between the alkali metal, zinc and the self-components of the tuyere composite brick will be hindered. Then, the CO and S in the gas react with zinc vapor to form high melting point ZnO and ZnS, and accumulate in the pores or cracks eventually filling the pores or cracks. At last, there will be some distinct layers in the brick, namely, kalsilite layer, ZnO layer, ZnS layer, gahnite layer and corundum layer. The expansion coefficient of different layers is different. When the furnace condition fluctuates (such as the edge gas flow becomes stronger), the non-uniform volume expansion between the foreign matters filled in the cracks or pores and the brick can lead to the generation of new cracks, which in turn loops over the above destruction process. The long-term accumulation eventually causes macro cracks. The fluctuations of the gas and liquid slag and iron will make the brick peel off along the macro cracks layer by layer.

## 4. Conclusions

- The main cause of the damage of the tuyere composite brick is found to be alkalis and zinc attack due to penetration of alkalis and zinc vapor through the pore or crack.
- Potassium is easier to permeate into the composite brick than zinc. Potassium is easy to react with Al-Si to form kalsilite or leucite while zinc is apt to react with the corundum in the composite brick to produce zinc-aluminum spinel. In addition, zinc can react with the blast furnace gas and the products are filled in the cracks of the composite brick.
- After more than nine years' use of the tuyere composite brick, the mineral phase of the residual tuyere composite brick changed from the original corundum, mullite and glass phase to zincite (ZnO), sphalerite (ZnS), gahnite ($ZnAl_2O_4$), kalsilite ($KAlSiO_4$), leucite ($KAlSi_2O_6$) and corundum ($Al_2O_3$).
- In the case of degradation of the raw materials, it can choose the micro corundum brick instead of the corundum-mullite or mullite brick to prolong the life of the tuyere composite brick.

**Acknowledgment**
This project was supported by the National Key R & D Program of China (2017YFB0304300 & 2017YFB0304302).

IOP Conf. Series: Materials Science and Engineering **668** (2019) 012004 doi:10.1088/1757-899X/668/1/012004

**References**


[1] Narita K, Onoye T, Satoh Y, Miyamoto M, Taniguchi K, Kamatani S, Sato T and Fukihara S 1981 *Trans. ISIJ* **21** 839-45

[2] Zou Z Q, Deng T and Song M S 2010 *Wisco Technology* **48** 1-4

[3] Gao Z X, Li H X, Shi G, Zhu R L and Jiang H 2009 *Microscopic Analysis of Lining Corrosion of Blast Furnace* (Beijing: Metallurgical Industry Press)

[4] Zhang S R and Yu Z J 2009 *Long Life Technology of Blast Furnace in WISCO* (Beijing: Metallurgical Industry Press)

[5] Naito M, Takeda K and Matsui Y 2015 *ISIJ Int.* **55** 7-35

[6] Zhou D D, Cheng S S, Wang Y S and Jiang X *ISIJ Int.* **55** 2519-24

[7] Yang X F, Chu M S, Shen F M and Zhang Z M 2009 *Acta Metall. Sin.* **22** 454-60

[8] Li N, Vainio E, Hupa L, Hupa M and Zabetta E C 2018 *Energ. Fuel* **32** 12971-80

[9] Sadik C, Amrani I E and Albizane A 2018 *Journal of Asian Ceramic Societies* **2** 83-96

[10] Li N, Vainio E, Hupa L, Hupa M and Zabetta E C 2017 *Energ. Fuel* **31** 10046-54

[11] Stjernberg J, Olivas-Ogaz M A, Antti M L, Ion J C and Lindblom B 2013 *Ceram. Int.* **39** 791-800

[12] Stjernberg J, Lindblom B, Wikström J, Antti M L and Odén M 2010 *Ceram. Int.* **36** 733-40

[13] Niu Q, Cheng S S, Xu W X, Niu W J, Li A F, Ma H, Zhang S and Cao T F 2018 *ISIJ Int.* **58** 667-76

[14] Dana K, Sinhamahapatra S, Tripathi H S and Ghosh A 2014 *Trans. Ind. Ceram. Soc.* **73** 1-14

[15] Li N, Gu H Z and Zhao H Z 2012 *Science of Refractories* (Beijing: Metallurgical Industry Press)

[16] Li X C 2003 *Study on Synthesis and Preparation of Corundum/Mullite/Gahnite-based Multiphase Materials* (Wuhan: Wuhan University of Science and Technology)

[17] Li X C 2003 *Journal of the Chinese Ceramic Society* **31** 1171-83

[18] Deng S Q 1991 *Blast Furnace Ironmaking Technology* (Beijing: Metallurgical Industry Press)

[19] Zhang X Z 1988 *Principles of Transfer in Metallurgy* (Beijing: Metallurgical Industry Press)

IOP Conf. Series: Materials Science and Engineering **668** (2019) 012005 doi:10.1088/1757-899X/668/1/012005

# Investigation of temperature and $CO_2$ concentration in converter gas flames by a combustion diagnostic system based on TDLAS

**Y Liu[1], J S Wang[1,2], Q G Xue[1], H B Zuo[1] and X F She[1]**

[1]State Key Laboratory of Advanced Metallurgy, University of Science and Technology Beijing, 30 Xueyuan Road, Haidian District, Beijing, 100083 P. R. China

E-mail: wangjingsong@ustb.edu.cn

**Abstract.** In this study, a novel combustion diagnostic system based on tunable diode laser absorption spectroscopy (TDLAS) was developed to simultaneously measure the temperature and $CO_2$ concentration in flames on a turbulent partly premixed burner. This system simulates harsh industrial combustion environment and enables *in situ* measurements of non-uniform temperature and $CO_2$ concentration distributions in converter gas-air flames. A mixture of CO, $CO_2$, and $N_2$ with the volume percentage ratio 80:15:5 was injected into the fuel stream to simulate effects of converter gas. Temperature and $CO_2$ concentration in flames along the axial direction at the burner centerline were simultaneously measured by online TDLAS. The flame temperature was compared with that obtained by thermocouple measurement. The $CO_2$ concentration was compared with those of samples taken from the flame. The results of online TDLAS measurements showed the same variation trend as temperature in the core of the flames measured using thermocouples and $CO_2$ concentration obtained from the analysis of samples. This work proved that non-uniform temperature and $CO_2$ concentration distributions in converter gas flames could be measured by this system and laid the foundation for further study about the effects of $CO_2$ and/or $N_2$ addition on converter gas/air flames.

## 1. Introduction

Converter gas [1] is generated during the process of oxygen converter steelmaking and contains approximately 80% carbon monoxide (CO), 15% carbon dioxide ($CO_2$) and 5% nitrogen ($N_2$). Converter gas is an important secondary energy resource and normally used as a fuel in hot-blast stoves, steel rolling heating furnaces, and power plants [2]. In order to get a primarily understanding of the combustion behavior of converter gas in harsh industrial environment, it is necessary to measure the fundamental parameters including temperature and species concentration in flames.

Tunable diode laser absorption spectroscopy (TDLAS) is a novel spectroscopic monitoring technique for temperature and species concentration measurement. Zhang *et al* studied two-dimensional temperature and $CO_2$ concentration distributions in methane co-flow diffusion flame [3]. Reconstruction technique based on TDLAS has been widely used for measuring temperature and species concentration distributions [4-7]. Ma *et al in situ* measured the non-uniform temperature, and $H_2O$ and $CO_2$ concentration distributions in a premixed methane–air laminar flame by TDLAS [8]. Sebastian Bürkle *et al* investigated the effects of oxy-fuel atmosphere on the temperature and $H_2O$, $CH_4$, OH and CO concentrations in methane flame [9]. TDLAS is useful for temperature and $CO_2$ concentration measurement in harsh ambient conditions.

IOP Conf. Series: Materials Science and Engineering **668** (2019) 012005 doi:10.1088/1757-899X/668/1/012005

The effects of $CO_2$ and/or $N_2$ addition on multi gas flames have been widely studied. Tu *et al* discussed MILD combustion and the flame characteristics of $CH_4/H_2$ diluted by $CO_2$ [10,11]. The effect of $CO_2$ addition on $CH_4$-air flames was a major subject investigated in combustion and energy [12-18]. Xi *et al* investigated the effects of $CO_2$ addition on temperature and soot formation in ethylene/air flame [19]. The effect of $CO_2$ and/or $N_2$ addition on the structure and temperature of some other hydrocarbon fuel flames has been discussed [20-22].

Based on the outcome of previous studies, a combustion diagnostic system based on TDLAS was developed for simultaneously measuring the temperature and $CO_2$ concentration in converter gas-air flames. The non-uniform temperature and $CO_2$ concentration in the flames along the axial direction at the burner centerline were simultaneously measured by online TDLAS. The online TDLAS was compared to thermocouple measurement and sampling analysis.

## 2. Experimental

### *2.1. Combustion diagnostic system based on TDLAS*
The simplified schematic of the combustion diagnostic system is shown in figure 1.

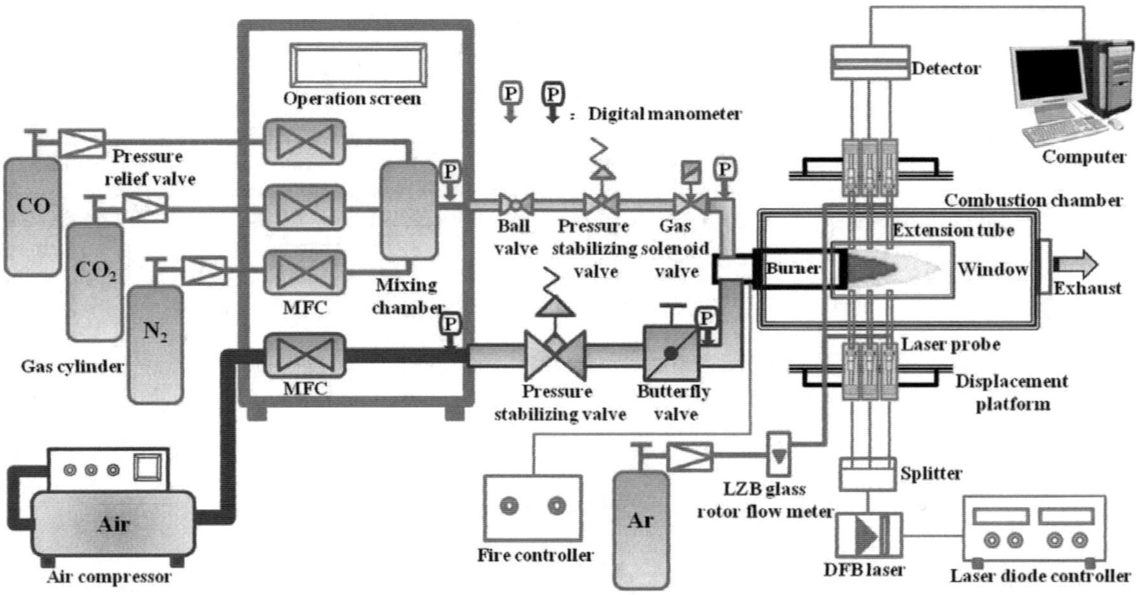


**Figure 1.** Schematic of the combustion diagnostic system.

This system consists of CO, $N_2$, and $CO_2$ gas cylinders, pressure relief valves, an air compressor, mass flow controllers (MFCs), a mixing chamber, digital manometers, an operation screen, a ball valve, pressure stabilizing valves, a gas solenoid valve, digital manometers, a butterfly valve, a partly premixed burner, a fire controller, a combustion chamber, laser diode controllers, DFB lasers, a splitter, laser probes, extension tubes, a detector, a computer, a displacement platform, an Ar gas cylinder, a pressure relief valve, and LZB gas flow meters.

CO, $N_2$, and $CO_2$ gas cylinders and air compressor provide fuel gas and air. The supplied CO, $N_2$, and $CO_2$ gases from Praxair have a purity of 99.9%, 99.999%, and 99.9%, respectively. The gas flow rates are controlled by mass flow controllers. The fuel gas and air are transferred through the ball valve, pressure stabilizing valves, the gas solenoid valve, digital manometers, and the butterfly valve into the partly premixed burner and ignited in the combustion chamber finally. The temperature and $CO_2$ concentration in flames are measured by online TDLAS and compared to thermocouples and sampling.

The laser diode controllers (ILX 3724C) exactly control the current and temperature. Two DFB lasers (NEL KELD1G5BAAA) generate two tunable continuous-waves. The splitter splits the light sources into three laser probes corresponding to three measurement positions. The light is emitted by laser probe and passes through the extension tube and flame. The detector receives the light and converts it into an electrical signal. The signal is transmitted to the data acquisition card in the computer and demodulated by LabView-based software. The extension tube is placed on the laser probe and connected to an Ar gas cylinder, a pressure relief valve and LZB gas flow meters, which is aimed to blow off the air inside the tube. The laser detection device is supported by a displacement platform, which is controlled to move along the axial and radial directions by UartAssist software in the ranges 0-500 mm and 0-400 mm.

This study utilized two-line scanned-wavelength direct absorption spectroscopy to measure the temperature and $CO_2$ concentration in converter gas flames. Two tunable continuous-waves near 1996 and 2004 nm were employed as the light sources. A spectral simulation of the selected absorption lines using the HITRAN database was plotted, as shown in figure 2. The strong absorbance of $CO_2$ and high temperature sensitivity were chosen to guarantee high signal-to-noise ratio (SNR) and minimal interference from ambient water vapor.

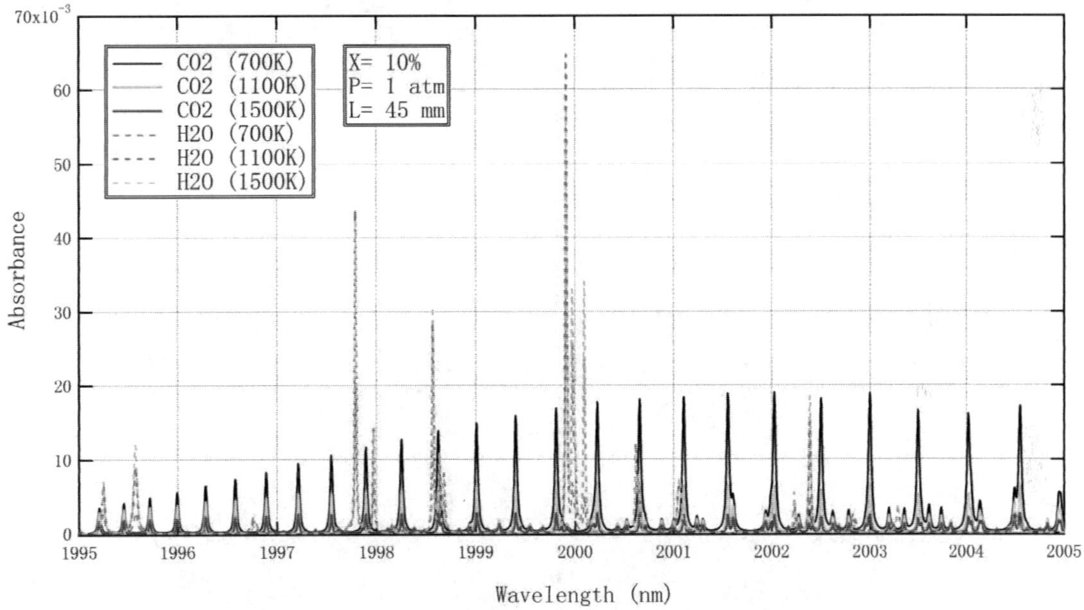


**Figure 2.** Simulation of the $CO_2$ and $H_2O$ absorption lines at different temperatures. The total pressure and path length were set at 1 atm and 45 mm, respectively.

## *2.2. Operating conditions and the analysis of data*

**Table 1.** Combustion conditions of the converter gas.

| | Converter gas pressure | Air pressure | CO volume flow rate | $CO_2$ volume flow rate | $N_2$ volume flow rate | Air volume flow rate |
|---|---|---|---|---|---|---|
| Unit | kPa | kPa | L/min | L/min | L/min | L/min |
| 1 | 0.88 | 0.32 | 33.4 | 6.2 | 2.1 | 302.4 |
| 2 | 0.88 | 0.31 | 33.4 | 6.2 | 2.1 | 302.3 |
| 3 | 0.88 | 0.32 | 33.4 | 6.2 | 2.1 | 302.4 |
| 4 | 0.88 | 0.33 | 33.4 | 6.2 | 2.1 | 302.5 |

**Table 1** lists the measurement conditions of the experiments. CO, $CO_2$, and $N_2$ with a volume percentage of 80:15:5 were injected into the fuel stream. The representative measurements (Plot 0) of the absorption spectra of $CO_2$ along with the corresponding Voigt-fitting profiles (Plot 1) and residuals (Plot 2) of the two lines are presented in Labview software, which is shown in figures 3 and 4.

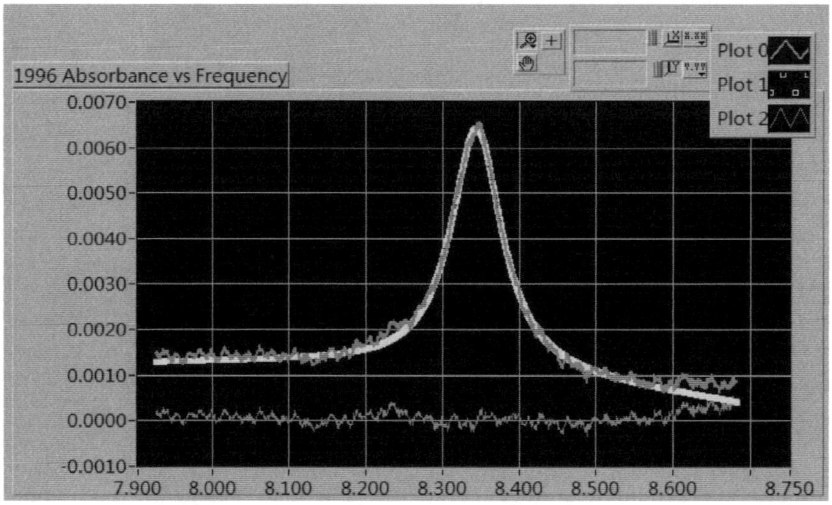


**Figure 3.** Measured absorption spectra of $CO_2$ near 1996 nm. The Voigt fit to the experimental data is plotted for comparison along with the residual.

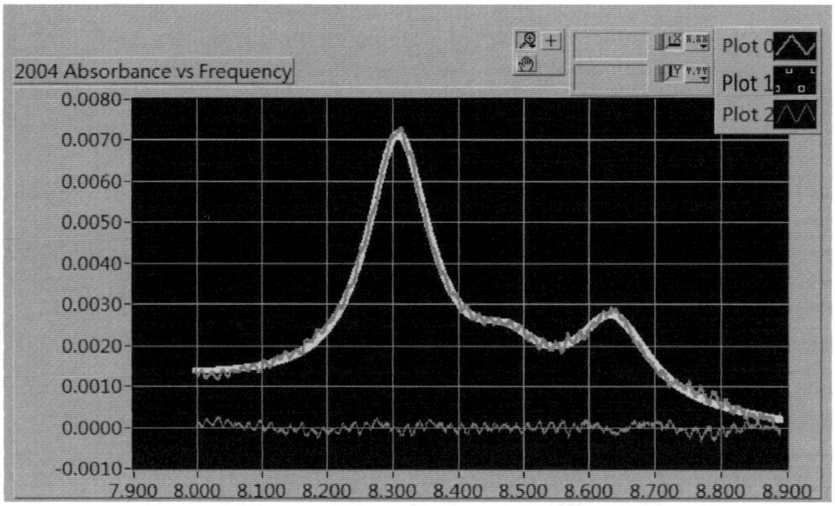


**Figure 4.** Measured absorption spectra of $CO_2$ near 2004 nm. The Voigt fit to the experimental data is plotted for comparison along with the residual.

Nine positions in total were selected for temperature and $CO_2$ concentration measurement. S-type platinum rhodium thermocouples (wrp-100) were used for temperature measurement in the core of the flames. The thermal response time of thermocouple is ~150 s, and the limit of intrinsic error is ±0.25%t. Six of the nine measurement positions were chosen for thermocouple measurement because of the long thermal response time. A black box, a sampling tube, a laser diode controller, a DFB laser, a laser probe, a detector and a computer were utilized for the detection of samples taken from the flames. The process of detection is shown in figure 5. The black box was completely evacuated before each measurement, and the whole process lasted for 180 s. Six of the nine measurement positions were chosen for the sample detection, because the sampling process needed too much time.

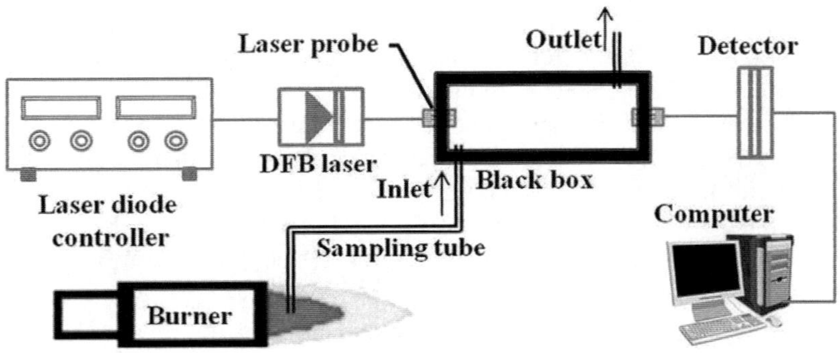


**Figure 5.** The process of sampling detection based on TDLAS.

## 3. Results and discussions

### *3.1. Temperature*

Figure 6 shows the online TDLAS temperature measurements and the temperature measured using thermocouples as a function of distance along the centerline axis, which are processed by software Origin Lab. In the entire flame area (ca. 80 mm along the axial direction), the TDLAS temperature remained constant between 600 K and 1400 K, while the thermocouple temperature remained constant between 800 K and 1500 K. The temperature uncertainty for all the points was ≤40 K. The variation rends of the results measured by the two methods agree well.

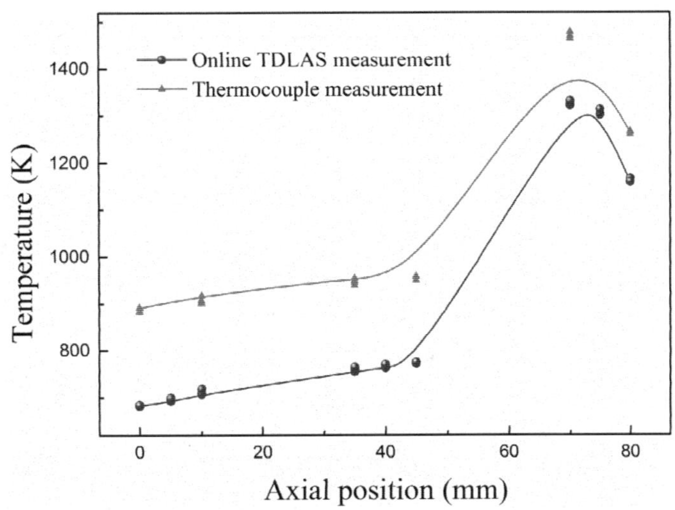


**Figure 6.** The two methods for temperature measurements in flames. Temperature measured by online TDLAS at 9 positions along the axial direction at the burner centerline is compared with thermocouple measurement at 6 positions.

The measurement area was divided into the flame end part (axis 0-10 mm along the center line), the flame center part (axis 35-45 mm along the centerline) and the flame tip part (axis 70-80 mm along the centerline). In the flame end part, the TDLAS temperature ranged 681-719 K, while the thermocouple

IOP Conf. Series: Materials Science and Engineering **668** (2019) 012005 doi:10.1088/1757-899X/668/1/012005

temperature ranged 882-917 K. In addition, the thermocouple temperature values were approximately 200 K higher than those of TDLAS. In the central part of the flame, the temperature measured by TDLAS ranged 753-774 K, while the thermocouple temperature ranged 938-958 K. The temperature measured by thermocouples was approximately 180 K higher than those of TDLAS. As the flame was kept at a distance from the burner exit, both the fuel and air gradually mixed well, leading to a rapid increase in the temperature, and the temperature fluctuations remained initially flat, which then increased with the increasing axial distance. In the flame tip part, the TDLAS temperature ranged from 1158 K to 1331 K, while thermocouple temperature ranged 1259-1478 K. The thermocouple temperatures were approximately 110 K higher than those measured using online TDLAS. Due to a good mixing between the fuel and air, the temperature of the flame rapidly increased to a maximum (1331 K for TDLAS and 1478 K for thermocouple) in the beginning. Then, the temperature decreased and temperature fluctuations increased as the horizontal flame turned upward under the action of updraft in the tip.

### *3.2. $CO_2$ concentrations*

The results of $CO_2$ concentrations along the axial direction at the centerline measured by online TDLAS and sampling are processed by software Origin Lab and shown in figure 7. In the detection area, the distribution of $CO_2$ concentrations measured using the online TDLAS remained constant within 15.6 mol%, whereas the results obtained from analyzing the samples were within 16.0 mol%. The results measured by analyzing the samples were the average concentrations in the detection device after sampling. In this section, mol% was defined as the mole percentage of the measured $CO_2$ concentration passing through the laser lights. The uncertainty in the $CO_2$ concentrations for all the points was ≤0.96 mol%.

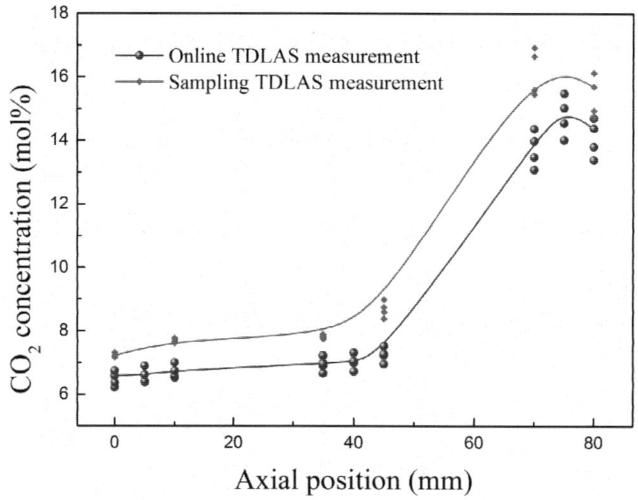


**Figure 7.** The two methods for $CO_2$ concentration measurements in flames. $CO_2$ concentration measured by online TDLAS at 9 positions along the axial direction at the burner centerline is compared with sampling TDLAS measurement at 6 positions.

The $CO_2$ concentrations measured using the online TDLAS ranged from 6.2 mol% to 15.6 mol%, while those obtained from the analysis of samples ranged 7.1-16.0 mol%. The results from the analysis of samples were 0.47 mol% to 0.96 mol% higher than those obtained using the online TDLAS. The $CO_2$ concentrations initially remained low, and then increased with the axial distance to a maximum. Finally, it decreased rapidly with the appearance of upward flame.

## 4. Conclusions

In conclusion, a novel combustion diagnostic system based on tunable diode laser absorption spectroscopy (TDLAS) was developed and successfully built to simultaneously measure the temperature and $CO_2$ concentration in converter gas/air flames on a turbulent partly premixed burner. The temperature and $CO_2$ concentration in converter gas flames along the axial direction at the burner centerline were simultaneously measured by online TDLAS. The flame temperature was compared with that measured using thermocouples. The $CO_2$ concentration was compared with those of samples taken from the flame. The results indicate that this newly developed system achieved *in situ* measurements of non-uniform temperature and $CO_2$ concentration in converter gas flames. The results of online TDLAS measurements showed the same variation trend as temperature in the core of the flames measured using thermocouples and $CO_2$ concentration obtained from the analysis of samples. The updraft in the tip of the horizontal flames affected the temperature and $CO_2$ concentration fluctuations. This study laid the foundation for further studying the effects of $CO_2$ and/or $N_2$ addition on converter gas/air flames.

## References

[1] Wang A H, Jiu-ju X P, Wang D and Zhou Q A 2007 *J. Iron Steel Res. Int.* **14** 22-6
[2] Wang Y, Lei X, Deng L, Wang X, Wang C and Che D 2014 *Thermal Power Generation* **43** 1002-3364(2014) 07-0001-09
[3] Zhang G, Wang G, Yan H, Wang Y and Liu X 2018 *OPTIK* **170** 166-77
[4] Deguchi Y, Kamimoto T and Kiyota Y 2015 *Flow Meas. Instrum.* **46** 312-8
[5] Wang F, Wu Q, Huang Q, Zhang H, Yan J and Cen K 2015 *Opt. Commun.* **346** 53-63
[6] Xia H, Xu Z, Kan R, He Y, Liu J and Zhang G 2015 *Infrared Phys. Techn.* **72** 170-8
[7] Xia H, Kan R, Xu Z, He Y, Liu J, Chen B, Yang C, Yao L, Wei M and Zhang G 2017 *Opt. Laser Eng.* **90** 10-8
[8] Ma L H, Lau L Y and Ren W 2017 *Appl. Phys. B-Lasers O* **123** 83
[9] Bürkle S, Dreizler A, Ebert V and Wagner S 2018 *Fuel* **212** 302-8
[10] Tu Y, Kai S, Hao L, Sheng C and Zheng C 2016 *Energ Fuel* **30** 21390-9
[11] Tu Y, Hao L and Yang W 2017 *Energ Fuel* **31** 33270-3280
[12] Guo H, Galizzi C and Escudié D 2009 *P. Combust. Inst. - Canadian Section Spring Technical Meeting* **12** 1913-9
[13] Min J, Baillot F, Wyzgolik A, Domingues E, Talbaut M, Patte-Rouland B and Galizzi C 2010 *Combust Sci. Technol.* **182** 1782-804
[14] Min J, Baillot F, Guo H, Domingues E, Talbaut M and Patte-Rouland B 2011 *P. Combust. Inst.* **33** 1071-8
[15] Oh J and Noh D 2014 *Fuel* **117** 79-86
[16] Shi B, Hu J and Ishizuka S 2015 *Combust Flame* **162** 420-30
[17] Erete J I, Hughes K J, Ma L, Fairweather M, Pourkashanian M and Williams A 2017 *J. Energy Inst.* **90** 191-200
[18] Zaidaoui H, Boushaki T, Sautet J C, Chauveau C, Sarh B and Gökalp I 2017 *Combust Sci. Technol.* **190** 5784-802
[19] Xi J, Wu Y, Gu Z, An H, Ye Y and Jie C 2017 *Environ. Prog. Sustain* **36** 476-82
[20] Liu F, Karataş A E, Gülder Ö L and Gu M 2015 *Combust Flame* **162** 2231-47
[21] Shi B, Peng W, Li B, Hu J, Wang N and Ishizuka S 2016 *P. Combust Inst.* **161** 1310-25
[22] Shmakov A G, Knyazkov D A, Bolshova T A, Dmitriev A M and Korobeinichev O P 2016 *Energ Fuel* **30** 32395-2406

IOP Conf. Series: Materials Science and Engineering **668** (2019) 012006 doi:10.1088/1757-899X/668/1/012006

# Measuring the dislocation density of VT1-0 titanium alloys with different content of hydrogen by x-ray diffraction method

**S P Xu[1,2], V N Kudiiarov[1], K Li[1] and V V Larionov[1]**

[1]Division for Experimental Physics, School of Nuclear Science & Engineering, National Research Tomsk Polytechnic University, Lenin Avenue 30, Tomsk, 634050, Russia

E-mail: isxusp@gmail.com

**Abstract.** In this study, the distributions of dislocation density with hydrogen concentration in titanium VT1-0 were obtained. The samples with different hydrogen concentrations was carried out by using the Sieverts method. The dislocation densities were obtained by using the full width at height medium calculations from the XRD results of Gaussian approximation fitting. For accurate calculation of the dislocation density in titanium alloys, the double-line separation phenomenon of the XRD results and the variation of the Burgers vector in different lattice directions were considered. The phenomenon of the double-line separation is more evident when the diffraction angle 2θ is larger than 40°. The mean value of dislocation density for the different hydrogen concentrations is about $10^{13} \sim 10^{14}$ m$^{-2}$. Moreover, as the increases of hydrogen content in the titanium alloy, the dislocation density also increases.

## 1. Introduction

Due to their high strength, modulus of elasticity and toughness, titanium and titanium alloys are used in a wide variety of aerospace, energy, and biomedical [1]. Recent advances in titanium alloy lead to understanding that studies of hydrogen-titanium system are important. Moreover, titanium alloys are considered for a variety of applications. In many cases, hydrogen changes the plasticity of most metals and alloys [3]. These properties are highly dependent on material texture, dislocation density, dislocation sliding, etc. Therefore, in order to predict the characteristics of materials and design materials with certain properties, the understanding of the deformation process from the microstructure parameters is crucial. The most well-known methods for measuring the presence of hydrogen and other gases in metals are the methods for measuring micro-hardness, x-ray diffraction (XRD), and photometric analysis [4]. These methods are laboratory methods, their use is difficult for production control and is ineffective for detecting local gas-saturated areas. Plastic deformation is the direct cause of dislocation motion, used the methods of statistical mechanics consistently with three-dimensional systems of curved dislocation lines also developed the mathematical foundations for calculating of the dislocation density [5]. In recent years, the XRD method has been used to calculate the dislocation density in materials [6], and in the method, the Burger vector influences the calculation. There are some studies on the Burger vector of materials different types [7]. The structure of titanium alloys is heterogeneous; the grain of the titanium is 10–15 μm or 50–100 μm in size. In this case, the dislocation density reaches $10^9$ m$^{-2}$. In particular, this refers to the study of the accumulation of energy in materials by creating boundaries. Information on the change in the structure of the grain boundaries leads to a change in the fraction of the resonance scattering of electrons on linear defects forming the boundary [8]. In the previous work, we have already discussed the electrical resistivity $\rho_d$ was calculated versus the density of defects $N_d$ using the model that includes the dilatation of the lattice and the existence of the quasi-stationary resonance electron states near the Fermi energy [9]. In this paper, we studied the dislocation density in the condition of different radiation for different hydrogen content.

## 2. Methods and materials

Titanium VT1-0 was studied in the following wt% composition: 0.18 Fe; 0.1 Si; 0.07 C; 0.12 O; 0.01 H; 0.04 N. For this research, we prepared samples with dimensions of 20 × 20 × 1 mm. The samples were cut from a sheet of titanium alloy VT1-0 by the method of electrical discharge machining (EDM). The surface of the specimens was mechanically polished to remove the oxide film. To remove defects and surface stresses, the samples were annealed in vacuum at 750°C for 60 minutes. To obtain different hydrogen concentration in titanium VT1-0, hydrogen permeation in the gaseous medium was carried out at different temperature (about 600°C) and hydrogen pressure in the chamber (about 1-2 atm) on the automated complex Gas Reaction Controller LPB [10]. After saturation, the hydrogen concentration was measured with the hydrogen analyzer RHEN602 (LECO™). The phase composition and structural parameters of the samples after saturation were determined by an XRD-7000 diffractometer using Cu-Kα radiation. The analysis of the phase composition was carried out using the PCPDFWIN and PDF-4 + databases and the full-profile analysis program (POWDER CELL 2.5). The concentration of hydrogen was changed in depth using a spectrum magnetic analyzer (SMA II, Germany) [11],[12]. Used the Philips SEM 515 scanning electron microscope to research the influences of hydrogenation for the surface state of titanium VT1-0; for the survey, the voltage is 20 keV and magnified 500 times.

## 3. Experimental results and discussion

In recent years, polycrystalline materials with ultrafine structure (grain size respectively less than 1 and 0.1 μm) and increased length of interfaces, have been intensively developed and investigated [13]. Among the methods for producing ultrafine-grained (UFG) metals and alloys, the most promising are the methods of intense plastic deformation (SPD) from the practical point, which make it possible to obtain the UFG structure in bulk metal blanks. The formation of such a structure using the SPD method leads to an increase in the energy of the interfaces, as a result of interaction with lattice defects. This changes the physical and mechanical properties of metals and alloys. And it is known that grains and phases largely determine the mechanical properties, the nature of the deformation behavior and the destruction of metallic materials.

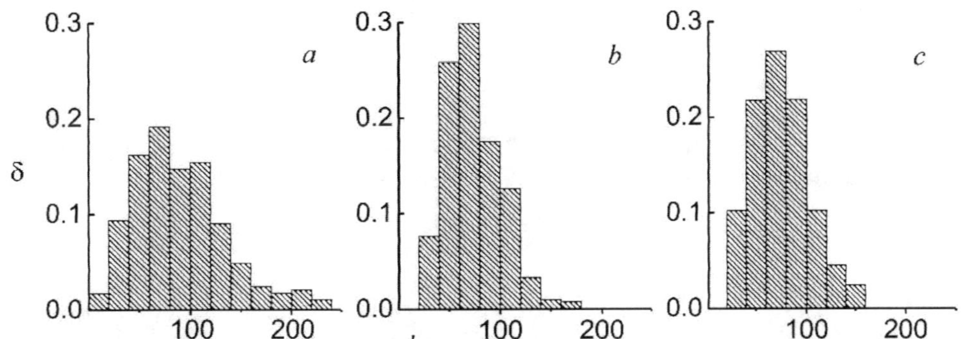


**Figure 1.** The distributions of grain size *a* – initial alloy VT1-0, *b* – $C_H$ =0.052 mass%, *c* – $C_H$ = 0.087 mass%.

By figure 1, we can see the distributions of grain size in titanium VT1-0 with the hydrogen content. It can be clearly seen that after hydrogenation, the larger grain size (50-120 μm) in titanium is significantly increased.

In order to study the dislocation density in materials by XRD, we need to think about the separation of double lines for the $K_{\alpha 1}$ and $K_{\alpha 2}$ x-rays. The separation of double lines is described by the function $\Delta\theta$ (deg)= $\tan\theta\cdot(\Delta\lambda/\lambda)\cdot(180/\pi)$. The Gaussian distribution was applied to the peak. In figure 2 can be observed that the peak splitting occurs, which is associated with strong deformations of the crystal. When the separation was observed at the peak, multiple peaks were attached to the specified curve, and the full width at height medium (FWHM) was determined for each fitting. Furthermore, the averaged values for all these FWHMs were used for the subsequent analysis of the dislocation density.

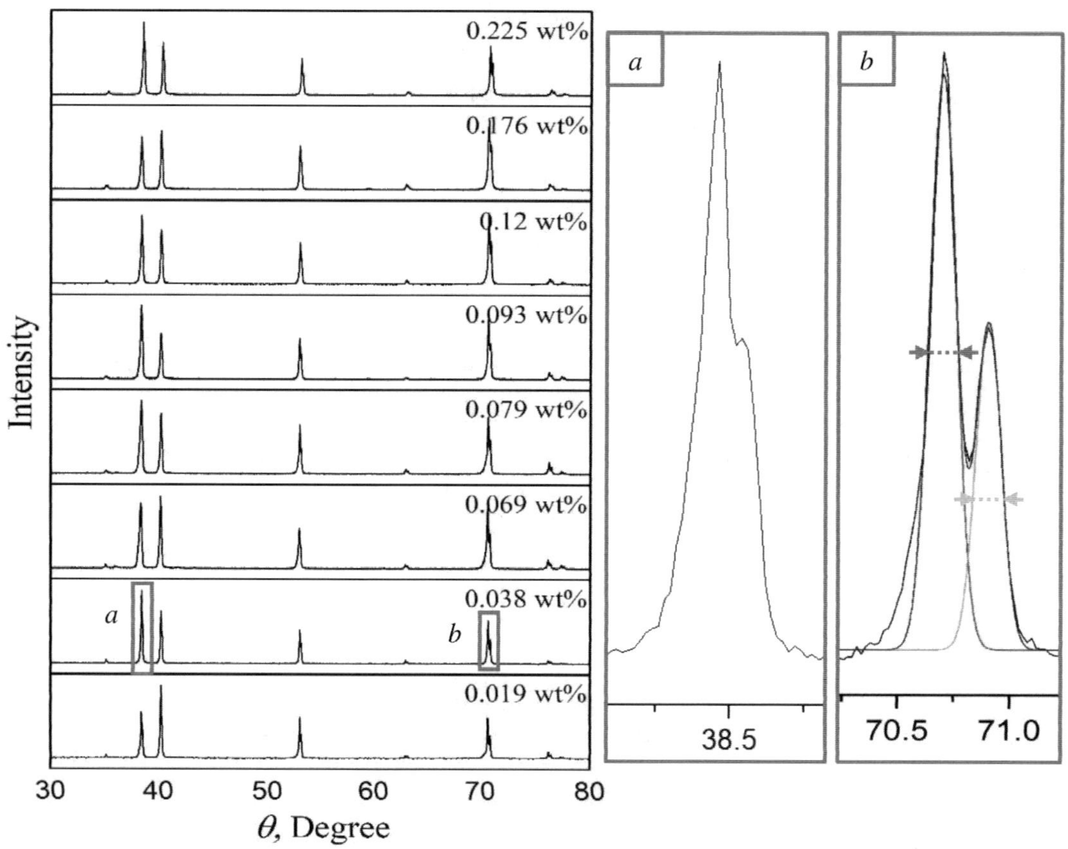


**Figure 2.** Phase transitions in samples of titanium alloy after saturation with hydrogen. And example of double line separation and full width at height medium for titanium with a hydrogen concentration of 0.038 wt%. *a* – for plane (002), *b* – for plane (103).

Dislocations increase the width of the XRD amplitude (FWHM) curve for three reasons: a) introduces the rotation of the lattice site; b) increases the deformation of the lattice; and c) in the strong damaged crystal, can form a walls-like between small crystals, which leads to a decrease in the size of the crystal. X-ray broadening analysis can be used as a method for calculating the dislocation density. In this work, the calculation of the dislocations density $N_d$ was carried out according to the formula [14]:

$$N_d = \frac{\pi(\beta_{adj}^{(hkl)})^2 \cot^2 \theta^{(hkl)}}{16(b^{(hkl)})^2}, m^{-2} \quad (1)$$

where $\beta$ is the broadening of X-ray lines caused by the lattice deformation, $\theta^{(hkl)}$ is the diffraction angle corresponding to the maximum of X-ray lines, $b^{(hkl)}$ is the Burgers vector.

By considering the Burger vector from different angles, discussed the result of dislocation density. The Burger vector is obtained in different directions by the model of pure titanium at a normal temperature. At room temperature ($T$ = 300K), the titanium alloys have the hexagonal close-packed structure, and the distance between the crystal surfaces can be obtained from the following forms:

$$b^{(hkl)} = d^{(hkl)} = \frac{a}{\sqrt{\frac{4}{3}(h^2 + k^2 + hk) + l^2 \frac{a^2}{c^2}}} \quad (2)$$

Taking into account the phenomenon of diffraction peak separation formed by the $K_{\alpha 1}$ and $K_{\alpha 2}$ x-rays in XRD results (figure 2(b)). Theoretically, the diffraction peak formed by irradiation of x-ray is only related to the atomic arrangement inside the crystal. If the double-line separation phenomenon is not considered, the full width at height medium of the diffraction peak will be larger, especially at a high diffraction angle (cf. figures 2(a) and 2(b)). For the same material, the dislocation density obtained from the diffraction peak formed by x-ray irradiation should be consistent. Therefore, in order to obtain an accurate dislocation density, we have averaged the dislocation density calculated from the diffraction peaks formed by the Kα1 and Kα2. The result of the dislocation density is shown in figure 3.

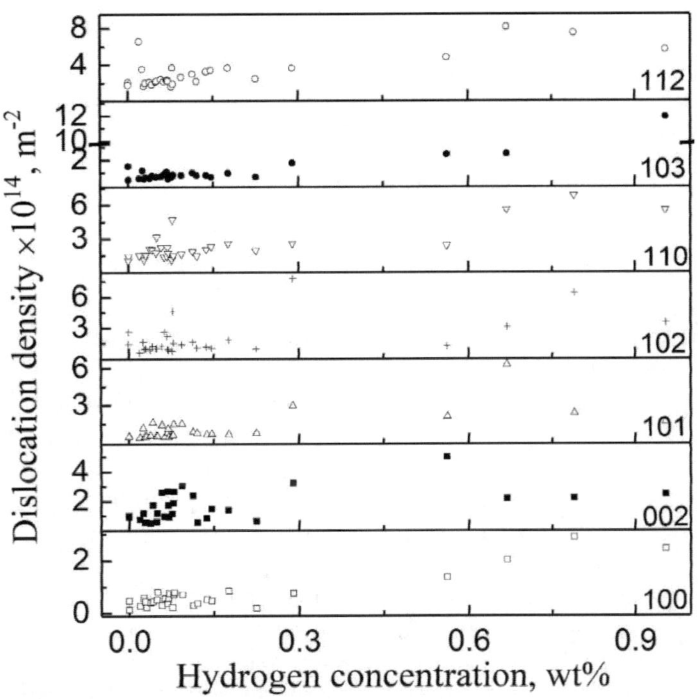


**Figure 3.** The change in the density of dislocations with hydrogen content for different orientations.

In the process of hydrogen sorption in titanium alloys, hydrides are formed. On the other hand, hydrogen in the form of atoms or molecules diffuses and migrates into titanium alloys. This will change the phase composition and the lattice structure of the titanium alloys. From the X-ray diffraction results, a change in the internal structure of the material is reflected as a change in the FWHM parameter. Theoretically, we know that grain size decreasing will broaden the diffraction peak. In addition, dislocations and crystal curvature also create conditions for the expansion of the diffraction peak [16]. Thus, under well-known conditions that affect the expansion of the diffraction peak, we obtain the dislocation densities by using equation (1). From the statistical theory, dislocations appear with a greater probability on the plane, where the slip will occur in the crystal. The titanium alloys have the hexagonal close-packed structure, this closest packing structure causes the dislocation lines to randomly appear on any crystal plane. Therefore, the effect on the broadening of the XRD diffraction peak is also random. Considered this random phenomenon, we used the Gaussian fitting when calculating the dislocation density. Therefore, from any crystal plane, we get the statistical results of the dislocation density in the crystal, which should be the same in theory. From figure 3, we can see that, on the planes (110) (102) (101) obtained the roughly consistent results. For the plane (103), due to the formation of the hydride at high concentrations, a peak of hydride formed on the plane, and broadened the diffraction peak. It leads to the anomalously large of the dislocation density. For the plane (112), due to appearing of the diffraction peaks in other orientations (200, 201) at this angle, it also leads to the calculated dislocation density being too large. For the orientations (100) and (002), because the two-line separation phenomenon is not obvious, the results are small by Gaussian fitting. So we chose the average of the dislocation density from the planes (110) (102) and (101). Table 1 shows us the results.

**Table 1.** The dislocation density for different hydrogen concentration.

| Hydrogen concentration, wt% | Dislocation density$\times 10^{14}$, $m^{-2}$ |
|---|---|
| 0.019 | 0.8537 |
| 0.030 | 1.0048 |
| 0.038 | 1.1783 |
| 0.063 | 1.4940 |
| 0.068 | 1.4795 |
| 0.093 | 1.5127 |
| 0.176 | 1.6855 |
| 0.562 | 1.9001 |
| 0.955 | 3.5647 |

## 4. Conclusion

When the X-rays diffraction peaks separation is considered for different orientations, the dislocation density value is approximately $10^{14}$ $m^{-2}$. The distribution of dislocations in the titanium alloy is not related to the lattice orientation. Theoretically, the number of dislocations in the material is not related to the orientation of the diffraction peak. All the factors in the material where the expansion of the diffraction peaks are involved, the dislocation is included. Therefore, whatever diffraction peak is selected, the result of the dislocation density should be similar. From the experimental results, when the diffraction angle exceeds $2\theta=40°$, the results are more consistent. However, for minor diffraction angles do not correspond to the theoretical result, given that the separation of two lines at small angles is not clear, and the approximation of the peaks at small angles, considering the two lines separation, is inaccurate. Consequently, taking into account the two-line separation, a diffraction peak with $2\theta > 40°$ is selected, and the dislocation density can be better approximated. Thus, can be concluded that when the hydrogen content exceeds 0.5 wt%, the dislocation density changes from $(0.1–3)\times 10^{14}$ $m^{-2}$ to $5\times 10^{14}$ $m^{-2}$, due to the formation of hydrides in the material with increasing hydrogen content.

**Acknowledgments**
This work was granted within the framework of realization of Strategic Programme on National

Research Tomsk Polytechnic University Competitiveness Enhancement in the Group of Top Level World Research and Academic Institutions. Project VIU-OEF-66/2019.

**References**

[1] Picu R C and Majorell A 2002 Mechanical behavior of Ti–6Al–4V at high and moderate temperatures—Part II: constitutive modeling *Mater. Sci. Eng., A* **326** 306-16
[2] Gurrappa I 2003 Characterization of titanium alloy Ti-6Al-4V for chemical, marine and industrial applications *Mater. Charact.* **51** 131-9
[3] Puls M P 2012 *The Effect of Hydrogen and Hydrides on the Integrity of Zirconium Alloy Components: Delayed Hydride Cracking* (Media Springer London Heidelberg New York Dordrecht)
[4] Patterson E A, Major M, Donner W, Durst K, Webber K G and Rödel J 2016 Temperature-dependent deformation and dislocation density in $SrTiO_3$ (001) single crystals *J. Am. Ceram. Soc.* **99** 3411-20
[5] Schulz K, Sudmanns M and Gumbsch P 2017 Dislocation-density based description of the deformation of a composite material *Modell. Simul. Mater. Sci. Eng.* **25** 064003
[6] Ungár T and Borbély A 1996 The effect of dislocation contrast on x-ray line broadening: a new approach to line profile analysis *Appl. Phys. Lett.* **69** 3173-5
[7] Dragomir I C, Li D S, Castello-Branco G A, Garmestani H, Snyder R L, Ribarik G and Ungar T 2005 Evolution of dislocation density and character in hot rolled titanium determined by X-ray diffraction *Mater. Charact.* **55** 66-74
[8] Petrushov S, Ershov V M, Gritsunova K A and Smelsky A 2013 The thin structure of cold-deformed mild steel *Journal of Donbas State Technical University* **39** 141-6 (in Russian)
[9] Larionov V, Xu S and Syrtanov M 2016 Measurements of hydrogenated titanium by electric methods *AIP Conference Proceedings* **1772** 040005
[10] Laptev R S, Kudiiarov V N, Bordulev Y S, Mikhaylov A A and Lider A M 2017 Gas-phase hydrogenation influence on defect behavior in titanium-based hydrogen-storage material *Progress in Natural Science: Materials International* **27** 105-11
[11] Larionov V V, Lider A M and Garanin G V 2015 Eddy current analysis for nuclear power materials *Advanced Materials Research* **1085** 335-9
[12] Larionov V V, Lider A M, Berezneeva E V and Kh M 2012 Krening, features of the electromagnetic control methods of stratified hydrogen content in construction materials *Appl. Phys.(Prikladnaya fizika in Russian)* **5** 20-4
[13] Panin A V, Kazachenok M S, Kozelskaya A I, Balokhonov R R, Romanova V A, Perevalova O B and Pochivalov Y I 2017 The effect of ultrasonic impact treatment on the deformation behavior of commercially pure titanium under uniaxial tension *Mater. Des* **117** 371-81
[14] Umansky I, Skakov Y, Ivanov A and Rastorguev L 1982 Crystallography, X-ray diffraction and electron microscopy *Metallurgy* **632** 351-73 (in Russian)
[15] Grobovetskaya G P, Nikitenkov N N, Michin I P, Duchkin I V, Stepanova E N, Sypchenko V S 2013 Diffusion of hydrogen in submicrocrystalline titanium *News of Tomsk Polytechnic University* **322** 55-9 (in Russian)
[16] Ayers J E 1994 The measurement of threading dislocation densities in semiconductor crystals by X-ray diffraction *J. Cryst. Growth* **135** 71-7

IOP Conf. Series: Materials Science and Engineering **668** (2019) 012007 doi:10.1088/1757-899X/668/1/012007

# Modelling and validation of fluid flow inside a dissipative ladle shroud and a continuous casting tundish

**J Zhang[1,3], Q Liu[1], Z Xin[1], B Lu[2], J Zhang[2] and J Li[2]**

[1]State Key Laboratory of Advanced Metallurgy, University of Science and Technology Beijing, Beijing 100083, China
[2]HBIS Tangshan Iron and Steel Co. LTD, Tangshan, Hebei 063000, China

E-mail: zjsustb@163.com

**Abstract.** Fluid flow plays a significant role in the continuous casting of molten steel. In this paper, two Reynolds Averaged Naiver-Stokes (RANS) models and a Large Eddy Simulation (LES) model were comparatively employed to characterize the fluid flow inside a dissipative ladle shroud and a tundish. LES model was proved to be powerful to characterize the turbulence structure inside the ladle shroud. The effect of meshing density on the computational accuracy was considered using the LES model. The fine-mesh model can capture multiscale vortices inside the ladle shroud; while, the coarse-mesh model disables the LES to obtain the detailed flow information. The experiment of particle image velocimetry (PIV) was used to verify the flow field obtained by the LES model inside the tundish. The PIV and the LES results agree well in terms of flow pattern and velocity vector.

## 1. Introduction

In the continuous casting process, the flow field of molten steel is of significance to the quality of casting blanks [1]. The turbulent flow inside the ladle shroud and tundish is associated with the air absorption, slag entrapment and inclusion removal, which could deteriorate the production. The method of combining physical and mathematical model has been widely used to study the flow field. For instance, the Reynolds Average Naiver-Stokes (RANS) model combined with hydraulic simulation is the most widely used strategy [1,2]. For example, Chattopadhyay *et al* [2] investigate the effect of submergence depth of ladle shroud on the fluid flow inside the tundish using a RANS model. Zhang *et al* [3] studied the flow behaviour inside a five-strand tundish with ladle changeover operation using a RANS and VOF model. RANS method has the advantages of low computation cost, easy convergence, and can obtain the main flow information. However, RANS is unable to give an accurate prediction of turbulence due to the time-averaged turbulence information at all scales of the whole flow field [4]. Direct Numerical Simulation (DNS) is capable of capturing all scale eddies, but requires very fine mesh and extremely high computational cost, which make it unfeasible to be applied in engineering applications. Large Eddy Simulation (LES) sits in between the RANS and DNS models and solve the large-scale eddies directly using the Naiver-Stokes equations with modelled small-scale eddies. A filter equation is used to differentiate the scale of the eddies and the small ones are solved using a subgrid model, which allows an in-depth understanding of the flow structure [5]. However, the LES is still less used to study the fluid flow in the tundish and the fundamental mechanism is still unclear.

This study focuses the modelling and validation of fluid flow inside a dissipative ladle shroud

(DLS) using both mathematical and physical modelling. Two RANS models and a LES model were examined. The numerical results were compared with physical observations. This study could be useful to better understanding the flow structure during the continuous casting process.

## 2. Experimental

### *2.1. Mathematical modelling*

*2.1.1. Assumption.* (1) Water, the fluid, was treated as incompressible continuum and its physical properties are fixed. (2) The flow was assumed to be isothermal in the ladle shroud. (3) The effect of argon injection and slag was neglected.

*2.1.2. Governing equations.* The flow inside the ladle shroud falls into turbulence. In RANS model, the instantaneous velocity and pressure are divided into the mean part and the fluctuating part. For example,

$$U_i = \bar{U}_i + U_i' \tag{1}$$

where $\bar{U}_i$ is the time-average velocity and $U_i'$ is the fluctuating velocity. The mean part is achieved by averaging over a time scale, while the fluctuating part is less considered. Thus, the equations of continuity and Naiver-Stokes are expressed as follows.

$$\frac{\partial \bar{U}_i}{\partial x_i} = 0 \tag{2}$$

$$\frac{\partial \bar{U}_i}{\partial t} + \frac{\partial \bar{U}_i \bar{U}_j}{\partial x_j} = -\frac{1}{\rho}\frac{\partial \bar{P}}{\partial x_i} + v_{\text{eff}} \nabla^2 \bar{U}_i - \frac{\partial (\overline{u_i' u_j'})}{\partial x_j} \tag{3}$$

In equation (3), $\overline{u_i' u_j'}$ indicates the Reynolds stress tensor, which can also be denoted as $R_{ij}$ and solved following the Boussinersq's hypothesis. The symbol of $v_{\text{eff}}$, effective viscosity, is composed of the molecular ($v_0$) and kinetic viscosity ($v_t$). The above equations are not closed to solve the five unknown variables and different turbulence models have been developed to close one or more of the equations.

The $k$-$\varepsilon$ model adopts the concept of Eddy-viscosity and solves the length and velocity scales in equation (3) using two equations. The kinetic viscosity is expressed as:

$$v_t = C_\mu \frac{k^2}{\varepsilon} \tag{4}$$

where $k$ and $\varepsilon$ are the turbulent kinetic energy and the dissipation rate respectively. These two values are solved by solving equations (5) and (6).

$$\frac{\partial k}{\partial t} + \bar{U}_j \frac{\partial \bar{U}_i}{\partial x_j} = v_t \frac{\partial \bar{U}_i}{\partial x_j}\left(\frac{\partial \bar{U}_i}{\partial x_j} + \frac{\partial \bar{U}_j}{\partial x_i}\right) - \varepsilon + \frac{\partial}{\partial x_j}\left[(v + \frac{v_t}{\sigma_k})\frac{\partial k}{\partial x_i}\right] \tag{5}$$

$$\frac{\partial \varepsilon}{\partial t} + \bar{U}_j \frac{\partial \varepsilon}{\partial x_j} = C_{\varepsilon 1}\frac{\varepsilon}{k} v_t \frac{\partial \bar{U}_i}{\partial x_j}\left(\frac{\partial \bar{U}_i}{\partial x_j} + \frac{\partial \bar{U}_j}{\partial x_i}\right) - C_{\varepsilon 2}\frac{\varepsilon^2}{k} + \frac{\partial}{\partial x_i}\left[(v + \frac{v_t}{\sigma_\varepsilon})\frac{\partial \varepsilon}{\partial x_i}\right] \tag{6}$$

Reynolds Stress Model (RSM) solves the transport equations for each of the terms in the Reynolds stress tensor, which shows clear advantage over the $k$-$\varepsilon$ model on flow with high degrees of anisotropy, such as curved flow and rotational flow.

$$\frac{DR_{ij}}{Dt} = P_{ij} + D_{ij} - \varepsilon_{ij} + \Pi_{ij} + \Omega_{ij} \tag{7}$$

Rather than averaging the time-dependent turbulence, LES model employs the concept of filtering, i.e., decomposing the turbulent eddies into large scales and small scales. The large-scale eddies are directly solved using the Naiver-Stokes equation while the contribution of small ones to turbulence is modelled using the so-called subgrid-scale (SGS) model. In FLUENT package, the spatial filtering is enabled based on the mesh size [5]. In the LES model, the Naiver-Stokes equation can be obtained by substituting the Reynolds stress tensor into the SGS stress tensor, which is defined as:

$$\tau_{ij} = \overline{u_i u_j} - \bar{u}_i \bar{u}_j \tag{8}$$

Boussinersq's hypothesis is used to again to model the SGS stress. However, the turbulent viscosity requires to be modelled and the Wall-Adapting Local Eddy-viscosity (WALE) model was used in the current study, which is believed to be reasonable and accurate in flows associated with complicate geometries [6]. The $\nu_t$ is given as [5]:

$$\nu_t = \rho L_s^2 \frac{(S_{ij}^d S_{ij}^d)^{3/2}}{(\bar{S}_{ij}\bar{S}_{ij})^{5/2} + (S_{ij}^d S_{ij}^d)^{5/4}} \tag{9}$$

where $L_s$ is the mixing length for sub-grid scales, defined as:

$$L_s = \min(\kappa d, C_w V^{1/3}) \tag{10}$$

where $d$ is the distance from the cell center to the closest wall; the von Kármán constant $\kappa$ is 0.418. $C_w$ is chosen to be 0.325, which has been confirmed to yield satisfactory results for a wide range of flow in FLUENT [5].

*2.1.3. Computational domain, boundary conditions and computational strategy*. The structures of the DLS and the small tundish are shown in figures 1 and 2 respectively. Their dimensions are listed in table 1. Water was selected as the fluid and its properties are shown in table 1 as well. The DLS comprises three enlarged chambers and a bell-shaped outlet. The effect of mesh density was considered, and the domain was meshed with coarse, fine and very fine grids as shown in figure 1.

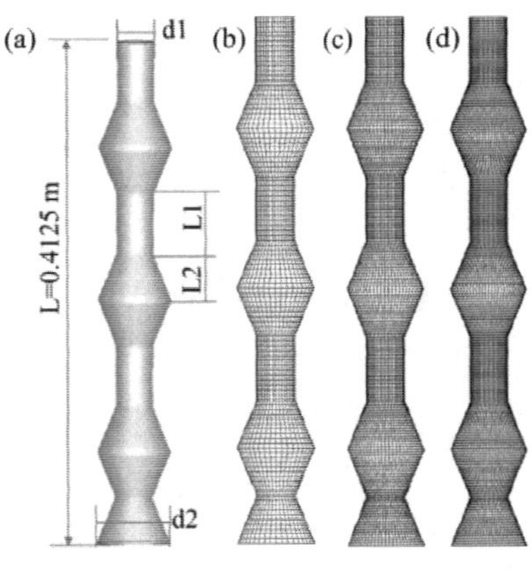


**Figure 1.** (a) The structure of a dissipative ladle shroud; (b) coarse mesh; (c) fine mesh; (d) very fine mesh.

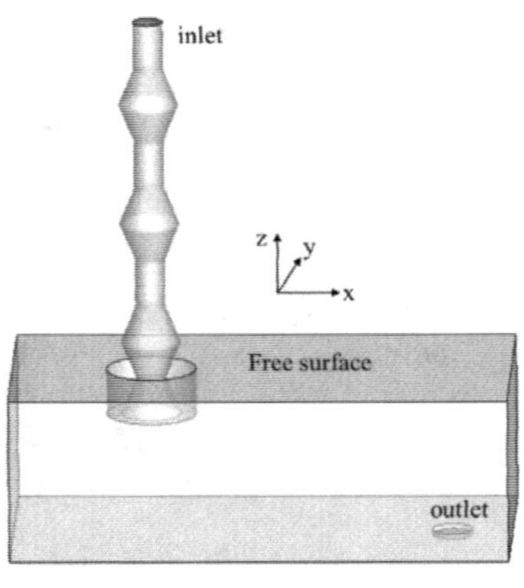


**Figure 2.** The structure of the ladle shroud and the tundish.

**Table 1.** The dimensions of the DLS and tundish, and fluid properties.

| Parameters | Values |
|---|---|

| | |
|---|---|
| Length of the domain, mm | 500.0 |
| Height of the domain, mm | 170.0 |
| Width of the domain, mm | 250.0 |
| Submergence depth of shroud, mm | 36.5 |
| Inlet diameter of shroud d1, mm | 30.0 |
| Outlet diameter of shroud d2, mm | 50.0 |
| Water density, kg/m$^3$ | 1000 |
| Water viscosity, kg/(m·s) | 0.001 |
| Flow volume rate, L/min | 30 |
| Whole Length of the shrouds (L), mm | 412.5 |
| Height of the trumpet section (h), mm | 200.0 |

The DLS is partly immersed into the tundish with a depth of 36.5 mm. The flow rate is 30 L/min. The tundish is a small scaled tundish with simplified structure, as the main purpose of the current study is to investigate the fluid flow inside the ladle shroud.

The boundary conditions of the domain are defined as follows.

- Inlet: a constant inflow velocity (0.708 m/s) profile was applied at the nozzle inlet;
- Outlet: a fixed pressure of 0 Pa (relative to the ambient) was applied at the outlet of the tundish;
- Top surface and walls: free-slip boundaries with zero normal velocity were used at the top surface. All walls of the domain were considered to be no-slip, and the wall boundary domain was handled using the Werner-Wengle formulation [7].

The governing equations were discretized with 2nd-order central differencing scheme for convection terms. Implicit Fractional Step Method was employed to realize the velocity-pressure coupling. The time integration was performed with 2nd order implicit scheme. The time-dependent LES modelling was started by a steady state calculation using the $k$-$\varepsilon$ two-equation model to help the convergence.

### 2.2. Physical modelling

A water model was established to investigate the fluid flow inside the domain with the same flow parameters and validate the mathematical results. A schematic illustration of the experiment set-up is shown in figure 3. A rotor was placed under the ladle shroud to measure the velocity magnitude. The LES results were also compared with measured results using PIV from the literature.

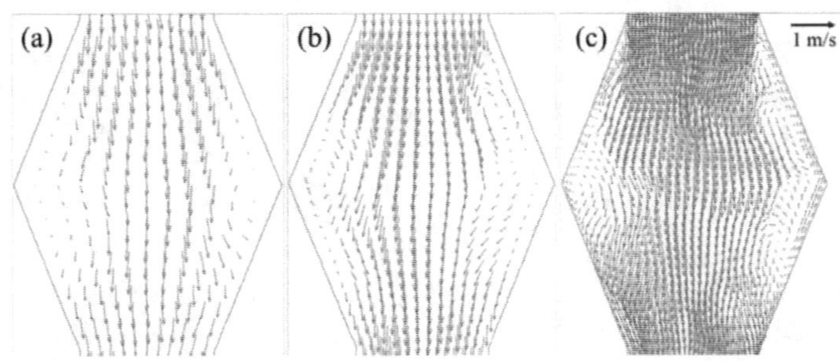


**Figure 3.** The velocity vectors inside the middle chamber of DLS with different mesh densities using LES: (a) coarse, (b) fine and (c) very fine.

## 3. Results and discussion

### 3.1. The effect of mesh density

The spatial mesh density is associated with the filtering operation in the LES model. It is shown in figure 3(a) that small-scale eddy is barely captured due to the large spatial size. The vector is apparently denser in figure 3(c) compared with the results in figure 3(b). However, figure (b) basically shows equivalent ability to show the morphology of the eddies.

Vorticity ($\zeta$) is one of the most important properties of fluid flow. The curl of fluid velocity is the vorticity of flow field, which is defined as:

$$\zeta = \nabla \times \vec{V} = \begin{vmatrix} \vec{i} & \vec{j} & \vec{k} \\ \dfrac{\partial}{\partial x} & \dfrac{\partial}{\partial y} & \dfrac{\partial}{\partial z} \\ u & v & w \end{vmatrix} \tag{1}$$

The unit of vorticity is $s^{-1}$. In the continuous casting process, the internal flow is mostly spiral due to the irregular of geometric structure, the viscous fluid of molten steel and the wall boundary layer. Vorticity is randomly distributed and composed of fluctuating vorticity of different scales. Thus, the vorticity magnitude is an important index to evaluate the fluid flow inside the DLS and the tundish. Figure 4 shows the contours of vorticity magnitude inside the DLS with different mesh densities. From the coarse to fine mesh, LES tends to capture finer turbulence scale, more obvious spin of eddy current, and larger vorticity. The maximum vorticity under the coarse grid is 150 $s^{-1}$, which can capture large vortexes with poor accuracy. The maximum vorticity under fine and very fine grids is 480 and 700 $s^{-1}$, respectively. Compared with these three vorticity contours, the fine grids can basically capture all turbulent structures.

In addition, the time consumptions to proceed 1.0 s calculation are 7 min, 18 min and 30 min for the coarse, fine and very fine meshes respectively. Thus, the fine mesh is selected to carry out the LES and RANS simulation considering a relatively acceptable computational cost.

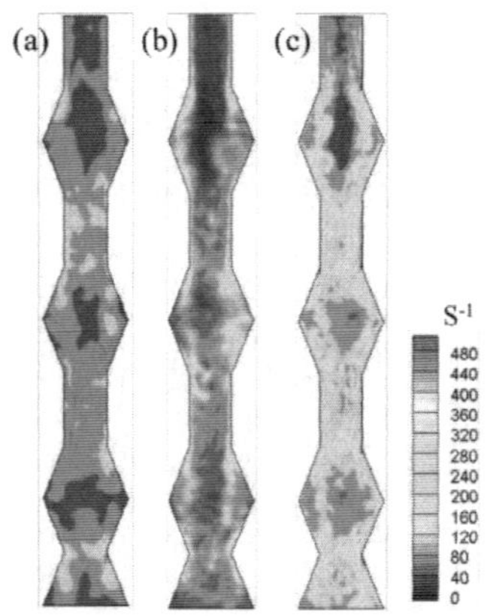


**Figure 4.** The contour of vorticity magnitude inside the DLS with different mesh density: (a) coarse, (b) fine and (c) very fine.

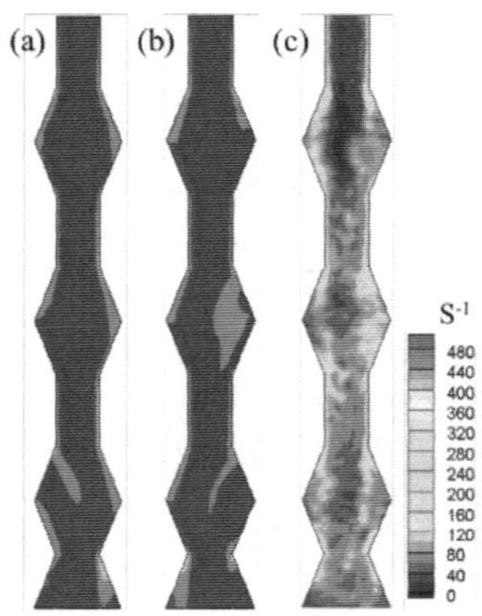


**Figure 5.** The contour of vorticity magnitude inside the DLS using different models: (a) $k$-$\varepsilon$, (b) RSM and (c) LES.

### *3.2. The results of different turbulent models*

Figure 5 shows the contour of vorticity magnitude inside the DLS using different models in the dissipative long nozzle calculated by SKE, RMS and LES models. It is noticed that the two RANS models can only capture a small number of vortexes, and the maximum vorticity value is less than 130 $s^{-1}$. This is mainly determined by the calculation method of Reynolds average, which results in the averaged vortexes with large vortexes and small vortexes. The LES results presented in figure 5(c) show that the model can capture vortices of different scales, and the maximum vorticity value captured is close to 600 $s^{-1}$.

Figure 6 shows two typical velocity vectors selected in the middle chamber. It can be seen that the instantaneous flow characteristics of RANS barely change at different times, and cannot show the random flow and eddy. These largely limit the RANS model to study the detailed flow structure inside the DLS with satisfactory accuracy. However, the computational efficiency of these two models is much higher than that of LES, and the convergence is relatively easy. Besides, the overall flow trend in the nozzle can be obtained.

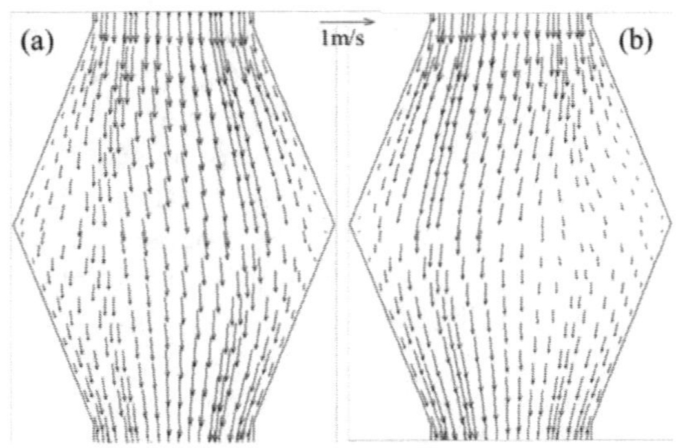


**Figure 6.** The velocity vectors obtained inside the middle chamber using (a) $k$-$\varepsilon$ model and (b) RSM.

These results confirm the advantages of LES in transient simulation. Compared with RANS model, LES is able to capture turbulent information of fluid more comprehensively, which is conducive to in-depth discussion of rotational motion in the domain. From the viewpoint of fluidynamics, the fluid flow in the DLS is a kind of non-equilibrium complex turbulence and LES has good adaptability to this kind of flow.

### *3.3. The fluid flow inside the tundish*

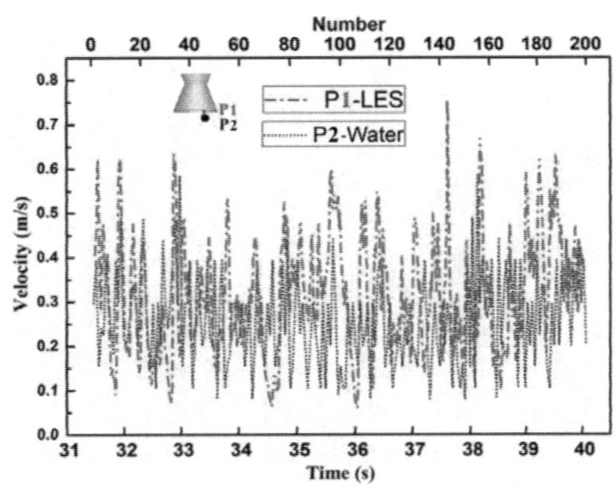


**Figure 7.** Comparison of velocity magnitude under ctors obtained inside the middle chamber using (a) $k$-$\varepsilon$ model and (b) RSM.

To validate the numerical results, the velocity magnitude is compared at two points under the DLS (P1 and P2), as shown in figure 7. P2 is 10 mm below the point P1 to facilitate the measuring. The rotor measured the distribution of velocity at the outlet of 200 groups, and the results show that the fluctuation range of velocity is 0.081-0.585 m/s; its average velocity at P2 is 0.269 m/s. LES findings show the fluctuation of P1 within 10 s. The velocity fluctuation range of LES results is large, ranging from 0.059 to 0.754 m/s, and its average velocity is 0.314 m/s. The average velocity measured at P2 is 14.3% lower than that at P1. This difference is mainly due to the fact that the measurement area and time step are much larger for the rotor than those of the LES. The rotor velocity measurement is 7 homogenized in a small area, so the measured value is relatively small as a whole. However, both of the measured results reflect that the velocity fluctuation at the outlet of the DLS is severe.

The LES model has also been validated by comparing simulation results with PIV-measured results of Ken Morales-Higa's work [8]. 2D PIV can capture the overall flow of fluid in a certain surface or region. It can be seen from the figure 8 that the two flow patterns are quite similar. After flowing out of the DLS outlet, the flow stream first shifts to the right-hand side, and then divides into two sub-streams near the bottom of the tundish, which flow to the left and right sides respectively. The left branch moves to the side wall of the tundish and goes up, creating a large circulation flow. The sub-stream on the right of the tundish quickly enter the whole pool and mix with the fluid in the tundish. In this flow mode, the velocity at the pouring region is within the range of 0.3 to 0.5 m/s, and it dissipates further after entering the whole tundish. Therefore, the comparison with the previous experimental results shows that the numerical simulation model basically matches well with the physical results. However, comparing the numerical and physical results, the location and size of the vortex cannot match completely, which is mainly caused by the complex and unpredictable nature of turbulence.

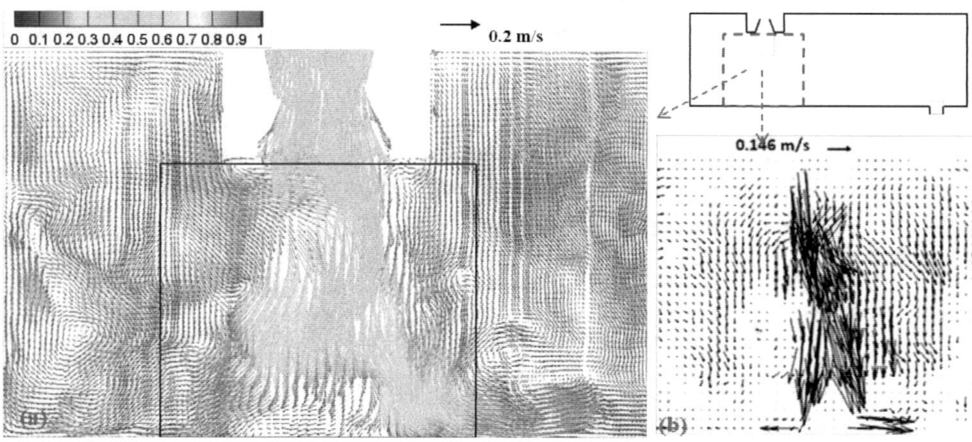


**Figure 8.** Comparison of the fluid flow in the pouring zone using (a) LES, and the measured results using PIV from [8].

## 4. Conclusions

Both numerical and physical simulations were conducted to investigate the fluid flow inside a DLS and a tundish. It is found that the LES simulation is sensitive the mesh density and a fine mesh is selected by comparing the vorticity magnitude and computational cost. The LES results successfully capture multiscale vortices inside the ladle shroud, while the two RANS models fail to capture the vortex due to the time-averaged operation in the simulation. The numerical and physical results match well in terms of flow pattern and velocity distribution; however, the velocity magnitudes are slightly different due to the complexity of turbulence. The LES model can be used to study the flow structure inside the ladle shroud and the tundish.

## Acknowledgments

The funding of Independent Subject of the State Key Laboratory of Advanced Metallurgy (No. 41619001) is highly acknowledged.

## References

[1] Chattopadhyay K, Isac M and Guthrie R 2010 *ISIJ Int.* **50** 331-48
[2] Chattopadhyay K, Isac M and Guthrie R 2011 *Ironmak. Steelmak.* **38** 398-400
[3] Zhang H, Fang Q, Luo R, Liu C, Wang Y and Ni H 2019 *Metall. Mater. Trans. B* **50** (online)
[4] Zhang J, Yang S, Li J, Yang W, Wang Y and Guo X 2015 *ISIJ Int.* **55** 1684-92
[5] FLUENT6.3-Manual ANSYS Inc Lebanon NH 2007
[6] Nicoud F and Ducros F 1999 *Flow Turbul. Combust.* **62** 183-200
[7] Piomelli U and Balaras E 2002 *Annu. Rev. Fluid Mech.* **34** 349-74
[8] Morales-Higa K Guthrie R Isac M and Morales R 2013 *Metall. Mater. Trans. B* **44** 63-79

IOP Conf. Series: Materials Science and Engineering **668** (2019) 012008 doi:10.1088/1757-899X/668/1/012008

# Study on the process of TC4 powders prepared by electrode induction melting gas atomization for laser 3D printing

**B Xie[1,2,3], Y Z Fan[1,2] and Q D Miao[1,2]**

[1]State Key Laboratory of Vanadium and Titanium Resources Comprehensive Utilization, Panzhihua 617000, Sichuan, China
[2]Pangang Group Research Institute Co., Ltd. Panzhihua 617000, Sichuan, China

E-mail: yunzangsuiyue@163.com

**Abstract.** This study employed TC4 rod as raw material to fabricate TC4 powders for laser 3D printing via electrode induction melting gas atomization (EIGA). The morphologies, phase compositions, particle size distributions, apparent densities and flowabilities of the powders were characterized by scanning electron microscopy (SEM), X-ray diffraction (XRD), laser particle size analyzer (LPS) and Hall flowmeter, respectively. Moreover, the effects of gas atomization pressure and melting temperature on the yield of TC4 powders for laser 3D printing were studied. The results showed that TC4 powders morphology was nearly regular spherical. The particle size of TC4 powders showed a single peak normal distribution, mainly distributed in the range of 15−180 μm. The powder was α'−Ti of a single phase solid solution. The optimum parameters were gas atomization pressure of 5MPa, melting temperature of 1750°C. Under the optimized condition, the average particle size D50 was 60.2 μm, the yield of printable TC4 powders was 35.6%, the flowability was 41.2 s/50g, the apparent density was 2.76 g/cm$^3$ and oxygen content was 800 ppm, which was in line with the ASTM test standard and was conformed to the requirement for laser 3D printing.

## 1. Introduction

3D printing technology is a kind of rapid prototyping technology. It has become research hotspot all over the world, because of high precision, high efficiency, materials saving and being suitable for manufacturing complex parts [1,2]. For metal laser 3D printing technology, the morphology, oxygen content, fluidity, apparent density and particle size of the powders directly determine the performance of the final product [3].

Ti-6Al-4V has drown increasing attention for 3D printing applications in aerospace, automobile and biomedical fields due to low density, high specific strength, good corrosion resistance, high heat resistance and good biocompatibility. Such as, engine fans, diffuser, cockpit window frames, aircraft structural beams, bone and joint, dental restoration materials, etc. [4-6]. TC4 powders produced in China has the defects of low sphericity, high oxygen and large particle size, the high-quality spherical TC4 powders still need to be imported, which seriously restricts the development of local 3D printing industry. Therefore, mastering the preparation technology of high-performance spherical TC4 powders is the key to the autonomy of 3D printing powders in China.

The preparation methods of spherical TC4 powders for laser 3D printing mainly include electrode induction melting gas atomization (EIGA) [7], induction plasma spheroidization (PS) [8], plasma rotating electrode process (PREP) [9] and plasma atomization (PA) [10]. Nevertheless, the particle size and composition of PS powders are affected by the original powders, and the powders have high oxygen content, which are not suitable for 3D printing in aerospace and biomedical fields [11]. The powders prepared by PREP have excellent sphericity and high purity. However, due to the limition of rotating speed, it is difficult to prepare the powders between 15 and 45 um [12]. The powders prepared by the PA have good spherical shape and good particle size, but the domestic technology has not yet been refined. The powders prepared by EIGA have the advantages of high purity, high sphericity, narrow particle size distribution, fast cooling rate and small environmental pollution [7]. Therefore, EIGA has become a major method for preparing TC4 powders. The particle size distribution primarily depends on the argon pressure, over-heat temperature and nozzle structure [13].

Laser 3D printing requires powder with a particle size of 15-45 μm which is defined as the yield of printable TC4 powders in the paper. At present, there is a problem that the yield of printable TC4 powders is low by EIGA, which results in a high price of printable TC4 powders. Therefore, the study focused on improving the yield of printable TC4 powders by studying argon pressure and melting temperature in the process of EIGA, in order to provide reference for the industrial production of laser 3D printing TC4 powders.

## 2. Experiment methods

The raw material in this study was TC4 rod with diameters of 45mm produced by Baoji Hi-Tech Zone Xinglong Titanium Industry Co., Ltd. The chemical composition of Ti-6Al-4V rod is shown in table 1. The end of the titanium rod was machined to an angle of 45°. The surface was cleaned with alcohol to remove grease. The atomizing gas was high purity argon (purity≥99.999%) supplied by Sichuan Pangang Messer Gas Products Co., Ltd.

**Table 1.** Chemical composition of Ti-6Al-4V rod (wt.%).

| C | H | O | N | Fe | Al | V | Ti |
|---|---|---|---|---|---|---|---|
| 0.041 | 0.011 | 0.048 | 0.003 | 0.026 | 6.21 | 3.87 | bal |

During the experiment, the TC4 rod rotated at low speed into the water-cooled copper induction coil for heating. The molten metal droplets were continuously dripped and atomized by the high-speed argon gas sprayed from the nozzle. Then, spray droplets were spheroidized and solidified into spherical particles. At last, TC4 powders were collected in the powder storage tank through the cyclone. Schematic of the atomizing processes was illustrated in figure 1.

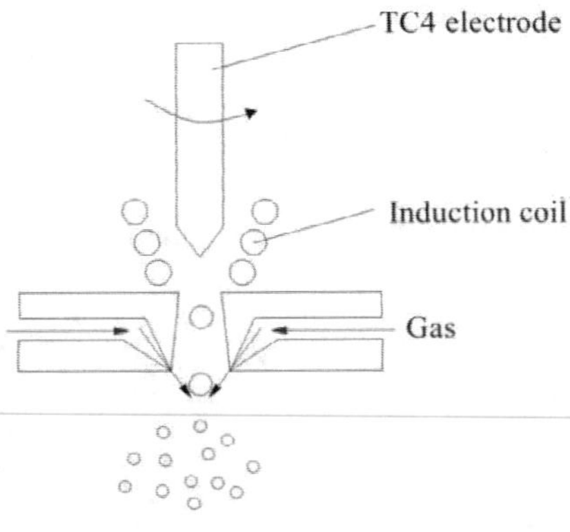


**Figure 1.** Schematic of TC4 powders prepared by EIGA.

The morphology and microstructure of TC4 powders were investigated using scanning electron microscopy (SEM, MLA650F, FEI, USA). The particle size distributions of TC4 powders were observed using a laser micron sizer (LMS, ZS90, Malver, UK). The phase constituents of TC4 powders were analyzed by X-ray powder diffraction (XRD, Empyrean, PANalytical B.V., The

Netherlands). The oxygen contents in the powders were recorded by using an oxygen and nitrogen analyzer (TC600, LECO, USA). The concentrations of metallic elements were determined by using an inductively coupled plasma atomic emission spectrometer (ICAP6300, Thermo, USA). The flowability and apparent density of TC4 powders were measured by using a Hall flowmeter(HY-102, Dandong Haoyu Technology Co., Ltd., China). All measurements were taken three times, and the average results were reported.

## 3. Results and discussion

### *3.1. Effect of atomizing gas pressure on yield of printable TC4 powders*

The EIGA method used argon gas to break molten metal stream into droplets and solidify into powders. The breakup process was mainly divided into three stages [14]: (1) The metal stream was torn into thin sheet by high-speed argon gas and left the atomizing gas center. (2)Thin sheets were torn into thin rod droplets under the action of the airflow field. (3) The thin rod droplets were torn into a large number of fine droplets due to surface tension and airflow field.

During the cooling process, the morphology of the powder was related to the spheroidization time and solidification time of the droplet. When the spheroidization time was less than the solidification time, the droplet had sufficient time to be spheroidized before being solidified, and finally the powder was spherical. when the spheroidization time was longer than the solidification time, and there was not enough time for the droplet to be spheroidized before being solidified. At last, irregular shape powder was obtained.

The atomizing gas pressure determines the gas flow rate Q and the flow velocity v. Generally, the larger the kinetic energy of the atomizing gas is, the better the breakup effect is. According to the principle of fluid dynamics, the kinetic energy of the atomizing gas ban be calculated using equation (1) [15].

$$\varepsilon=\frac{1}{2}mv_g^{\ 2}=\frac{1}{2}\frac{\rho_g dV}{dt}\left(\frac{Q}{S}\right)^2=\frac{1}{2}\rho_g\times Q\times\left(\frac{4Q}{\pi d^2}\right)^2=\frac{8\rho_g Q^3}{\pi^2 d^4} \tag{1}$$

Where m is the mass flow rate of the gas, $v_g$ is the speed at which the gas stream contacts the solution, $Q$ is the gas flow rate, $\rho_g$ is the density of the gas, d is the internal diameter of tuyere. From the equation (1), it was clearly that with a fixed gas nozzle design and metal composition, the particle size was dominated by atomizing gas pressure. In other words, a higher atomizing gas pressure could result in a higher yield of printable TC4 powders.

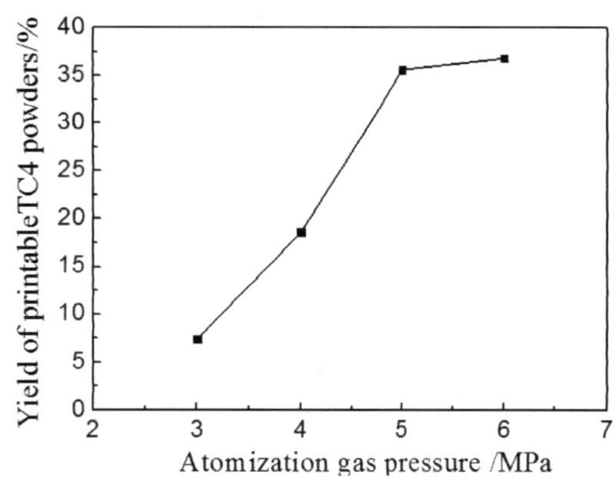


**Figure 2.** Yield of printable TC4 powders under different atomization gas pressure.

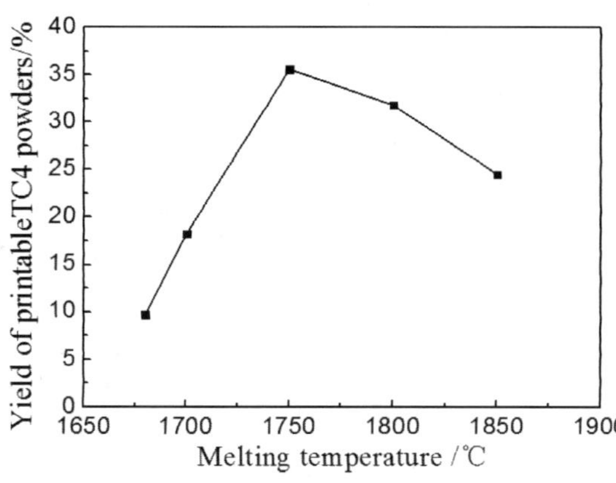


**Figure 3.** Yield of printable TC4 powders under different melting temperature.

It can be seen from figure 2 that the pressure of the atomizing gas increased, the yield of printable

TC4 powders increased. When the pressure of the atomizing gas was 5 MPa and 6 MPa, the yield of printable TC4 powders was 35.6% and 36.8% respectively. When the pressure of the atomizing gas exceeded 5 MPa, the atomizing gas flow rate was greater than the design speed of the atomizing nozzle, the yield of printable TC4 powders didn't increase significantly.

### *3.2. Effects of melting temperature on yield of printable TC4 powders*

The relationship between the surface tension of metal melt and the temperature can be expressed by the Ramsay-Sheilds formula [16].

$$\sigma = K(T_C - T - 6)V_m^{-2/3} \tag{2}$$

Where $\sigma$ is the surface tension of metal melt, K is the empirical constant, Tc is the critical temperature when the surface tension is zero, $T$ is the actual temperature of the alloy liquid, and $V_m$ is the molar volume of the alloy liquid. The relationship between viscosity of metal melt and temperature is explained in equation (3) [17].

$$\eta = Ae^{E/(kT)} \tag{3}$$

Where $\eta$ is the viscosity of metal melt, A is a constant, E is the activation energy, k is the Boltzmann constant, and $T$ is the actual temperature of the alloy liquid.

It could be seen from equations (2) and (3) that as melting temperature increased, the surface tension and viscosity of the alloy liquid decreased. Therefore, molten metal stream was easy to break into small droplets in the breakup stage. The yield of printable powders increased with increasing melting temperature up to 35.6%. After that, the yield began to reduce gradually as shown in figure 3. This was attributed to the powders adhered to each other and agglomerated during the slow solidification causing by higher temperature. The powders could be easily oxidized under high temperature conditions. When the temperature was below 1750°C, the solidification time of the particles was less than spheroidization time, and the particles were solidified before being spheroidized, the TC4 powders were irregular in shape as shown in figure 4(b).

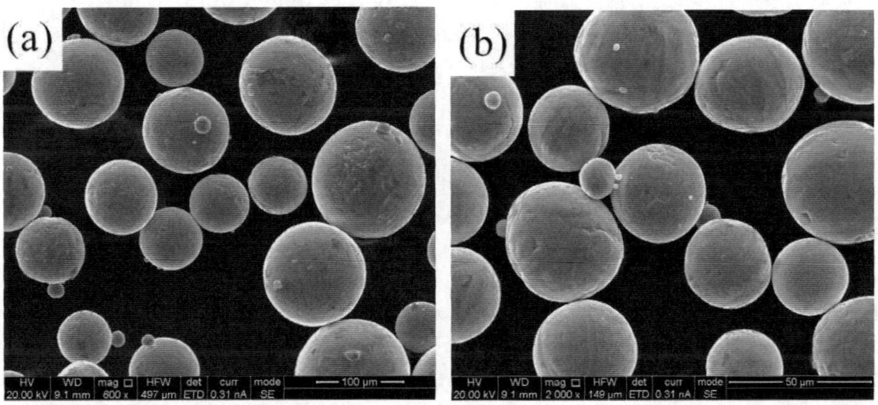


**Figure 4.** SEM morphology of TC4 powders under different melting temperature. (a) 1750°C and (b) 1700°C.

### *3.3. The optimum process parameters*

Considering the yield of printable TC4 powders and cost of argon, the optimum process parameters were found to be melting temperature of 1750°C and atomizing gas pressure of 5 MPa. The yield of printable TC4 powders was 35.6%.

Figure 4(a) shows the micrograph of TC4 powders. It was apparent that most of the TC4 powders had good sphericity and smooth surface, indicating that the TC4 droplet had been fully spheroidized before solidification. Some small satellite balls adhered to the surface of the spherical particles. This was mainly attributed to the much higher cooling rate for small particles as compared to the large particles. The small particles which first solidified adhered to the surface of the unsolidified large particles by the air current to form satellite balls. The satellite particles had a negative influence on the free-flowing of the powders, which were thus not desired for 3D printing.

Figure 5 shows the XRD diffraction pattern of the TC4 powders. According to the standard titanium card, the TC4 alloy powders prepared by the EIGA method was a close-packed hexagonal phase. After being crushed by high-speed argon gas, the TC4 droplets were spheroidized and solidified. During rapid cooling, the β-phase of the body-centered cubic was transformed into a close-packed cell through a non-diffusion phase transition process. The alpha phase of the hexagonal structure finally produced a metastable state HCP-α' phase.

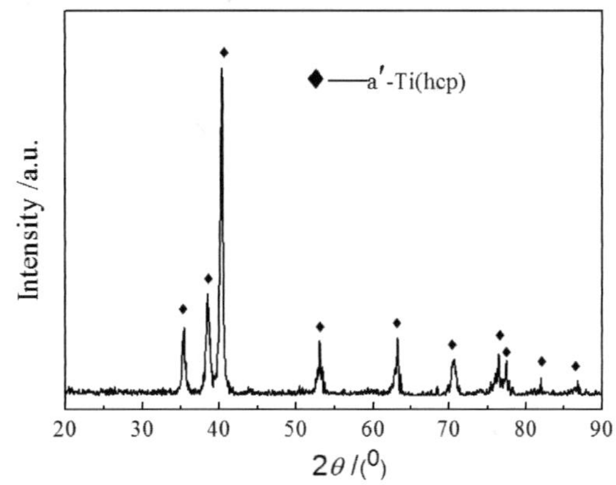


**Figure 5.** XRD pattern of TC4 powders.

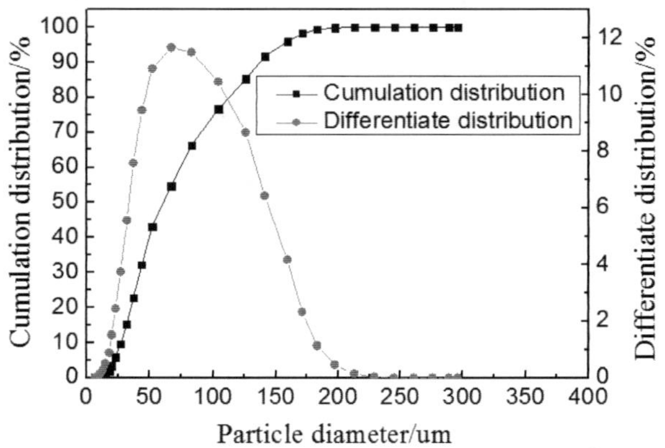


**Figure 6.** The particle size distributions of TC4 powders.

It can be seen from figure 6 that TC4 powders mainly distributed in the range of 15−180 μm. The particle size differential distribution curve of TC4 powders was a single peak and was approximately normal distribution, the average particle size D50 was 60.2 μm.

**Table 2.** Properties of TC4 powders prepared by EIGA.

| Particle size | Oxygen content | Nitrogen content | Apparent density | Flowability |
|---|---|---|---|---|
| 15～45 μm | 800 ppm | 70 ppm | 2.76 g/cm$^3$ | 41.2 s/50g |

Properties of TC4 powders were shown in table 2, the contents of oxygen and nitrogen were 800 ppm and 70 ppm respectively. The combination of high sphericity and non-satellite particles contributed to improve its flowability. The flowability was 41.2 s/50g, the apparent density was 2.76 g/cm$^3$, which was in line with the ASTM test standard and was conformed to the requirement for laser 3D printing.

## 4. Conclusion

- As the pressure of the atomizing gas increased, the yield of the printable TC4 powders increased. The yield of printable TC4 powders first increased and then decreased with increasing melting temperature. The optimum parameters were gas atomization pressure of 5 MPa, melting temperature of 1750°C.
- Under the optimized condition, TC4 powders morphology was nearly regular spherical, which surface adhered a small amount of satellite balls. The powder was α′−Ti of a single phase solid solution, the average particle size D50 was 60.2 μm, the yield of printable TC4 powders was

35.6%.

- The flowability was 41.2 s/50g, the apparent density was 2.76 g/cm$^3$ and oxygen content was 800 ppm, which was in line with the ASTM test standard and was conformed to the requirement for laser 3D printing.

**Acknowledgments**
This work was financially supported by Key Research and Development Project of Sichuan Province (2018GZ0145).

**References**

[1] Herzog D, Seyda V, Wycisk E and Emmelmann C 2016 *Acta. Mater.* **117** 371-92
[2] Clayton J, Millington-Smith D and Armstrong B 2015 *JOM.* **67** 544-8
[3] Gu H, Gong H, Dilip J, Pal D, Hicks H, Doak H and Stucker B 2018 *Int. J. Powder. Metall.* **51** 35-42
[4] Harun W, Manam N, Kamariah M, Sharif S, Zulkifly A, Ahmad I, Miura H and Kamariah M 2018 *Powder. Technol.* **331** 74-97
[5] Cao F, Zhang T, Ryder M and Lados D 2018 *JOM.* **70** 1-9
[6] Seyda V, Herzog D and Emmelmann C 2017 *J. Laser. Appl.* **29** 022311-1-8
[7] Chen G, Tan P, Zhao S, He W and Tang H 2016 *Key. Eng. Mat.* **704** 287-92
[8] Wei W, Wang L, Chen T, Duan X and Li W 2017 *Adv. Powder. Technol.* **28** 2431-7
[9] Yin J, Chen G, Zhao S, Ge Y, Li Z, Yang P, Han W, Wang J, Tang H and Cao P 2017 *J. Alloy. Compd.* **731** 222-8
[10] Chen G, Zhao S, Tan P, Wang J, Xiang C and Tang H 2018 *Powder. Technol.* **333** 38-46
[11] Sun P, Fang Z, Xia Y, Zhang Y and Zhou C 2016 *Powder. Technol.* **301** 331-5
[12] Sun P, Fang Z, Zhang Y and Xia Y 2017 *JOM.* **69** 1853-60
[13] Wei M, Chen S, Liang J and Liu C 2017 *Vacuum.* **143** 185-94
[14] Jalaal M and Mehravaran K 2012 *Int. J. Multiphas. Flow.* **47** 115-32
[15] Su C, Chou J, Liu S and Chiang C 2010 *Mater. Trans.* **51** 1594-601
[16] Cherevko A 2009 *High. Temp.* **47** 920-3
[17] Koike T 1993 *J. Appl. Polym. Sci.* **47** 387-94

IOP Conf. Series: Materials Science and Engineering **668** (2019) 012009 doi:10.1088/1757-899X/668/1/012009

# Research of unpolluted deoxidization technology with applied electric field

**J X Jia[1,2,3], Q T Guo[1,2], D G Li[1,2], X W Liao[1,2] and C L Peng[1,2]**

[1]Ansteel Group Iron and Steel Research Institute, Wuyi Road, Anshan, Liaoning, China
[2]State Key Laboratory of Metal for Marine Equipment and Application, Wuyi Road, Anshan, Liaoning, China

E-mail: jjxman@sina.com

**Abstract.** A commercial experiment system of unpolluted deoxidization with applied electric field is designed and fabricated based on laboratory and pilot research. A simulation model for the electrochemical deoxidization system between molten steel and slag with direct current electric field is found and calculated. A series of research work has been carried out in order to explore the influence of the various technology parameters on the electrochemical deoxidization efficiency. The effects and influences of the electric field parameters, the electrode parameters, the properties of the molten slag and molten steel on the distribution of the current density in the electrochemical deoxidization system are analyzed according to the simulation experiments and the commercial experiments. The simulation results are verified based on experiments. The result shows that the oxygen transfers from the molten slag into the molten steel without electric field because of the high oxidizing slag. And the electrochemical deoxidization can be carried out effectively with the application of electric field. The maximum deoxidization rate can reach to $11.3\times10^{-6}\cdot min^{-1}$ and the deoxidization rate can be improved by increasing oxygen ion current. Increasing both of the mass transfer coefficient and the reaction interface area can increase the oxygen ion current between the steel and slag.

## 1. Introduction

In refining process, the deoxidizer is added into the molten steel directly with the traditional deoxidization methods [1]. It results in the formation of nonmetallic inclusions in the steel and the quality of the product can be affected. In 1970s, in order to prevent the molten metal from being unpolluted, the deoxidizer of $H_2$, CO and other active metals--all these deoxidizers and their products of $H_2O$, $CO_2$ and oxide could not be contacted with molten metal by the methods of electrolytic and galvanic. Thus the oxygen in molten metal has been decreased without pollution [2,3]. A concept of controlled oxygen flow metallurgy technology was proposed by Zhou G Z in 1980s. Based on the electrochemistry, the oxygen ion moving direction could be controlled and then the oxygen could be transfered out of the molten steel. Unpolluted deoxidization technique had been studied with applied electric field between slag and molten steel or Cu melt by Lu, Zhang and Li [4-6], the oxygen content in molten steel or Cu melt was decreased rapidly and effectively with this method. The kinetic model with electrochemical characteristic representing oxygen ion immigration had been worked out, and the theoretical results were in good agreement with experiments. The electrochemical research on oxygen transfer process of CaO-MgO-$Al_2O_3$ slag and Steel was carried out by Li [7], and the results showed that under the experimental conditions, the oxides in the slag decomposed in the electric field. The electrode electrochemical reactions were controlled by the voltage parameters and ion concentration of the slag and steel.

In Iron & Steel Research Institute of Ansteel, the laboratory equipment of the deoxidization experiment with DC field with 500 kg molten steel was exploited in 2010. A series of unpolluted deoxidization experiments were carried out [8,9]. In order to research the deoxidization effect continuously, the commercial research equipment of the deoxidization experiment was set up based on 100t ladle. The experimental processes and the results are introduced in this paper.

## 2. Experimental

### *2.1. Theory*

Molten slag is a kind of electrolyte that includes oxygen ion, various positive metal ions and mult-ion groups, which can be used as a passage for the oxygen ion elimination. The moving direction and reaction velocity can be controlled by the applying DC field. The following steps occur during the process.

① Oxygen ion is transfered from the molten steel to the interface

② Cathode reaction in the steel-slag interface: $[O] + 2e \rightarrow (O^{2-})$

③ Oxygen ion is transfered into the slag with the application of DC field

④ Anode reaction in the interface of the slag and anode: $(O^{2-}) \rightarrow 1/2O_2 + 2e$

Based on the electrochemistry theory, the positive charge is accumulated on the steel-slag interface with the step ②. With step ④, the negative charge is accumulated on the slag-anode interface. So the electric field is formed between the two interfaces and the transport of the oxygen ion will be prevented. Thus the deoxidization process is prevented. In order to maintain the deoxidization process, an electric field should be imposed to eliminate the original electric field between the two interfaces.

### *2.2. Equipment*

The equipment system includes: 100t ladle, DC source, anode (graphite, diameters of the two anodes are 400 mm and 150 mm respectively), cathode (inner is graphite and the diameter is 80 mm, exterior is the insulated material that prevents the erosion from the steel and slag, and diameter is 150 mm), clamps and elevators for the electrodes, control room and stirring system, as shown in figure 1.

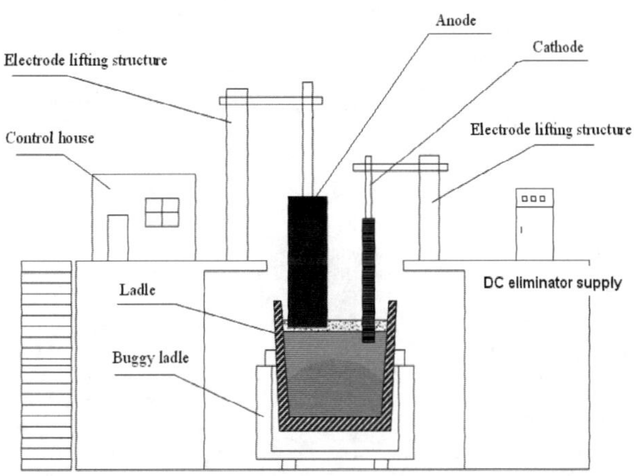


**Figure 1.** Schematic of applied electrical field steel refining equipment.

## 3. Simulation

IOP Conf. Series: Materials Science and Engineering **668** (2019) 012009 doi:10.1088/1757-899X/668/1/012009

### *3.1. Model*

Based on the experiment system, a simulation model is set up by the finite element software. Molten steel and slag are the two main parts, one anode in the slag and another cathode in the steel, as shown in figure 2. The diameter of the anode and cathode are 400 mm and 150 mm respectively. The slag thickness is 100 mm and anode is inserted in the slag with 50 mm depth. The cathode is inserted in the steel with the depth of 100 mm.

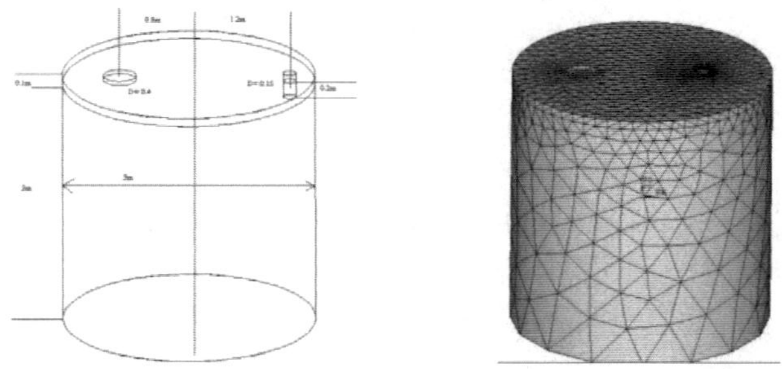

**Figure 2.** Schematic of simulation model and net mesh.

Simulation parameters are: 5 V imposes in the anode and 0v in the cathode. Others are shown in table 1.

**Table 1.** Simulation parameter.

| | Height | Diameter | Resistivity | Inductive capacity |
|---|---|---|---|---|
| Molten steel | 3 m | 3 m | $1.4*10^{-6}$ $\Omega\cdot$m | 1000 |
| Molten slag | 100 mm | 3 m | 0.2 $\Omega\cdot$m | 6 |

### *3.2. Simulation results*

The current density in the steel and slag interface is shown in figure 3. It can be seen that the maximum current density under the anode is 258 A/m$^2$ when the anode diameter is 400 mm and the current in this area is almost 32 A. The total current is higher than 50 A in the steel-slag system when adding the current around the anode. When the anode diameter is 150 mm, the maximum current density is 53.5 A/m$^2$, which is lower than that with the bigger anode. So in practice, a bigger anode should be adopted to achieve a better result.

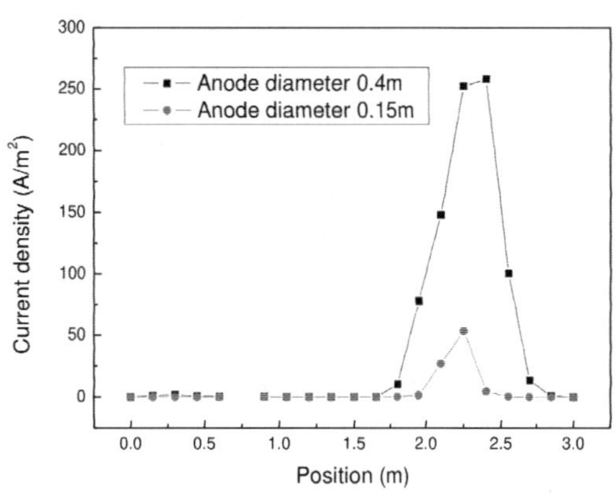


**Figure 3.** Comparison of current density at the steel-slag interface of different simulation.

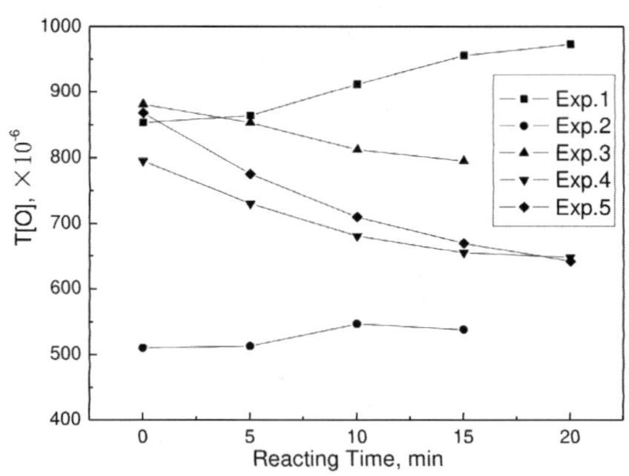


**Figure 4.** The result of experiments.

## 4. Experiment

### *4.1. Experimental scheme*

Q235B is selected as the experimental steel. Tapping temperature is higher than 1710°C, and the slag is pushed off when tapping in order to control the slag content strictly. Lime and refining slag are added into the ladle in the final stage of tapping. Oxygen content is checked after 2 minutes of large amount Argon bottom blowing. Then the electrodes are descended by the elevators respectively. Anode is immerged in the molten slag and cathode is immerged in the molten steel. An electric field is provided by a DC source. The experimental scheme is shown in table 2.

**Table 2.** Experimental scheme.

| No. | Condition | DC field | Reaction, min |
|---|---|---|---|
| Exp.1 | no electrodes | None | 20 |
| Exp.2 | anode diameter:Φ400mm, distance to the steel surface:50mm | None | 15 |
| Exp.3 | anode diameter:Φ150mm, distance to the steel surface:50mm | 3~5V, 30~50A | 15 |
| Exp.4 | anode diameter:Φ400mm, distance to the steel surface:50mm | 3~5V, 30~50A | 20 |
| Exp.5 | anode diameter:Φ400mm, distance to the steel surface:30mm | 3~5V, 50~80A. | 20 |

### *4.2. Results and discussion*

In the experiments, the effects of the electrode diameter, the current and the immersion depth are researched respectively.

*4.2.1. Results and analysis without DC field.* In order to validate the deoxidization effect, the contrast experiments 1 and 2 are carried out firstly. Figure 4 shows the results. It can be seen that the oxygen content increases from $853\times10^{-6}$ to $972\times10^{-6}$ in 20 mins without electrode and DC field in exp 1. The oxygen content increases from $510\times10^{-6}$ to $538\times10^{-6}$ in 15 mins with electrode while without DC field in exp. 2.

Table 3 shows the composition and content of the steel before and after the contrast test. The slag content and components before the test are shown in table 4.

**Table 3.** Content and composition of the steel (wt,%).

| Stage | C | Si | Mn | P | S |
|---|---|---|---|---|---|
| Initial | 0.0252 | <0.01 | 0.020 | 0.0050 | 0.026 |
| End | 0.0248 | <0.01 | 0.020 | 0.0062 | 0.027 |

**Table 4.** Content and composition of the slag (wt,%).

| CaO | $SiO_2$ | MgO | $Al_2O_3$ | MnO | $P_2O_5$ | TFe | FeO | $Fe_2O_3$ |
|---|---|---|---|---|---|---|---|---|
| 38.00 | 12.15 | 6.80 | 11.40 | 1.17 | 0.99 | 21.75 | 19.04 | 9.94 |

It can be seen that the oxidizability of the slag is very high because the total content of FeO and MnO is higher than 20%. The $a_{\mathrm{FeO}}$ is used to characterize the oxidizability of the slag and $Fe_2O_3$ is another supporter for the slag oxidizability except for FeO. According to the oxygen contain:

$$w(\mathrm{Fe_tO}) = w(\mathrm{FeO}) + 1.35w(\mathrm{Fe_2O_3}) \tag{1}$$

There is a balance for Ferous oxidizing:

$$K_{FeO}^{\Theta} = a_{(FeO)} /(a_{[Fe]} a_{[O]}) = \exp(2.303 * (\frac{6320}{T} - 2.734)) \quad (2)$$

In the molten steel, $a_{[Fe]} \approx 1$, so

$$a_{[O]} = a_{(FeO)} / K_{FeO}^{\Theta} \quad (3)$$

The oxygen active coefficient can be calculated with the interaction coefficient in the molten steel.

$$\lg f_{[O]} = e_O^C w[C] + e_O^O w[O] + e_O^{Si} w[Si] + e_O^{Mn} w[Mn] + e_O^P w[P] + e_O^S w[S] \quad (4)$$

where, $e_O^C = -0.45$, $e_O^O = -0.20$ (or-0.30), $e_O^{Si} = -0.131$, $e_O^{Mn} = -0.021$, $e_O^P = 0.07$ (or 0.006), $e_O^S = -0.18$ (or -0.133) [10] (Annotation: the interaction coefficients are various from the different materials.)

The steel components are substituted into equation (4), then $f_{[O]}$ can be calculated.

$$f_{[O]} = 0.95 \quad (5)$$

When temperature is 1600°C, $K_{FeO}^{\Theta}$ equals to 4.369. From figure 5, $a_{Fe_tO}$ is 0.7 according to the slag composition in table 4, as shown the red area in figure 5. Then the oxygen content in the steel should be:

$$w[O]_e^* = \frac{a_{(FeO)}}{f_{[O]} \cdot K_{FeO}^{\Theta}} = 0.168\% \quad (6)$$

It is much higher than the total oxygen that of molten steel. So the oxygen ions will transport into the steel if there is no DC field. In exp. 2 the oxygen content increases slowly. It exhibits that the electric field balance between steel and slag is destroyed because of the immersion of the electrodes. More oxygen ions move into the slag and react with the electrodes and the oxygen transport ratio is reduced.

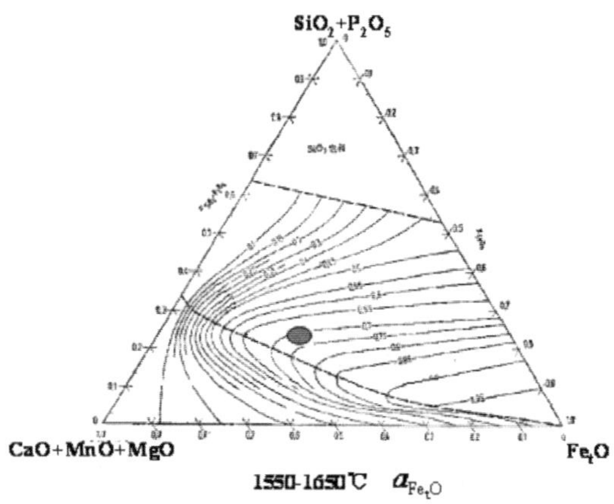


**Figure 5.** $Fe_tO$ activity in the slag [11].

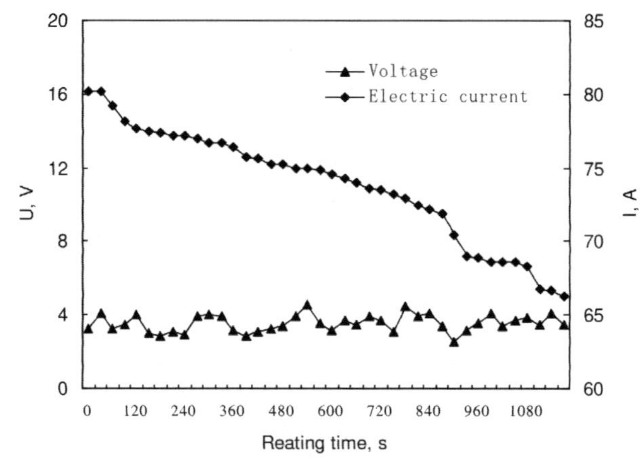


**Figure 6.** The transformation of applied electric field in exp. 5 during the reaction.

*4.2.2. Analysis of the deoxidization experiment with DC field.* The effect of the various diameter and immersion depth of the anode on the deoxidization are tested. The results are shown in figure 4. In exp. 3, the anode diameter is 150 mm and the distance to the steel and slag interface is 50 mm. The oxygen content decreases from $881\times10^{-6}$ to $795\times10^{-6}$ in 15 mins and the deoxidization rate is $5.73\times10^{-6}\cdot\text{min}^{-1}$. In exp. 4, the anode diameter is 400 mm and the distance is 50 mm also. The oxygen content decreases from $795\times10^{-6}$ to $650\times10^{-6}$ in 20 mins. The deoxidization rate is $7.25\times10^{-6}\cdot\text{min}^{-1}$. So the deoxidization rate can be improved by increasing the reaction interface area. The difference between exp. 4 and exp. 5 is the immersion depth of anode. In exp. 5, the distance between anode end to the interface of steel and slag is 30 mm. With DC field, the oxygen content decreases from $868\times10^{-6}$ to $642\times10^{-6}$ in 20 mins. The deoxidization rate is $11.30\times10^{-6}\cdot\text{min}^{-1}$.

The deoxidization experiments with DC field between the steel and slag were also carried on by Lu [6]. A kinetic model to describe the deoxidization process had been set up. As following:

$$O_m = O_{m,0} - \int_0^t \frac{I_{O^{2-}} M_O \times 10^6}{2FM_{melt}} d\xi \tag{7}$$

Where,

$$I_{O^{2-}} = I^0_{O^{2-}} - exp\left(-\frac{\alpha\rho_{melt}A}{M_{melt}}t\right) \tag{8}$$

The following equation can be obtained.

$$O_m = (O_0 - O_C)\cdot exp\left(-\frac{\alpha\rho_{melt}A}{M_{melt}}t\right) + O_C \tag{9}$$

Where, $I_{O^{2-}}$ is the oxygen ion current, $t$ is time, $M_O$ is the mole mass of oxygen, $F$ is faraday constant, $M_{melt}$ is the weigh of the melt, R is gas factor, $O_C$ is the balance oxygen concentration in the melt with the graphite, $\alpha$ is mass transfer coefficient, $\rho_{melt}$ is the melt density, $A$ is the area between the steel and slag, $O_m$ is the oxygen content in sometime and $O_0$ is the initial oxygen content in the steel.

From the kinetic model we can see that the deoxidization rate is mainly depend on the oxygen ion current between the steel and slag, the oxygen ion current is mainly influenced by mass transfer coefficient and reaction interface area of the steel and slag. It means that the oxygen ion current can be increased by increasing the mass transfer coefficient and the reaction interface area. Thus the reaction rate can be improved.

Compared exp. 3 and exp. 4, the deoxidization kinetic condition is improved because of increasing the reaction interface area and then the deoxidization rate can be improved. From exp.4 and exp.5, it can be seen that there will be some influences with the increase of the immersion depth, such as decrease of the slag resistance, increase of the oxygen ion current and then the deoxidization rate increases. It also can be obtained that deoxidization rate in the initial stage is higher than the terminal stage. In the initial 5 mins, the oxygen content decreases $93\times10^{-6}$ and the deoxidization rate is $18.6\times10^{-6}\cdot\text{min}^{-1}$, while in the terminal 5 mins, $28\times10^{-6}$ oxygen is deoxidized and the deoxidization rate is $5.6\times10^{-6}\cdot\text{min}^{-1}$ in exp. 5.

The relationship between conductivity and viscosity of molten slag had been studied by Bockris [12] in 1980s. He found the electric conductivity of molten slag not only depend on ion concentration but also on the interaction force of plus or minus ion. The relationship between conductivity and viscosity expressed as follow equation:

$$K^{\text{n}} \cdot \eta = \text{C} \tag{10}$$

$K$ is conductivity of molten slag, $\eta$ is viscosity of molten slag, n is an index (>1), C is a constant.

The relationship between ion conductivity and diffusion coefficient had been studied by Goto [13] in 1980s. He found the relationship following the principle of Nernst-Einstein equation.

$$D_i = \frac{RT}{F^2 z_i^2 c_i} K \tag{11}$$

$D_i$ is the diffusion coefficient of ion, $z_i$ is the electrovalence of ion, $c_i$ is the concentration of ion, $K$ is conductivity of molten slag.

The electric conductivity is mainly influenced by viscosity and diffusion coefficient according to their researches. At the beginning of experiments the mass transfer coefficient is relatively higher because of the high temperature of the steel and slag and lower viscosity of the slag. With the proceeding of the reaction, the viscosity of slag increases, the oxygen ion transports more difficult.

Figure 6 shows the variation of the DC field in the reaction process. It can be seen that the voltage is relatively stable in the process except for the fluctuation due to the slag flow. The system temperature decreases, the viscosity becomes higher and the resistance of the slag increases. All these lead to the reduction of the current. At the same time, the oxygen ion current decreases while the oxygen content of the steel and the oxygen ion transportation decrease during the process. Finally the system current also reduces.

## 5. Conclusions

- Unpolluted deoxidization can be carried out by applying a DC field between the steel and slag interface. The oxygen content in the molten steel can be decreased from $868\times10^{-6}$ to $642\times10^{-6}$ in 20 mins on the condition of voltage is 3~5 V, current is 50~80 A. The deoxidization rate reaches to $11.3\times10^{-6}\cdot min^{-1}$ and the initial ratio is higher than that of terminal one.
- Without DC field, oxygen ion must be transported into the steel due to the high oxidizability of the slag. The balance of the electric field between the steel and slag can be broken by the immersed electrode and the transfer rate of the oxygen ion can be restricted.
- The oxygen ion current can be improved by increasing both of the mass transfer coefficient and the reaction area between the steel and slag, and the deoxidization rate can be improved by optimizing the kinetic conditions.

## References

[1] Qu Y 1992 *Principles of Steelmaking* (Beijing: Metallurgical Industry Press) pp 188-202
[2] Shi Y, Uday P and Kuo C 1994 Deoxidization of molten metals by short circuiting yttria-stabilized zirconia electrolyte cell *Elec Soc* **141** 467-74
[3] Iwase M 1981 Electronically driven transport of oxygen from liquid iron to $CO+CO_2$ gas mixtures through stabilized zirconia *Metal Trans* **12B** 517-24
[4] Zhou G Z 2006 *To Explore New Directions of Metallurgical Process- Controlled Oxygen Flow Metallurgy Wei Shoukun Academy 100th Birthday Festschrift* (Beijing: Science Press) pp 171-6
[5] Gao Y M, Jiang Y, Zhang H, Gao X M and Zhou G Z 2007 Metallurgy with controlled oxygen flow *J. Wuhan Uni. Sci. & Tech. (Natrual Science Edition)* **5** 449-53
[6] Lu X G, Ding W Z, Li F S, Li L F and Zhou G Z 2002 Electrochemistry of oxygen iron transport in slag *Trans. Nonferrous Met. Soc. China* **2** 326-9
[7] Li Z P, Wang J, Wang Z M, Zhou A L, Li J C and Zhang J Y 2018 Electrochemical study on oxygen transfer process of $CaO-MgO-Al_2O_3$ slag and liquid IF steel *Shanghai Metals* **1** 45-50
[8] Jia J X, Liao X W, Zhao G, Cao D, Guo Q T and Zhu X L 2012 Research on pollution free hot metal deoxidization method by externally applied electric field *Steelmaking* **4** 47-50

[9] Guo Q T, Jia J X, Liao X W, Li G B, Cao D and Li D G 2014 Simulation study on deoxidization method with no pollution under direct current electric field *Angang Tec* **2** 24-8
[10] Li W C 2001 *Physical Chemistry of Metallurgy and Materials* (Beijing: Metallurgical Industry Press) pp 158, 521-523
[11] German Steel Engineers Association (Wang J and Mao Y W Translation) *Slag Atlas* (Beijing: Metallurgical Industry Press) p 171
[12] Bockris J O N and Reddy A K N 1977 *Modem Electrochem* (Plenum Publishing Corp.) pp 384-5
[13] Goto K S, Sasabe M and Kawakami M 1977 Relation between trace diffusivity and electrical conductivity on multi-component oxide slags at 900°C to 1600°C *Trans Iron Steel Inst Jpn* **17** 212 *ISU* 2

IOP Conf. Series: Materials Science and Engineering **668** (2019) 012010 doi:10.1088/1757-899X/668/1/012010

# Effect of two-step austempering process on carbidic ductile iron

**V Dakre[1], A Likhite[1,2], S U Pathak[1] and D R Peshwe[1]**

[1]Department of Metallurgical & Materials Engineering, Visvesvaraya National Institute of Technology (V.N.I.T.), Nagpur, Maharashtra, 440010, India

[2]E-mail:alikhite@gmail.com

**Abstract.** The two-step austempering process when applied to ductile iron, results in significant improvement in mechanical properties due to the strengthening of the ausferitic matrix.The present work was undertaken to investigate the effect of the two-step austempering process on carbidic ductile iron. Carbidic ductile iron is a new member to the ductile iron family, it is produced by alloying the ductile iron composition with chromium.In the present work, the carbidic ductile iron thus produced, was austempered by both conventional and two-step austempering route. For both processes, all specimens were austenitized at 975°C and soaked for 180 min. In the conventional process, the specimanwas quenched in salt bath held at to 350°C, 400°C, and 450°C and soaked at respective temperatures for 180 min. Whereas, in the two-step austempering process, specimenwere first quenched in salt bath heldat 250°C. After stabilization at 250°C, the salt bath temperature was raised to 350°C, 400°C, and 450°C. After the respective temperature was attained, the specimens were soaked for 180 min. The austempered samples obtained by both processes were analyzed with the help of SEM, XRD, and TEM. Two-step process resulted in finer ausferitic matrix than the conventional process. The XRD analysis revealed an increase in carbon content of austenite and volume fraction of Carbon Stabilized Austenite (CSA) in sample austempered by a two-step process. The TEM analysis indicated that the lath width of austenite and ferrite was small in samples austempered by two-step process than the conventional process, due to a high degree of super-cooling.

## 1. Introduction

Austempered Ductile Iron (ADI) is known to be very useful in many engineering applications as automobile parts (camshafts). The ADI has unique property of high strength at very low cost [1]. The properties of ADI are attributed to its unique structure comprising of acicular ferrite and carbon stabilized austenite called as "ausferrite" (2). ADI is produced by austempering of ductile iron.

Carbidic austempered ductile iron (CADI) is the newest member in the cast iron family. As the name suggests, it has carbides embedded in the ausferitic matrix. The presence of carbides improves the hardness and wear resistance of alloy but on the other hand reduces the toughness [1, 2].

The carbide can be introduced in matrix ADI by various methods [1, 3-6]. In this investigation, chromium was added into the melt to obtained carbide in ADI matrix.

Putatunda introduced the concept of two-step austempering process [7, 8]. Many researchers have studied and compared the two-step and conventional austempering processes. They reported many advantages of two-step process over the conventional process [9-16].

The aim of present study was to study the effect of two-step austempering, on the microstructure and structural characterization using XRD and TEM analysis of CADI.

## 2. Experimental details

### *2.1. Metal casting and heat treatment*

The carbidic ductile iron was produced in a medium frequency induction furnace. The molten metal was treated with 5% Mg-Fe-Si alloy using the sandwich technique. The molten metal was inoculated with foundry grade ferrosilicon with nominally 75% silicon. After spheroidising treatment, the molten metal was poured into mould. The casting was produced in cylindrical (φ20 mm, L = 300 mm) form. The chemical composition (wt. %) of the ductile iron alloyed with chromium is given in table 1.

**Table 1.** Chemical composition of alloy

| Element | C | Si | Mn | Cu | Ni | Cr | CE* |
|---|---|---|---|---|---|---|---|
| Wt.% | 2.42 | 1.88 | 0.86 | 0.57 | 0.57 | 3.63 | 3.1 |

*CE (carbon equivalent) = %C + 1/3 %Si

The carbidic ductile iron was austempered by both conventional and two-step austempering route. All specimens were austenitized at 975°C for 180 min. In the two-step process, specimens were quenched to 250°C. After stabilization at 250°C, the salt bath temperature was raised to 350°C, 400°C, and 450°C. After the respective temperature reached, the specimens were soaked for 180 min. In the conventional process, specimen were quenched at 350°C, 400°C, and 450°C and soaked for 180 min.

The SEM analysis was carried out on JEOL 6380A make machine. X-ray diffraction analysis was carried on ''Xpert-Pro'' diffracto-meter. For X-ray, Cu $K_\alpha$ radiation at 40 mA and 45 kV was used. The austempered specimens were examined over an angular range of 20° to 110°. The scan speed is 0.03° per sec. From XRD analysis volume fraction of CSA, the carbon content of austenite and ferrite cell size were estimated as described in previous literature [12,17-21].

## 3. Results and discussions

### *3.1. Microstructural analysis*

The figure 1 shows the microstructure of the alloy in the as cast condition.

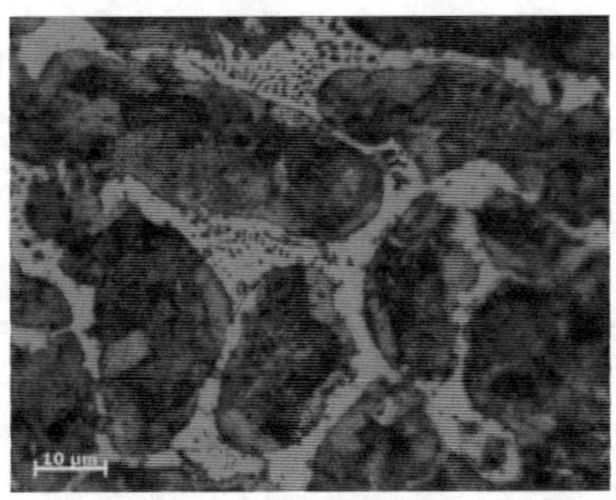


**Figure 1.** As as-cast carbide ductile iron (2% Nital)

The SEM images of conventionally and two-step austempered samples of the alloy at three different temperatures are shown in figures 2 and 3 respectively.

In the as-cast condition and in samples austempered by either of the processes, the coarse eutectic carbides, the carbides having globular and blocky morphologies were also observed. The similar observations have been also reported by other researchers, after the conventional austempering of CADI [2, 22].

The ausferitic matrix gets coarser, with increasing austempering temperature in both processes. For given austempering temperature, a fine ausferitic matrix was observed in samples process by two-step austempering route than conventional austempering route.

**Figure 2.** SEM images of conventionally austempered samples at (A) 350°C for 180 min, (B) 400°C for 180 min, and (C) 450°C for 180 min.

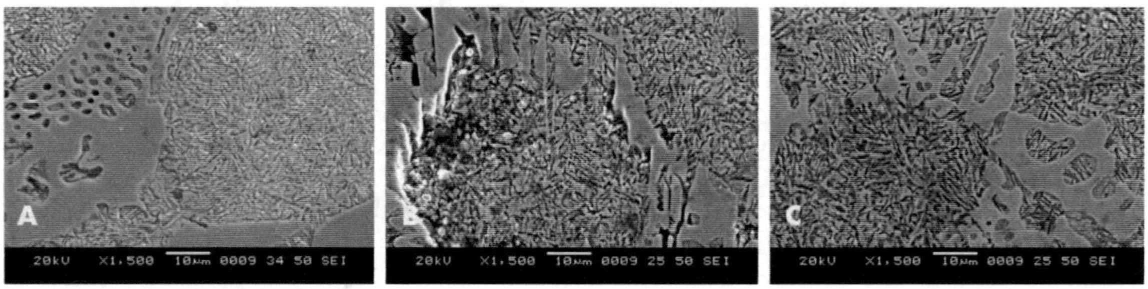


**Figure 3.** SEM Images of Two-Step Austempered Samples at (A) 350°C for 180 min, (B) 400°C for 180 min, and (C) 450°C for 180 min.

*3.2. XRD analysis*

The representative XRD plots for samples austempered by conventional and two-step process austempered at 400°C are given in figures 4 and 5 respectively. Along with austenite and ferrite peaks, some peaks corresponding to carbide like $Fe_3C_7$ and $Cr_7C_3$ were also observed in both processes. The austenite peak observed at (111) at 43.4° in conventional austempering, shifted slightly toward left in sample austempered by two-step austempered indicating a change in the lattice parameter of austenite. The lattice parameter of austenite changes due to the presence of carbon in austenite [20]. The intensity of austenite peak (111) was more in samples austempered by two-step than that of the conventional austempering process. This suggests an increased volume fraction of CSA in two-step than that of conventional austempering process.

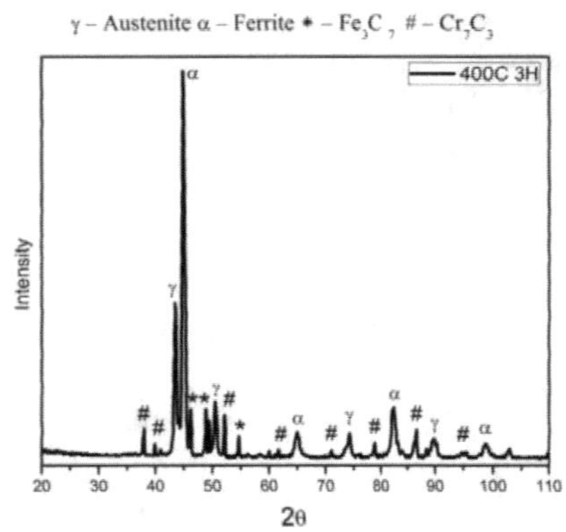


**Figure 4.** XRD plot for conventionally austempered samples at 400°C for 180 min.

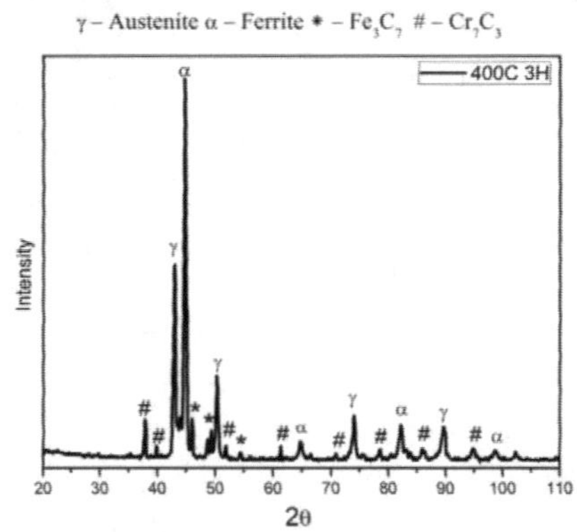


**Figure 5.** XRD plot for two-step austempered samples at 400°C for 180 min.

The comparison of ferrite cell size, the carbon content of austenite and volume fraction CSA for both austempering is given in figures 6-8 respectively. From figure 6, it clear that the two-step process results in smaller ferrite cell size when compared to that of conventionally austempered samples. It is also observed that the ferrite cell size increases with austempering temperature in both processes. Both observations are in correlation with the observation in SEM images.

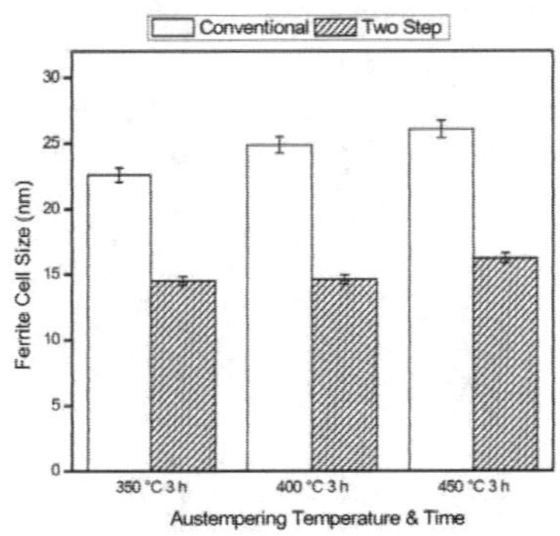


**Figure 6.** Comparison of Ferrite Cell Size for Conventional and Two-Step Austempering.

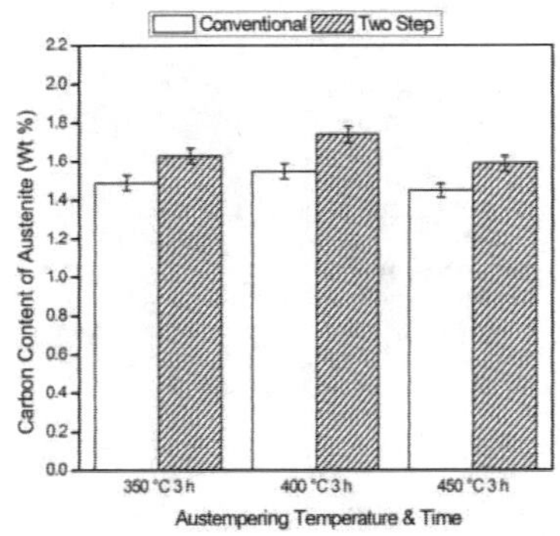


**Figure 7.** Comparison of Carbon Content of Austenite for Conventional and Two-Step Austempering.

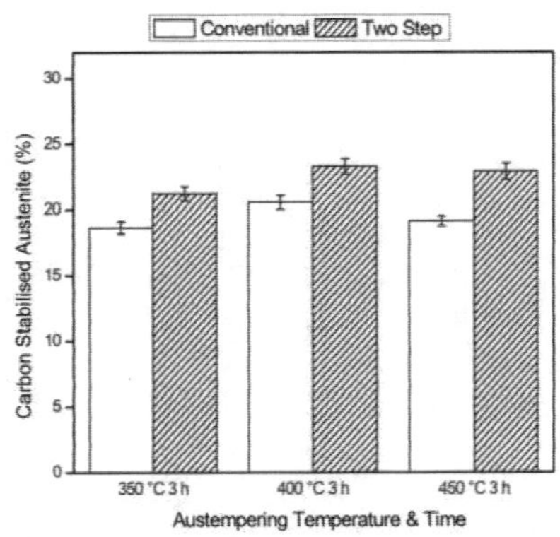


**Figure 8.** Comparison of volume fraction of CSA for conventional and two-step austempering.

The carbon content of austenite and volume fraction of CSA was increased in samples austempered by two-step process than that of conventionally austempered samples (figure 7 and figure 8).

*3.3. TEM analysis*

TEM analysis was carried out on JEOL JEM 2100. The accelerating voltage was 200 kV.

The TEM images are shown in figure 9 (Conventional austempering treatment) and figure 10 (Two step austempering treatment).

It was observed that in case of conventional austempering, average lath width of ferrite was 306 nm and that of austenite was 315 nm. As against that the two step austempering process produced ferrite with an average lath width of 248 nm and austenite with a width of 260 nm.

In a conventional process, less number of ferrite nuclei gets nucleated compared to that of two-step process due to higher quenching temperature. As number of ferrite nuclei is less, they can grow much bigger in given austenite grain and vice versa. In both process, austenitization temperature is 975°C and hence the prior to quenching condition is same, i.e. size of austenite grain and austenite carbon [23-25]. Thus, in this case, austempering or quenching temperature plays an important role in final microstructure. Hence, the two-step austempering process produces the finer ausferitic matrix than the conventional process.

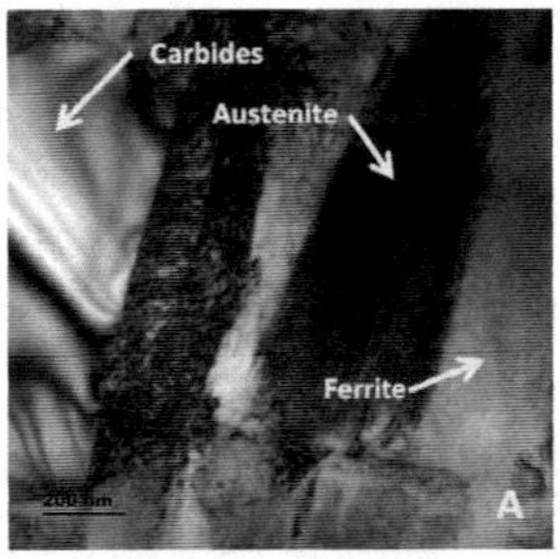


**Figure 9.** TEM analysis for conventionally austempered samples at 400°C for 180 min.

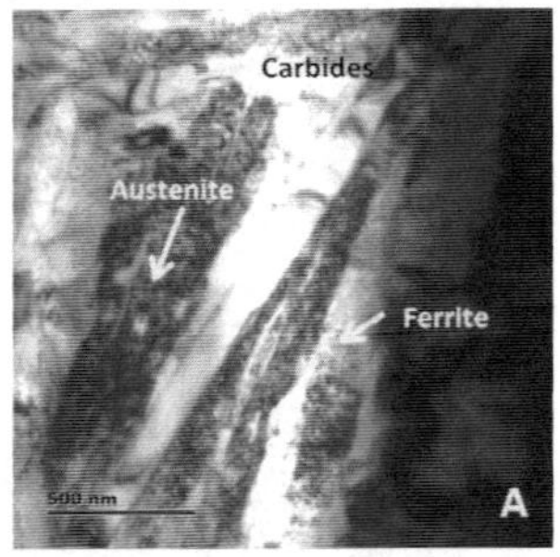


**Figure 10.** TEM analysis for two-step austempered samples at 400°C for 180 min.

Further, large number of ferrite nucleation result in more carbon atom rejected from ferrite to get loaded into surrounding austenite in sample processed by two-step route. This result in higher carbon content in austenite, in samples austempered by two-step process than conventional process. Higher carbon content in austenite increases the amount of carbon-stabilized austenite (CSA). Hence, the CSA volume fraction and its carbon content is more in sample austempered by two-step process than that in case of the conventional process.

## 4. Observations and conclusions

The carbides having globular and blocky morphologies were observed in samples austempered by conventional and two-step austempering process. The XRD analysis confirms the presence of iron carbide ($Fe_3C_7$) and chromium carbide ($Cr_7C_3$). The area fraction of carbide is reduced after austempering from 22–28% to 15–18%.

The TEM analysis shows the ferrite and austenite lamellar structure in ausferite matrix of CADI along with blocky carbide. This lamellar structure was found in both samples austempered by both routes. SADP analysis confirms the presence of carbides.

The fine ausferrtiic matrix was observed in SEM images of samples austempered by two-step route when compared to the SEM images of samples austempered by conventional route. The lath width of ferrite and austenite is smaller in sample austempered by two-step process than that of conventional process. The lower quenching temperature in two-step favors the high nucleation rate of ferrite. Consequently, a large number of ferrite gets nucleated and the fine ausferite matrix was obtained.

## References

[1] Hayrynen K L and Brandenberg K R 2003 *Trans. Am. Foundry Soc.* **111** 845-50
[2] Laino S, Sikora J A and Dommarco R C C 2008 *Wear* **265** 1-7
[3] Caldera M, Rivera G L, Boeri R E and Sikora J A2005*Mater. Sci. Technol.* **21** 1187-91
[4] Rao P P and Putatunda S K 2003 *Mater. Sci. Eng. A***346** 254-65
[5] Bartosiewicz L, Singh I, Alberts F A, Krause A R and Putatunda S K, 1995 *J. Mater. Eng. Perform.* **4** 90-101
[6] Liu H H, Wang J, Shen B L, Yang H S, Gao S J and Huang S J 2007 *Mater. Des.* **28** 1059
[7] Putatunda S K 2001*Mater. Sci. Eng. A***315** 70-80
[8] Rao P P and Putatunda S K 1998 *Metall. Mater. Trans. A***29** 3005-16
[9] Yang J and Putatunda S K 2005 *Mater. Sci. Eng. A* **406** 217-28
[10] Ravishankar K S, Udupa K R and Rao P P 2008 Development of austempered ductile iron for

high tensile and fracture toughness by two step austempering process The Proceedings of the World Foundry Congress(Chinnai, India) pp 35-40
[11] Ibrahim K M, Nofal A A and Ibrahim M M 2008 Effect of alloying additions and two-step austmpering on the microstructure and mechanical properties of ductile iron*The 9th International Conference on Mechanical Design and Production (MDP-9)*(Cairo, Egypt)pp 8-10
[12] Dakre V, Peshwe D R, Pathak S U and Likhite A 2017 *Trans. Indian Inst. Met.* **70** 2381
[13] Yang J and Putatunda S K 2004 *Mater. Des.* **25** 219-
[14] Dakre V, Peshwe D R, Pathak S U and Likhite A, 2019*Trans. Indian Inst. Met.***72** 911-7
[15] Yang J and Putatunda S K 2004 *Mater. Sci. Eng. A* **382** 265-9
[16] Yang J and Putatunda S K 2005 *Mater. Sci. Eng. A***393** 254-8
[17] Dakre V, Peshwe D R, Pathak S U and Likhite A, 2018 *Int. J. Miner. Metall. Mater.***25** 770-8
[18] Dakre V S, Peshwe D R, Pathak S U and Likhite A A2018 *IOP Conf. Ser. Mater. Sci. Eng.* **346** 1-10.
[19] Parhad P*et al* 2012 *Trans. Indian Inst. Met.* **65** 449-58
[20] Roberts C S 1953 *Trans. AIME* **197** 203-204
[21] Cullity B.D. (1978) *Elements of X-Ray Diffraction* 2$^{nd}$ ed. Addison-Wesley Reading MA (Massachusetts - Menlo Park) pp 100-105 pp 277-9
[22] Nili Ahmadabadi M 1998 *Metall. Mater. Trans. A* **29** 2297-6
[23] Rundman K B 2004 It's about austenite and carbon, mate - A story of the physical metallurgy of ADI - part II, *The 35$^{th}$ Australian Foundry Institute National Conference* (Adelade, Australia)
[24] Minkoff I 1983 *The Physical Metallurgy of Cast Iron* (New York:John Wiley & Sons Ltd) pp 290-2
[25] Elliott R1988*Cast Iron Technology* 1$^{st}$ edi.(London:Butterworth & Co. Ltd.) pp 142-3

IOP Conf. Series: Materials Science and Engineering **668** (2019) 012011 doi:10.1088/1757-899X/668/1/012011

# Ceria doped $TiO_2$ as photocatalyst for water treatment under visible light

**V O Ndabankulu[1], S Maddila[1] and S B Jonnalagadda[1,2]**

[1]School of Chemistry & Physics, University of KwaZulu-Natal, Westville Campus, Chiltern Hills, Durban-4000, South Africa

[2]E-mail: jonnalagaddas@ukzn.ac.za

**Abstract.** The photocatalytic activity of bare titania and ceria supported mesoporous titania ($CeO_2/TiO_2$) catalysts were assessed for the degradation of caffeine by visible light irradiation in absence of the oxidants. Different wt% metal loaded (bare, 0.1, 0.5 and 1.0) materials were synthesised by a sol-gel method using anionic surfactants. Various instrumentation techniques were used to characterise the prepared photo-catalysts such as P-XRD, BET, TEM, SEM-EDX, FT-IR, Raman, UV-DRS and photoluminescence spectroscopy. 0.5% $CeO_2$ doped titania exhibited excellent efficiency in photo degradation of caffeine in aqueous solution relative to all the other prepared heterogeneous catalysts. All the materials showed good efficiency and were recyclable without loss of catalytic activity up to three times. Analysis of degradation products was carried by Liquid chromatograph-mass spectrometry. Two reaction products 6-amino-1,3-dimethyl-2,4-dioxo-1,2,3,4-tetradropymidin-5-ly)-methylcarbamic acid (ATCA) and *N*-methyl-*N*-(methylcarbomoyl)-2-oxoacetamide (MMO) were positively identified.

## 1. Introduction

The persistence of organic pollutants in water systems is one of the major concerns for the environmental protection and the entire world. Pharmaceutical compounds are a group of organic pollutants that have become an environmental topic of discussion. Wastewater treatment plants (WWTP) and sewage treatment plants (STP) have been noticed as main causes of environment discharge. Due to they were not designed to deal with this type of contaminants. As a result, pharmaceutical compounds have been perceived in surface waters and ground waters in sediments [1-4]. Pharmaceuticals may reach the aquatic environment through different pathways, i.e. from domestic wastewater from urban areas, effluent from pharmaceutical manufacturing industries, hospital disposal, and the disposal of unused medicine [5,6]. The metabolites originating from the human consumption of medicines entering the water bodies and reservoirs is an issue of concern for environmental health and sustainability. Appropriate eco-friendly technologies for effective removal or degradation of such organic chemicals is vital.

In recent years, heterogeneous photo-catalysts have been widely investigated for water purification and degradation of emerging contaminants in wastewater [7,8]. Particularly, titania ($TiO_2$) is one of the most promising and developing technology semiconductors that has been applied in the degradation of organic contaminates in aqueous system. Some of the attractive qualities of $TiO_2$ are its high activity, strong oxidizing power, chemical stability, robustness against photo-corrosion, low toxicity, eco-friendliness and cost-effectiveness [9,10]. The water molecules and the hydroxides effectively act as photo-excited hole traps on the catalyst surface and generate $OH^{\cdot}$ radicals, which are powerful oxidants in degrading organic compounds [11,12]. UV and visible light are used as the energy source of photons to excite the semiconductor across the band gap. Then it leads to the excitation of an electron from the valence band to the conduction band, thereby generating an electron-hole pair. Further, the combination of the photo-generated electrons and holes lead to the release of heat or photon. Sometimes, there is a possibility of the relocation of the photo-generation holes to move into the surface of the semiconductor [13,14]. There are some limitations with using $TiO_2$ as a photo-catalyst, such as the large band gap energy (3.2eV) for anatase. This limits the use of $TiO_2$ in the ultraviolet region (< 380 nm), which is only about 5 % of the solar spectrum [15].

Various techniques including surfactants in preparation and loading the metal ions are employed to decrease the bandgap of the semiconductors to be sensitive to visible light. Rare earth metals having empty 5d and incomplete 4d orbitals are ideal metals to serve as dopants and they promote photocatalysis [16,17]. Ceria has a good redox pair $Ce^{3+}/Ce^{4+}$, which plays a vital role as a catalyst, which led to its application as photo-catalyst in degradation of dye [18]. Ryu *et al* has showed that doping $TiO_2$ with Ce increased its photocatalytic activity in degradation of 2-mercaptobenzothiazole (MBT) [19].

In this present work, we report ceria doped on $TiO_2$ photo catalysts in degradation of caffeine using visible light. Ceria used as a metal dopant to reduce the band gap energy of anatase into visible region. Caffeine used as a model compound, which among the pharmaceuticals are the mostly found in wastewater treatment plant effluents.

## 2. Experimental section

### *2.1. Catalyst synthesis*

This sol-gel procedure was adopted for the preparation of the ceria doped on titania ($CeO_2/TiO_2$) (wt% 0.1, 0.5, and 1.0) and bare $TiO_2$. For a model synthesis, 2 g of Pluronic F-127 was dissolved in absolute ethanol (60 mL) into a closed 400 mL Pyrex beaker. Then it was stirred for 4 h. Six mL of titanium (IV) isopropoxide was added dropwise to the solution and stirring was continued for 2 h, followed by dropwise addition of deionised water (36 mL). The mixture was continuously for 24 h till it turned from clear to white gel followed by aging it 12 h in the dark at room temperature. Next, the gel was dried in an oven at 80°C for overnight and the yellowish solid obtained was then calcined in the presence of air, at 450°C for 4 h to obtain the white powder of undoped $TiO_2$. Different wt% of the ceria doped on titania materials were synthesised following the similar protocol, but using 36 ml of water with requisite amounts cerium acetylacetonate hydrate precursor dissolved in it. The resultant catalyst materials were denoted as bare $TiO_2$, 0.1 wt% $CeO_2/TiO_2$, 0.5 wt% $CeO_2/TiO_2$ and 1.0 wt% $CeO_2/TiO_2$. Materials and methods section is provided in the supporting information (SI-I).

### *2.2. Photocatalysis experiments*

Photo-degradation experiments of caffeine were carried on lab scale using a 32 W / 480 Osram dulux F378 daylight – Compact fluorescence lamp. The lamp was fitted into a cylindrical Pyrex jacket used to cover the lamp. Reaction was conducted in a 1 L Pyrex beaker, and the vessel was set on a magnetic stirrer together with a lamp in the jacket. 300 ml of caffeine solutions containing 5, 10 or 15 mg $L^{-1}$ were used in the experiments. In a typical run, 50 mg of catalyst was added to the caffeine solution. Then the solution was ultra-sonicated for 10 min and stirred in the dark for 60 min before light illumination. After 1 h, the solution was an irradiated with 32 W day light fluorescence lamp for 2 h. To monitor the progress of the reaction, 4 ml aliquots of reaction mixture was withdrawn at 20 min intervals. Samples were centrifuged using a Mikro 120 micro centrifuge for 2 min at 140 rpm and filtered using 0.45 micron. The extent of degradation of caffeine was monitored spectrophotometrically by recording the absorbance values at its absorption maxima 271 nm as function of reaction progress.

### *2.3. Products analysis*

The product analysis was done using Shimadzu, LC-MS TOF mode. System was equipped with an electrospray ionization source ESI. An injection volume was keep at 10 µL. Isocratic elution using a reverse phase, in which mobile phase was made of 65% v/v acetonitrile and 35% v/v methanol. The chromatogram was recorded for 15 min run and the product peaks appeared between retention times of 3 and 4 minutes. Samples were analyzed was done by after 20 minutes starting of the caffeine degradation reaction.

## 3. Results and discussion

### *3.1. Catalyst characterisation*

*3.1.1. BET analysis.* Figure 1 shows the $N_2$ isotherms of the synthesized materials. The BET surface areas of the material are tabulated in table 1. All the catalyst materials were observed to be type IV with H1 hysteresis loop [19]. Type IV isotherms are usually associated with mesoporous materials and H1 hysteresis loop is caused by agglomerates in the materials. Further, it is also characteristic of material with high pore size uniformity. The calculated BET surface area of all four catalysts exhibited the similar pattern, which is a commonly observed trend in doped materials, when the metal dopant possess larger surface area than $TiO_2$.

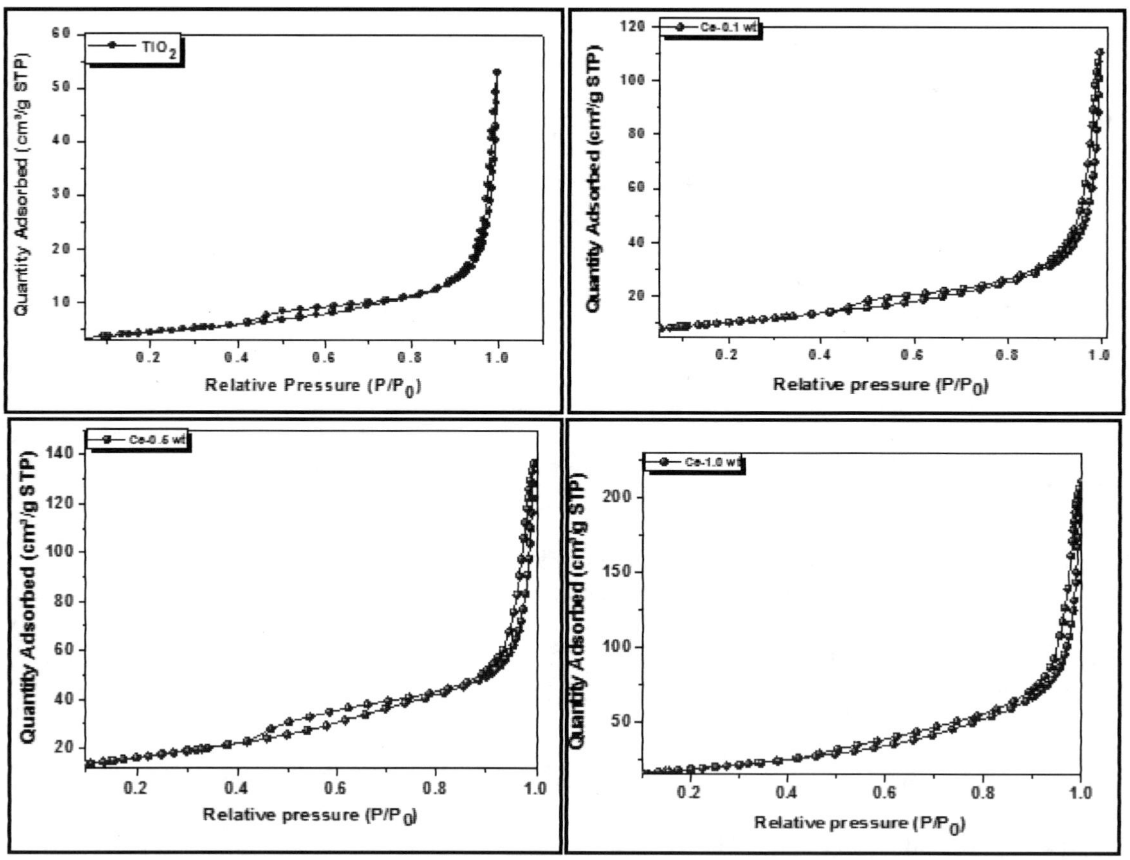


**Figure 1.** $N_2$ sorption/desorption isotherms of (a) bare $TiO_2$ (b) 0.1% Ce/$TiO_2$ (c) 0.5% Ce/$TiO_2$ and (d) 1.0% Ce/$TiO_2$ as prepared material.

**Table 1.** Summary of the textural characterizations, XRD and band gap analysis for the prepared materials.

| Sample Wt% | BET Surface area/ $m^2g^{-1}$ | BJH adsorption pore volume/ $cm^3g^{-1}$ | BJH adsorption pore size/ nm | FWHM (101) / rad | Anatase crystal size (101) / nm | Band gap energy / eV | EDX / Wt. % |
|---|---|---|---|---|---|---|---|
| Bare $TiO_2$ | 15.955 | 0.0841 | 18.012 | 0.01556 | 8.92 | 3.14 | -- |
| 0.1% Ce/$TiO_2$ | 36.754 | 0.1762 | 17.232 | 0.01829 | 7.58 | 3.12 | 0.098 |
| 0.5% Ce/$TiO_2$ | 57.884 | 0.2229 | 12.613 | 0.01902 | 7.29 | 3.05 | 0.481 |
| 1.0% Ce/$TiO_2$ | 66.048 | 0.3342 | 16.475 | 0.01831 | 7.54 | 2.88 | 1.025 |

*3.1.2. SEM & TEM analysis.* SEM and TEM micrograph of $CeO_2/TiO_2$ (wt% 0.1, 0.5, and 1.0) and bare $TiO_2$ are shown in the figures 2 and 3 below. An examination of the micrographs indicates that particles have irregular shapes and agglomerated. XRD spectra shows that ceria doping decreases the crystalline size by inhibiting growth, but this does not affect the morphology of the material. The presence of ceria was established by EDX analysis of the catalyst material. The results suggest that ceria is well dispersed throughout the $TiO_2$ surface. EDX revealed the presence of Ti, O and Ce.

**Figure 2.** Scanning electron micrographs of (a) bare $TiO_2$ (b) 0.1% Ce/$TiO_2$ (c) 0.5% Ce/$TiO_2$ and (d) 1.0% Ce/$TiO_2$ as prepared material.

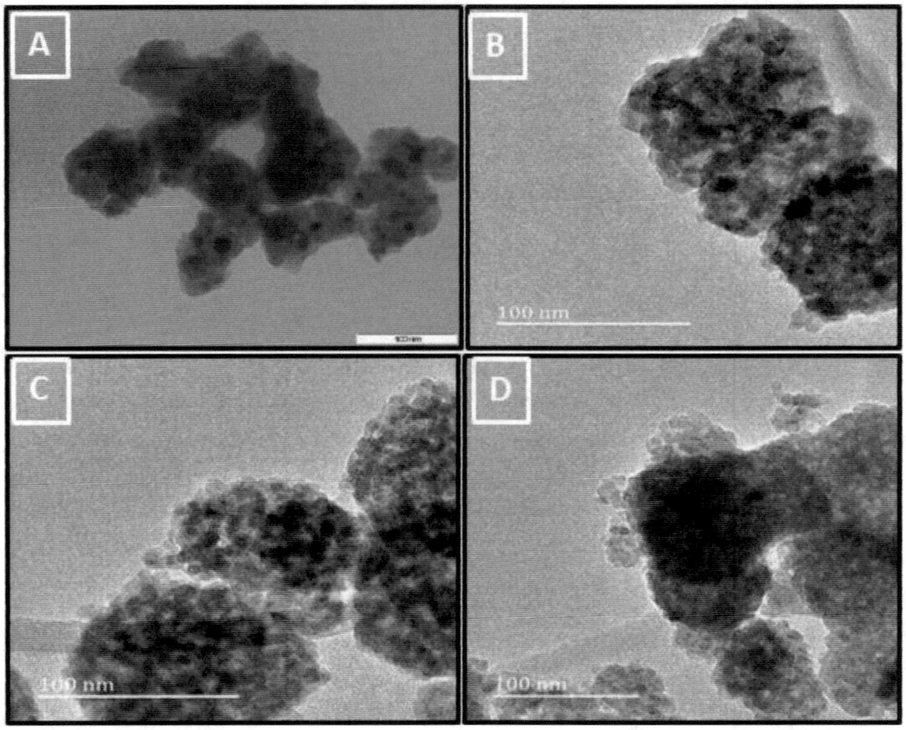


**Figure 3.** Transmission electron micrographs of (a) bare $TiO_2$ (b) 0.1% $Ce/TiO_2$ (c) 0.5% $Ce/TiO_2$ and (d) 1.0% $Ce/TiO_2$ as prepared material.

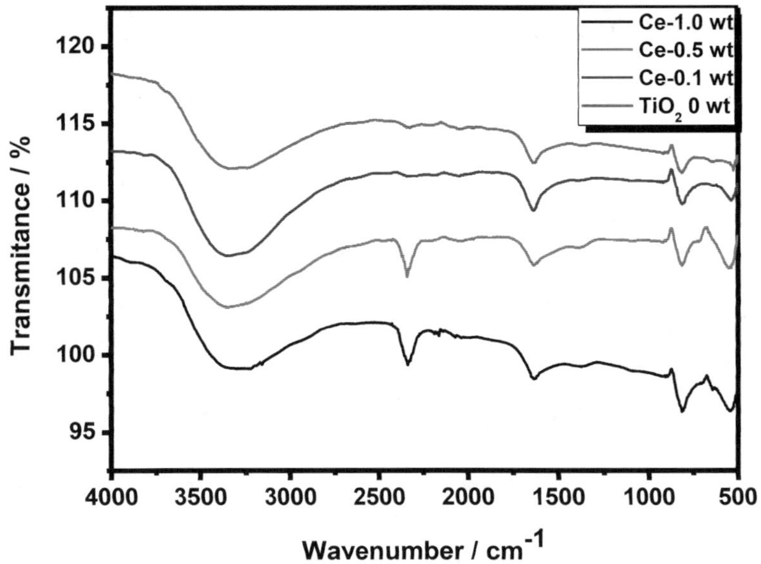


**Figure 4.** IR Spectra of the prepared material.

*3.1.3. FTIR analysis.* Figure 4 displays the FTIR spectrum of $CeO_2$ doped $TiO_2$ samples and bare $TiO_2$. The absorption bands found in the low range of 500–800 $cm^{-1}$ for correspond to Ti–O and Ti–O–Ce bands [20,21]. The broad absorption peaks appeared at 3300 $cm^{-1}$ due to the physisorbed OH group at the surface of the material [4]. The Ti–O–Ti bands are seen in the region of 380–1130 $cm^{-1}$ [20].

The bands that appears in the range of 2200–2400 $cm^{-1}$ are assumed to be due to Ce–O bands. There are no peaks attributed to organic compounds because of the fact that they were removed when the material was calcined. The peaks appeared at 1627 $cm^{-1}$ at 3388 $cm^{-1}$ are the characteristic of $H_2O$ bending and vibration of hydroxyl groups [22-25].

*3.1.4. UV-DRS analysis.* Figure 5 illustrates the optical absorption spectra of the synthesized catalyst materials. The results obtained indicate significant red shift in all doped samples. The red shift is caused by $CeO_2$ energy levels in the band gap [16], which is to the transitions of the dopant ceria electrons between the conducting and valance bands of titania [17]. The optical band gap of the prepared material was calculated from the Tauc plot in figure 5 using this equation.

$$(\alpha h\nu)^2 = A\,(h\nu - E_g),$$

where hv the energy of phone and α denotes the absorption coefficient.

$$\alpha = \left(\frac{4\pi k}{\lambda}\right);$$

where k = absorption index/ absorbance; λ = wavelength and A = constant related to the material.

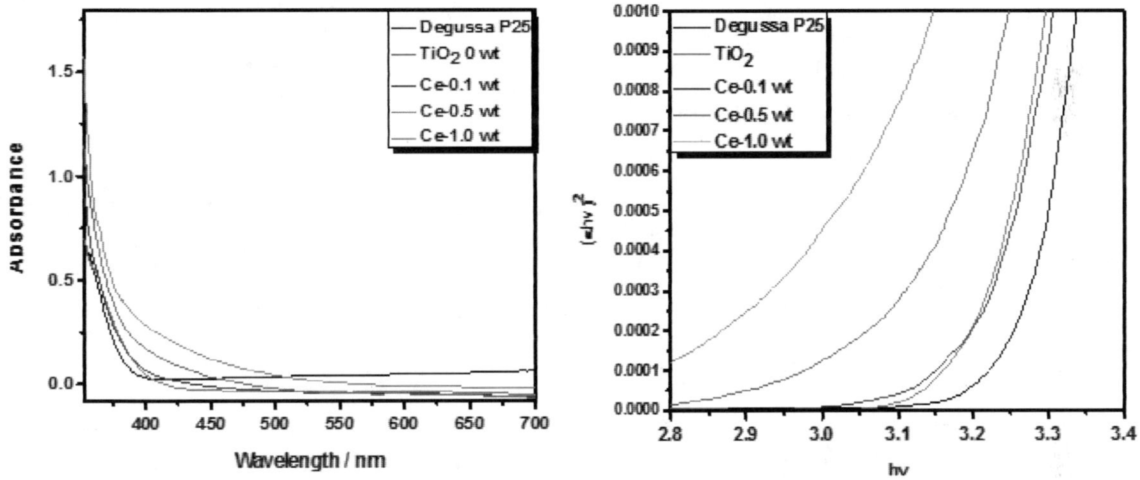


**Figure 5.** UV/ Vis – DRS and Tauc plot of the prepared material.

The calculated values from the Tauc plot display that all the catalyst materials had a reduced band gap compared to bare $TiO_2$ or commercial material, Degussa P25. It was seen that the band gap reduced with the increase in the metal loading of the dopant. Reduction in band gap suggest that $e^-h^+$ can be generated using visible light. The band gaps of different catalyst materials are summarised in table 1.

*3.1.5. XRD and Raman spectra analysis.* Figure 6 represents the powder XRD patterns and the summary of Raman spectra of bare $TiO_2$, and $CeO_2$ doped $TiO_2$ catalysts materials. Figure 6(a) indicates that all the materials peaks characteristic of polycrystalline nature. It suggests that all the samples possess high crystallinity with anatase structure only (2θ = 25.2, 37.76, 48.02, 54.05, 55.03, 62.80, 68.85, 70.19 and 75.07) corresponds to the JCPDS file no: 21-1272. No ceria diffraction peaks observed in the spectra, which could be due to either $CeO_2$ is incorporated in the crystal structure of $TiO_2$ or it may be that the wt% of dopant is too minor. An observation of the diffraction patterns, indicate that doping of ceria on titania caused a marginal shift only suggest that no significant structural changes. The average crystallinity size for the anatase diffraction peak (101) was calculated using Debye–Scherrer equation [26].

$$D = \frac{K\lambda}{\beta \mathrm{Cos}\theta}$$

where D = crystallite size in nm; k = Scherrer's constant ≈ 0.9; λ = wavelength of the X-ray radiation (Cu$K\alpha$ = 0.15406 nm); $\beta$ = full width at half-maximum of the diffraction peak; and θ = the diffraction angle.

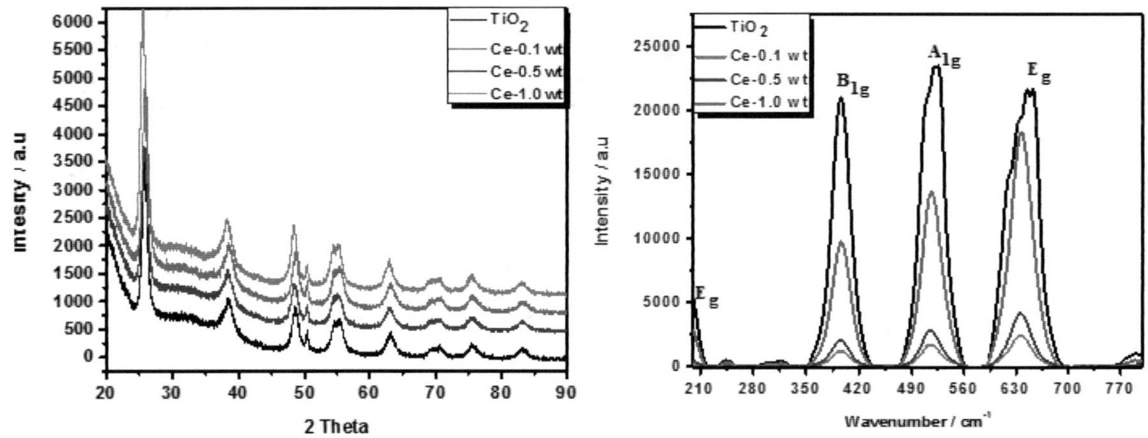


**Figure 6.** (a) X-ray diffraction and (b) Raman spectra.

As shown in table 1, there is a decrease in crystalline size with addition of cerium dopant from 8.92 nm to 7.29 nm with the exception of 7.54 for ceria (1.0 wt%). This result indicate the presence of ceria, to an extent inhibit the growth of crystalline anatase phase.

Raman spectra was applied to study the physical modifications of all the prepared materials. According to symmetry group analysis anatase $TiO_2$ has six mode of vibrations. Four of these vibration modes are indicated in figure 6(b). 146 ($E_g$), 197 ($E_g$), 399 ($B_{1g}$), 515 (A1g), 519 ($B_{1g}$ is superimposed with the $A_{1g}$), and 639 $cm^{-1}$ ($E_g$) and the other three modes are IR active and one mode is inactive [27-29].

*3.1.6. Photoluminescence analysis.* Photoluminescence (PL) is a tool employed in investigation of $e^-h^+$ pair relationship of the semiconductor and asses the ion trapping, migration and charge carrier transfers [30]. The exciton emission suggests that the catalyst materials possess huge exciton binding energies with extended exciton radiative period [31-33]. PL study is used to understand the oxygen vacancy, surface state and defects of metal doping using ceria. From the results obtained in the figure 7, it is clearly seen that with increase in the weight composition of ceria, there is a decrease in the emission intensity compared with bare $TiO_2$. The decrease in emission intensity was attributed to oxygen vacancy or defects caused by the metal, which lead to an increase in optical properties. While, the three emission peaks at 376, 378, and 382 nm are attributed to the exciton recombination, the band at 485 nm corresponds to the self-trapped excitons. The longer wavelength peaks are attributed to the oxygen vacancies in the lattice of titania [23]. The signal seen around 530 nm is assumed to be caused by an increase oxygen vacancies on the surface of the catalyst [34]. Therefore, 0.5% $CeO_2/TiO_2$ showed higher intensity compared to other catalysts.

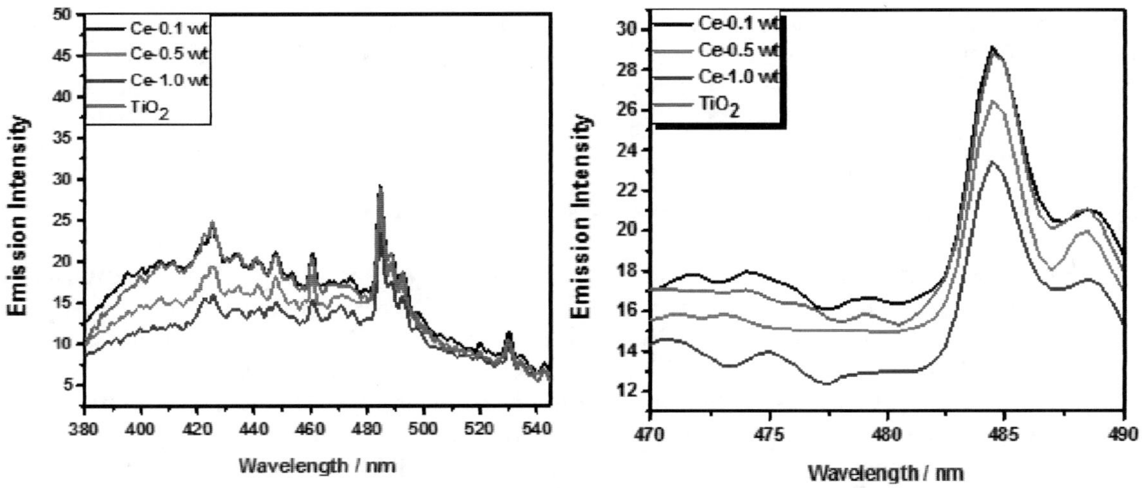


**Figure 7.** Photoluminescence spectra of the prepared material.

*3.1.7. Photocatalytic degradation of caffeine.* Figure 8 shows degradation of CAF as a function of time using various synthesized materials and a commercial catalyst Degussa P25 under visible light after equilibrating in the dark for 60 min. Degradation of CAF was first done using a blank, in which light was used to degrade CAF without the catalyst.

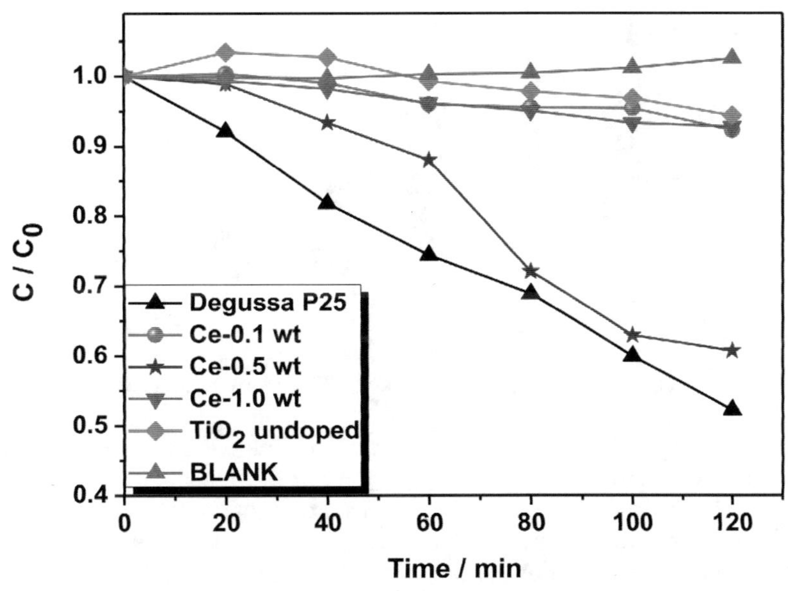


**Figure 8.** Photocatalytic degradation of caffeine.

Assessing the obtained results it is clear that BET surface area does not play much role in photo degradation of CAF. Although, 1.0% $CeO_2/TiO_2$ photocatalyst has the highest surface area and it had lower activity compared to 0.5% $CeO_2/TiO_2$, which possessed smaller surface area. The Tauc plot data shows that gap energy decreased with an increase in metal dopant. Further, there was no photo-degradation observed in absence of any catalyst (figure 8). This meant that visible light alone had no effect in degrading CAF. Better photocatalytic performance was observed with the commercial catalyst P25, which is assumed to comprise both the $TiO_2$ phases (anatase and rutile). As shown in the figure, bare $TiO_2$ had low photocatalytic activity. However, ceria doped $TiO_2$ demonstrated photo catalytic activity. This also suggests that though the prepared materials had reduced band gap compared to P25, their lower efficiency is because of high electron-hole recombination. The increase in photocatalytic activity of ceria doped titania catalyst material compared to the bare titania and it can be also attributed to the redox pair of ceria. It has been reported to act as an effective electron scavenger, for majority of electrons from $TiO_2$, this leads to efficient adsorption of $O_2$ to generate superoxide ($O^{2-}$) radicals, which increases photocatalytic activity [6,35].

*3.1.8. Effect of pH.* The variation in the solution pH results from the changes in $H^+$ and $OH^-$ ion concentrations. Hydroxyl radicals are widely known to play a vital role in photocatalytic degradation. Hence a change in pH to alkaline medium is expected to increase $OH^-$ ions in the reaction medium, which will then be transferred into hydroxyl radicals [36,37]. However, in caffeine is a weak base, it is more soluble in low pH aqueous medium in high pH medium. In low pH solutions, caffeine exists completely as a protonated salt, thus making it more hydrophilic.

In the aim of optimising the efficiency of Ce-0.5 wt. catalyst, degradation of caffeine was performed in different pH conditions and the results are displayed in figure 9. From the obtained results, it can be seen that degradation of caffeine is favoured in acidic conditions.

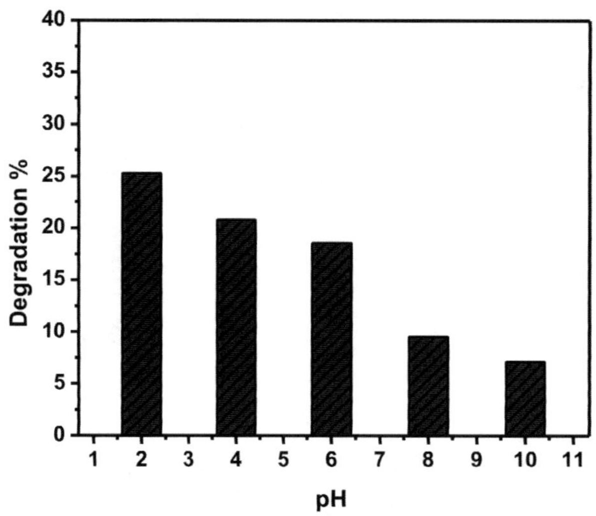


**Figure 9.** Effect of pH in degradation of caffeine over 120 min.

*3.1.9. Detection of degradation products.* Pure caffeine and its degradation products were analysed by LC-MS TOF by injecting the standard solution of caffeine and its reaction products under similar conditions for 5 min runs. The pure caffeine was characteristic of a long intensity at m/z 194. Samples of reaction mixture collected at 20 min intervals were collected and investigated. One intermediate and two reaction products, 6-amino-1,3-dimethyl-2,4-dioxo-1,2,3,4-tetradropymidin-5-ly)-methylcarbamic acid (ATCA) and *N*-methyl-*N*-(methylcarbomoyl)-2-oxoacetamide (MMO) were positively identified. among the degradation products (Supporting information, SI-II). Scheme 1 illustrates the probable mechanistic scheme based the identified intermediate and reaction products.

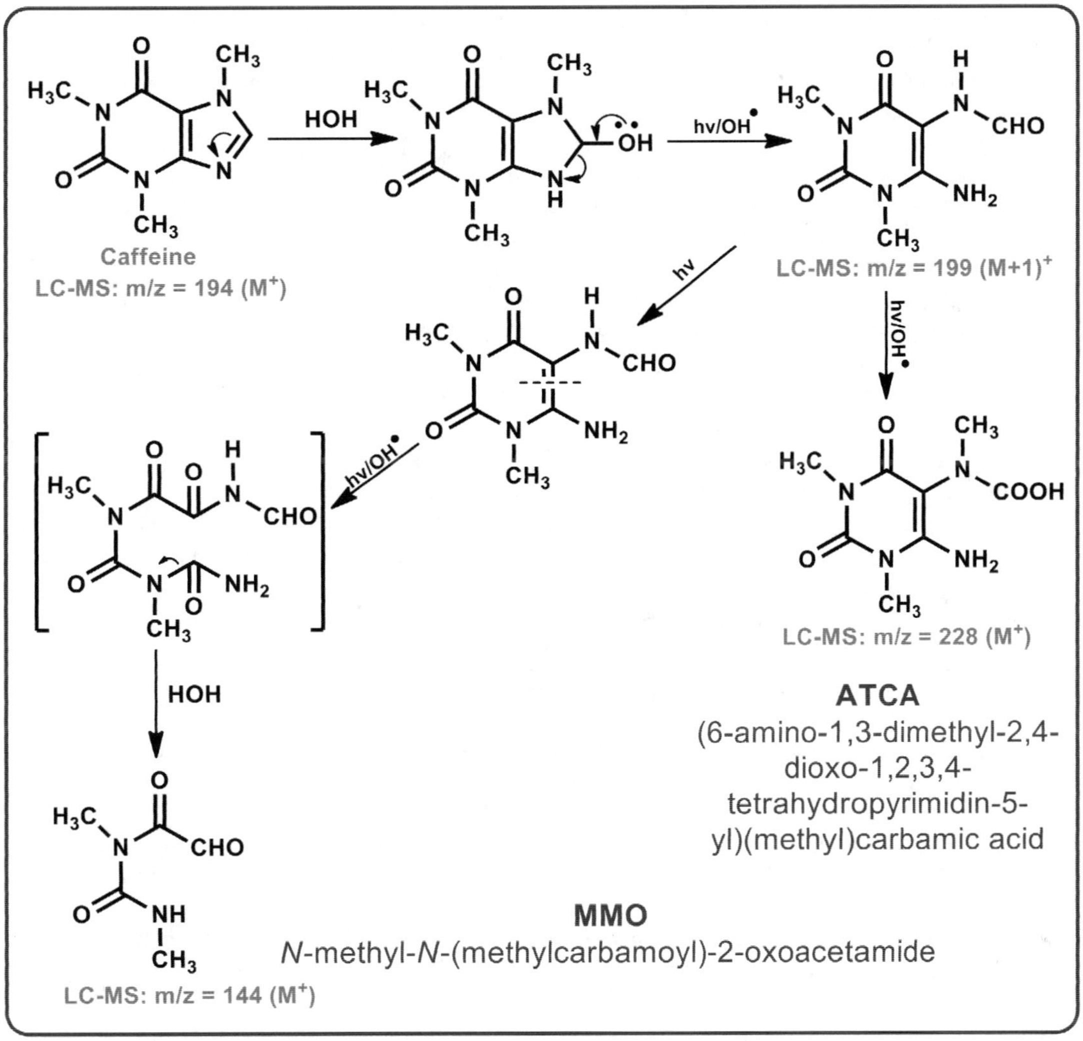


**Scheme 1.** Plausible reaction mechanism of photocatalytic degradation of caffeine under visible light using Ce/$TiO_2$.

## 4. Conclusion

An efficient ceria doped $TiO_2$ materials were synthesised by using sol-gel procedure. It was seen from nitrogen isotherms that all as prepared materials were mesoporous and exhibited type IV isotherms. P-XRD showed that all ceria doped $TiO_2$ were polycrystalline in nature and they were all anatase phase. Synthesized materials were showed excellent photocatalytic activity by the degradation of caffeine in the presence of visible light. An increase in pH showed a decrease in catalytic activity of the as prepared materials.


## Acknowledgments

National Research Foundation (NRF) of South Africa and University of KwaZulu-Natal, Durban are duly acknowledged respectively for the financial support and research facilities.


## References

[1] Emnet P, Gaw S, Northcott G, Storey B and Graham L 2015 Personal care products and steroid hormones in the antarctic coastal environment associated with two antarctic research

stations, mcmurdo station and scott base *Environ. Res.* **136** 331-42

[2] Pereira A M P T, Silva L J G, Meisel L M, Lino C M and Pena A 2015 Environmental impact of pharmaceuticals from Portuguese wastewaters: Geographical and seasonal occurrence, removal and risk assessment *Environ. Res.* **136** 108-19

[3] Mumford S L, Kim S, Chen Z, Gore-Langton R E, Barr D B and Louis G M B 2015 Persistent organic pollutants and semen quality: The LIFE study *Chemosphere* **135** 427-35

[4] Wong C S and MacLeod S L 2009 JEM spotlight: Recent advances in analysis of pharmaceuticals in the aquatic environment *J. Environ. Monit.* **11** 923-36

[5] Frederic O and Yves P 2014 *Chemosphere* **115** 31-9

[6] Magureanu M, Piroi D, Mandache N B, David V, Medvedovici A and Parvulescu V I 2010 Degradation of pharmaceutical compound pentoxifylline in water by non-thermal plasma treatment *Water Res.* **44** 3445-53

[7] Djurisic A B, Leung Y H and Ng A M C 2014 Strategies for improving the efficiency of semiconductor metal oxide photocatalysis *Mater. Horiz.* **1** 400-10

[8] Fujishima A, Rao T N and Tryk D A 2000 Titanium dioxide photocatalysis *J. Photochem. & Photobiol. C: Photochem. Rev.* **1** 1-21

[9] Gaya U I and Abdullah A H 2008 Heterogeneous photocatalytic degradation of organic contaminants over titanium dioxide: A review of fundamentals, progress and problems *J. Photochem. & Photobiol. C: Photochem. Rev.* **9** 1-12

[10] Marques R R N *et al* 2013 Photocatalytic degradation of caffeine: Developing solutions for emerging pollutants *Catal. Today* **209** 108-15

[11] Sreethawong T, Ngamsinlapasathian S and Yoshikawa S 2014 Positive role of incorporating P-25 $TiO_2$ to mesoporous-assembled $TiO_2$ thin films for improving photocatalytic dye degradation efficiency *J. Colloid and Interf. Sci.* **430** 184-92

[12] Vijayalakshmi R and Rajendran V 2012 Arch. synthesis and characterization of nano-$TiO_2$ via different methods *App. Sci. Res.* **4** 1183-90

[13] Zhang J, Wu Y, Xing M, Leghari S A K and Sajjad S 2010 Development of modified N doped $TiO_2$ photocatalyst with metals, nonmetals and metal oxides *Energy & Environ. Sci.* **3** 715-26

[14] Quiñones D H *et al* 2015 Boron doped $TiO_2$ catalysts for photocatalytic ozonation of aqueous mixtures of common pesticides: Diuron, o-phenylphenol, MCPA and terbuthylazine *Appl. Catal. B: Environ* **178** 74-81

[15] Xu H *et al* 2013 Anatase $TiO_2$ single crystals exposed with high-reactive {111} facets toward efficient $H_2$ evolution *Chem. Mater.* **25** 405-11

[16] Myilsamy M, Murugesan V and Mahalakshmi M 2015 Indium and cerium co-doped mesoporous $TiO_2$ nanocomposites with enhanced visible light photocatalytic activity *Appl. Catal. A: Gen.* **492** 212-22

[17] Tong T *et al* 2007 Preparation of Ce–$TiO_2$ catalysts by controlled hydrolysis of titanium alkoxide based on esterification reaction and study on its photocatalytic activity *J. Colloid and Interf. Sci.* **315** 382-8

[18] Silva A M T, Silva C G, Drazić G and Faria J L 2009 Ce-doped $TiO_2$ for photocatalytic degradation of chlorophenol *Catal. Today* **144** 13-8

[19] Ryu S-Y, Kim D S, Jeon J-D and Kwak S-Y 2010 Pore size distribution analysis of mesoporous $TiO_2$ spheres by $^1H$ nuclear magnetic resonance (NMR) cryoporometry *The J. Phy. Chem. C* **114** 17440-5

[20] Tripathi A K *et al* 2013 Study of structural transformation in $TiO_2$ nanoparticles and its optical properties *J. Alloys and Comp.* **549** 114-20

[21] Park A T, Koh J H, Seo J A and Kim J H 2011 Formation of mesoporous $TiO_2$ with large surface areas, interconnectivity and hierarchical pores for dye-sensitized solar cells *J. Mater. Chem.* **21** 17872-80

[22] Rey A *et al* 2015 Influence of structural properties on the ativity of $WO_3$ catalysts for visible

light photocatalytic ozonation *Chem. Eng. Sci.* **126** 80-90
[23] Zhao Y *et al* 2007 Synthesis and optical properties of $TiO_2$ nanoparticles *Mater. Lett.* **61** 79-83
[24] Zeng T *et al* 2014 Hydrothermal steam induced crystallization synthesis of anatase $TiO_2$ nanoparticles with high photovoltaic response *Mater. Lett.* **119** 43-6
[25] Liu Q *et al* 2016 Enhancement of photodegradation of Ce, N, and P Tri-doped $TiO_2$/AC by microwave radiation with visible light response for naphthalene *J. Taiw. Institu. Chem. Eng.* **68** 506-13
[26] Burda C *et al* 2003 Enhanced nitrogen doping in $TiO_2$ nanoparticles *Nano Lett.* **3** 1049-51
[27] Gajović A, Stubičar M, Ivanda M and Furić K 2001 Raman spectroscopy of ball-milled $TiO_2$ *J. Mol. Struct.* **563–564** 315-20
[28] Sekiya T *et al* 2001 Raman spectroscopy and phase transition of anatase $TiO_2$ under high pressure *J. Phy. & Chem. Sol.* **62** 717-21
[29] Su W *et al* 2008 Surface phases of $TiO_2$ nanoparticles studied by UV raman spectroscopy and FT-IR spectroscopy *The J. Phy. Chem. C* **112** 7710-6
[30] Kernazhitsky L *et al* 2015 Photoluminescence of Cr-doped $TiO_2$ induced by intense UV laser excitation *J. Lumines.* **166** 253-8
[31] Oseghe E O, Ndungu P G and Jonnalagadda S B 2015 Photocatalytic degradation of 4-chloro-2-methylphenoxyacetic acid using W-doped $TiO_2$ *J. Photochem. & Photobiol. A: Chem.* **312** 96-106
[32] Reyes-Coronado D *et al* 2008 Phase-pure $TiO_2$ nanoparticles: Anatase, brookite and rutile *Nanotechnol.* **19** 145605
[33] Mathew S *et al* 2012 UV-visible photoluminescence of $TiO_2$ nanoparticles prepared by hydrothermal method *J. Fluores.* **22** 1563-9
[34] Khan M M *et al* 2014 Band gap engineered $TiO_2$ nanoparticles for visible light induced photoelectrochemical and photocatalytic studies *J. Mater. Chem. A* **2** 637-44
[35] Li F B *et al* 2005 Enhanced photocatalytic activity of $Ce^{3+}$–$TiO_2$ for 2-mercaptobenzothiazole degradation in aqueous suspension for odour control *Appl. Catal. A: Gen.* **285** 181-9
[36] Zhang J and Nosaka Y 2014 Mechanism of the OH radical generation in photocatalysis with $TiO_2$ of different crystalline types *The J. Phy. Chem. C* **118** 10824-32
[37] Nosaka Y and Nosaka A 2016 Understanding hydroxyl radical (•OH) generation processes in photocatalysis *ACS Energy Lett.* **1** 356-9

IOP Conf. Series: Materials Science and Engineering **668** (2019) 012012 doi:10.1088/1757-899X/668/1/012012

# Application of TRIZ theory in steel rolling production

**J B Li**[1,2]

[1]Room 307, Fuhai International Port, 17 Daliushu Road, Haidian District, Beijing, China

E-mail: 2396411055@qq.com

**Abstract.** In the face of two problems encountered in the process of steel rolling production, the TRIZ theory was applied to analyse and solve the problems, and unprecedented solutions were put forward, both of which were applied for patents and authorized and published. It is expected that through this paper, innovative thinking will be stimulated and the problems encountered in iron and steel production can be solved quickly.

## 1. Introduction of the TRIZ theory

The former Soviet scientist Genrich S. Altshuler through the analysis and study of 2.5 million invention patents, established a comprehensive theoretical system composed of a variety of methods and algorithms for solving technological issues and realizing innovation and development, known as the Theory of Inventive Problem Solving (TRIZ) theory.[1] The theory was developed by summarizing the laws and patterns followed by the development and evolution of various technologies in multi-disciplinary fields, and the innovative principles and rules for solving various technical and physical constraints.[2]

The tools included in the TRIZ theory include: 40 principles of invention, conflict matrix, 76 standard solutions, Algorithm for Inventive-Problem Solving (ARIZ), Anticipatory Failure Determination (AFD), substance-field analysis, Innovation Situation Questionnaire (ISQ), Directed Evolution (DE), Ideal Final Result (IFR), trimming, 8 evolution types, effective knowledge base and so on.[3][4] The international research on TRIZ has been pushed from the field of engineering technology to the fields of management, strategy, finance, information and so on. For example, Valeri Souchkov has developed the method of TRIZ application in the field of management, Sergei Ikovenko has successfully applied TRIZ to patent avoidance design and the development of enterprise patent strategy.

## 2. Application practice of TRIZ theory

The following two examples illustrate the application of TRIZ theory to solve the problem in the iron and steel rolling process.

### *2.1. Invention of a steel antirust device*

We were informed that a black box had been installed behind the rolling production line. When the steel passed through the black box, the surface of the steel was smooth and no rust could be seen. The content of the black box was not revealed and was kept as a technical secret. We initiated a thought-process, yet could not come up with a solution.

After taking part in the technical innovation method and TRIZ theory training, we employed the principles of the TRIZ theory in the production process of H-section steel using the field analysis and knowledge base. We developed a new device to prevent the corrosion of steel surface by adding a field and a substance, and the patent was declared.[5]

According to the theory, 'a substance itself could not achieve a certain effect, only after contact with a certain "field" produced the effect on another object or underwent the corresponding reaction'. As far as the field of science is concerned, temperature field, mechanical field, sound field, gravitational field, magnetic field, electric field and so on are the concrete forms of substance field. Substance-field analysis refers to the analysis of the interactions between Substance 1 and Substance 2. Figure 1 represents the existing substance-field model analyzed in the steel rolling process.

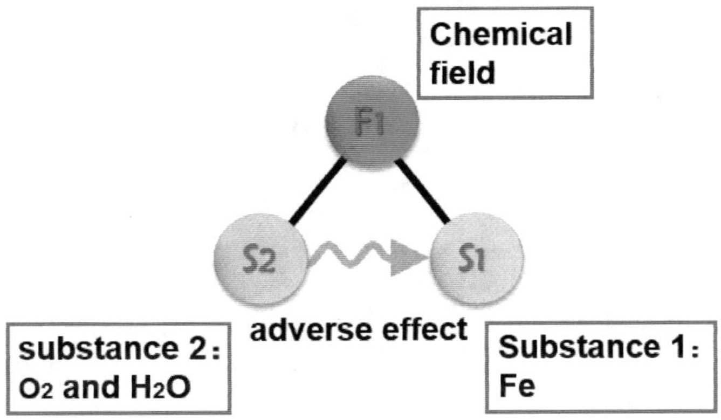


**Figure 1.** Existing substance-field model.

To address this problem, we considered changing the field model:

Change 1: As shown in figure 2, a new field F2 was added in the process.

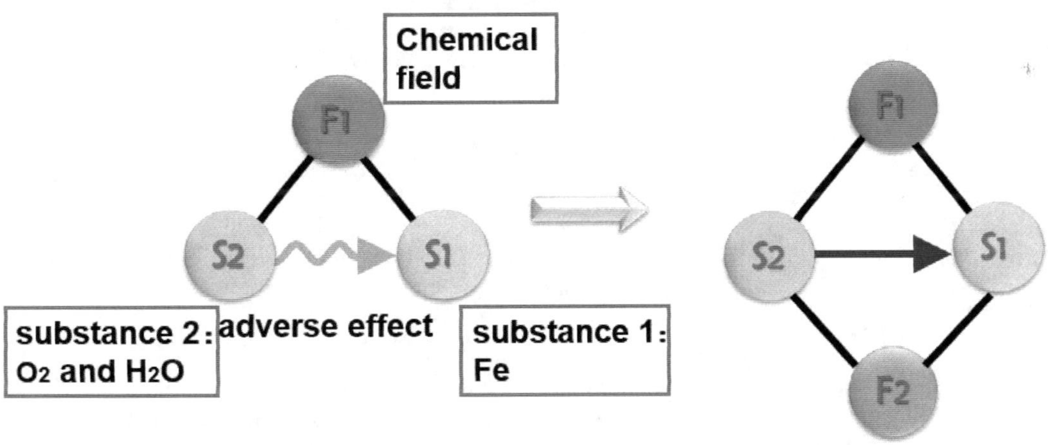


**Figure 2.** Addition of a new field F2 in the substance-field model.

Change 2: As shown in figure 3, a new substance S3 was added to minimize S2's harmful effects on S1.

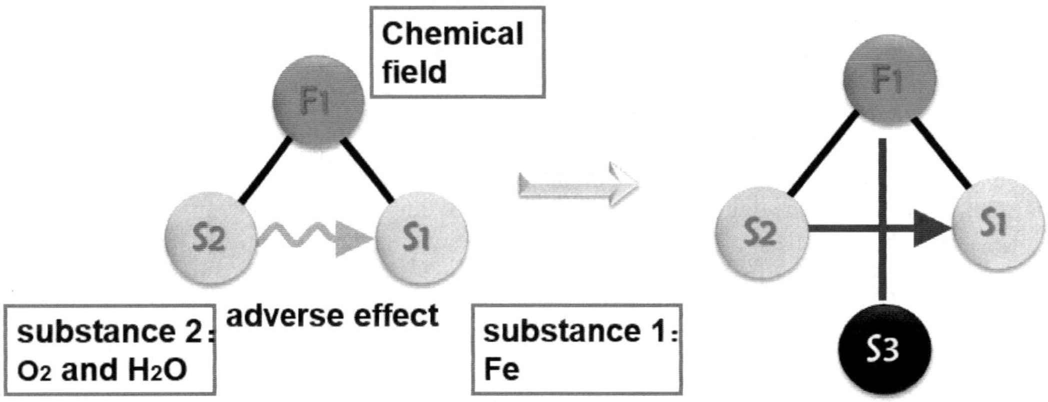


**Figure 3.** Addition of a new substance S3 in the substance-field model.

The results are presented in figure 4.

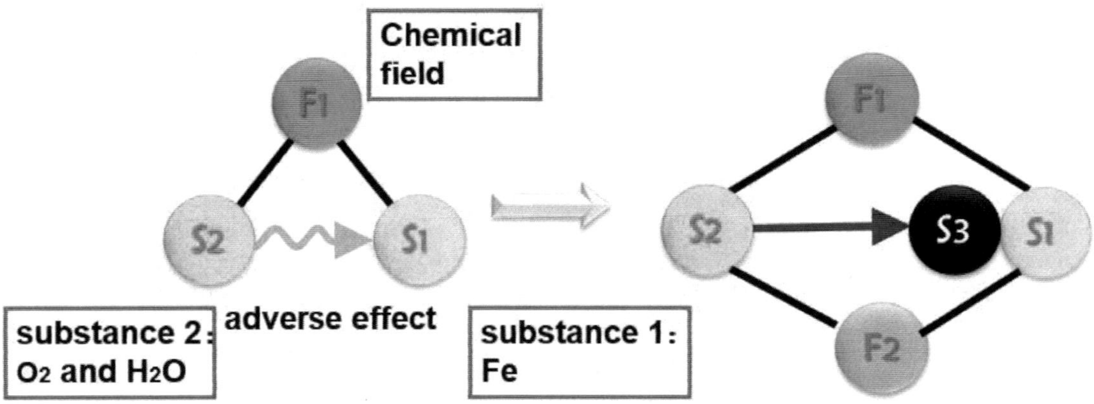


**Figure 4.** Addition of a new field F2 and a new substance S3 in the substance-field model.

*2.1.1. Choice of the added new substance S3.* The gases $H_2$ and $N_2$, graphite powder, paint and others are considered to be substances. $N_2$ is a colorless and tasteless gas, which is not lively and non-flammable. It is one of the main components of air, accounting for 78.08% (volume fraction) of the total atmosphere, and its density is lesser than that of air. Due to the chemical inertia of $N_2$, it is often used as a protective gas to prevent certain objects from being oxidized when exposed to air. We decided to choose $N_2$.

*2.1.2. Choice of the added new field F2.* The fundamental cause of steel rust is due to the reaction between Fe and $O_2$ in water and air. Hence, there is a necessity to prevent the formation of the rust ($Fe_2O_3$, $Fe_3O_4$ and FeO), by increasing the electromagnetic field to move the positively charged Fe ion and the negatively charged O ion in the opposite direction. Due to the difference in the major study in this case, the application effect knowledge base was taken into account.

After screening the various types of electromagnetic radiations such as microwave, infrared ray, ultraviolet ray and so on, considering the difficulties like the effect, reflection and so on, the choice of microwave was finally decided. The developed system is shown in figure 5.

Microwave has the ability to penetrate through the glass, ceramics and other substances, but cannot penetrate through the metal. Therefore, taking advantage of this fact, the microwave emitted on the web of H-section steel was reflected alternately in the upper and lower webs, so that the general H-beam could be treated by microwave.

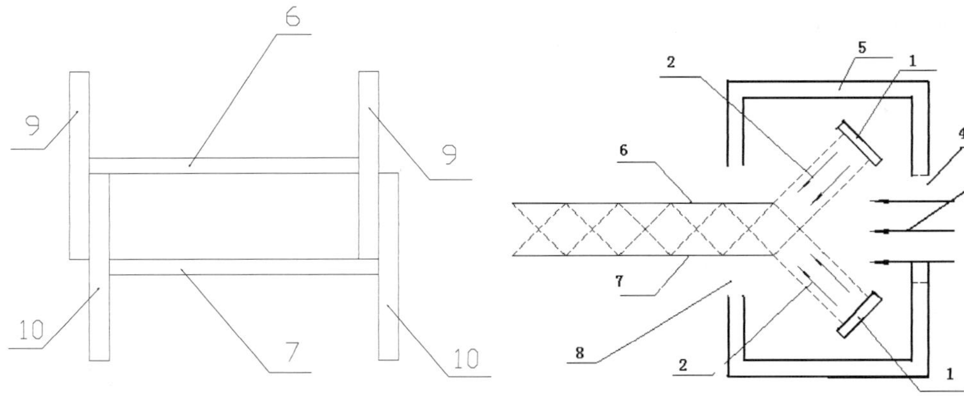


1.Magnetron, 2. Microwave, 3. Nitrogen, 4. Blowing holes, 5. Box, 6. Upper h-beam web, 7. Lower h-beam web, 8. Opening, 9. Upper h-beam flange, 10. Lower h-beam flange

**Figure 5.** Steel rust prevention device.

The formation of patent for the utility type model is shown in table 1.

**Table 1.** Patent formation of steel rust prevention device.

| Patent type | Name | Patent number | Application date | Date of announcement of authorization |
|---|---|---|---|---|
| Utility model | A steel antirust device | 200920280716.6 | 2009-12-9 | 2010-10-27 |

### *2.2. Invention of a new universal mill*

The technical agreement of forklift gate frame channel steel was made with a forklift truck company. A kind of forklift gate frame steel with high dimensional precision and good surface quality was developed, which was much better than the conventional production in the export forklift, as shown in figure 6.

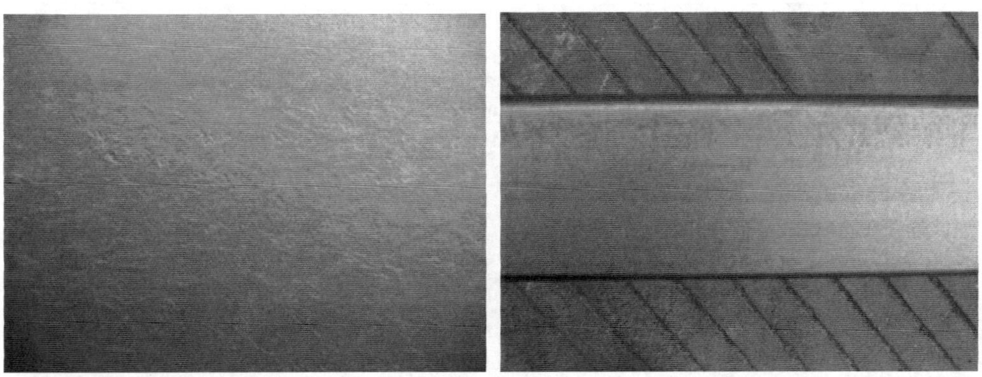

**Figure 6.** Surface comparison of channel steel between two forklift gantry frames.

The precision and surface quality directly affcts the product grade. Although this fact is clearly known, it was not easy to solve it. Therefore, the TRIZ theory was used and analysis was done. The form of ordinary universal rolling mill is shown in figure 7.

In production, to improve the precision and performance of section steel, the rolling of section steel needs to be done with a controlled rolling, controlled cooling and high strength materials.[6-8] As a result, a higher rolling force is required.[9] However, if the rolling force increases, it may lead to roll breaking and bring great losses to the production and management.[10]

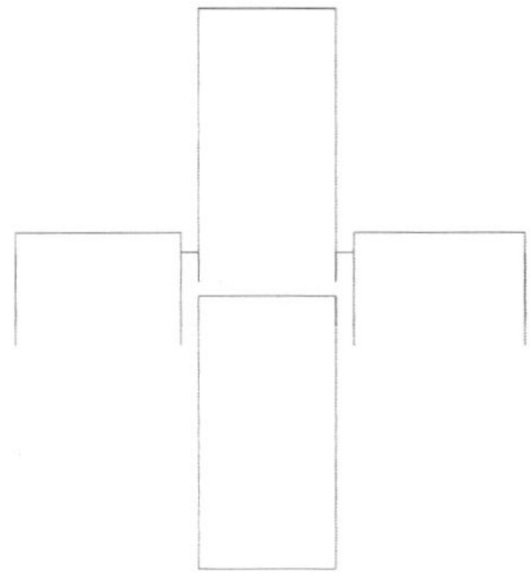

**Figure 7.** Ordinary universal rolling mill.

Therefore, the physical contradiction existing in the rolling of profile steel is that the rolling force is expected to be both, high and low.[11]

On the basis of a great deal of research, analysis and summary of patents, Altshuler abstracted 40 important principles of invention in TRIZ theory with universal use, which is shown in table 2.

**Table 2.** Important principles of invention in TRIZ theory.

| | | | |
|---|---|---|---|
| 1. Fragmentation | 11. Cushion in advance | 21. Rushing through | 31. Porous materials |
| 2. Extraction | 12. Equipotentiality | 22. Convert a harm into a benefit | 32. Change the color |
| 3. Local conditions | 13. Inversion | 23. Feedback | 33. Homogeneity |
| 4. Asymmetry | 14. Spheroidality | 24. Mediator | 34. Rejecting regenerating parts |
| 5. Consolidation | 15. Dynamicity | 25. Self-service | 35. Transform the physical/chemical state |
| 6. Universality | 16. Partial or excessive actions | 26. Copying | 36. Phase transformation |
| 7. Nesting | 17. Shift to a new dimension | 27. Disposable objects | 37. Thermal expansion |
| 8. Anti-weight | 18. Mechanical vibration | 28. Replacement of mechanical system | 38. Strengthen oxidation |
| 9. Prior counteraction | 19. Periodic action | 29. Pneumatics or hydraulic construction | 39. Inert environment |
| 10. Prior action | 20. Continuity of useful action | 30. Flexible "shells" or thin films | 40. Composite materials |

The solving physical contradiction guide table in TRIZ theory was used. According to the power which is the general engineering parameter of 15, the corresponding principle of invention is 35,03,13,10,17,19,28. Through comparative screening, 35 this scheme was considered to be more feasible. 35 is the 35th of the 40 principles of invention-transform the physical/chemical state. Applying this principle, we changed each horizontal roll of the universal mill into one horizontal work roll and one horizontal support roll, and changed each vertical roll into one working vertical roll and one supporting vertical roll. In this way, the universal mill improved the rolling precision and offered a higher rolling force. We were able to produce the hot rolled section steel with a thin size and a higher precision to reduce the residual stress in the section steel, which met the user's requirements better, and promoted energy saving and environmental protection. Besides the breakage of the roll was very difficult using the developed strategy.[12] This formed a brand-new universal rolling mill scheme shown in figure 8 (four more rollers than figure 7).

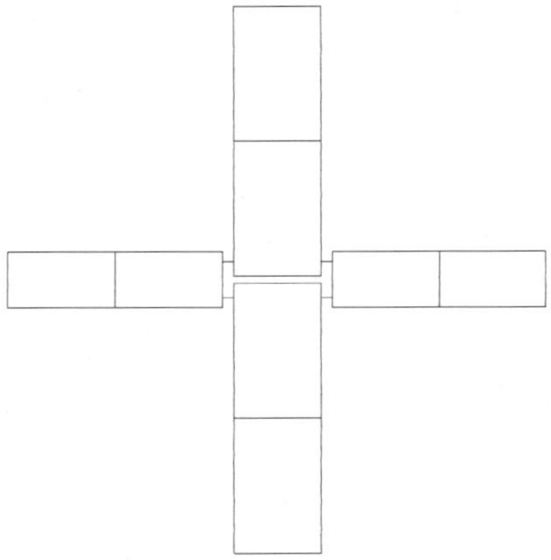

**Figure 8.** Developed new universal rolling mill.

The patent formation of the scheme is shown in table 3.

**Table 3.** Patent formation of the universal steel rolling type.

| Patent type | Name | Patent number | Application date | Date of announcement of authorization |
|---|---|---|---|---|
| Utility model | A universal steel rolling type | 201020678371.2 | 2010-12-24 | 2011-09-07 |

## 3. Conclusion

With the application of the TRIZ theory in our study, three major conclusions were pointed out as follows:

### *3.1. Stimulate innovative thinking*

The application of TRIZ theory to solve problems made it possible to do things that could not be done before, and made difficult problems easier and clearer.

### *3.2. Efficient problem solving*

The problems encountered in the iron and steel production were solved quickly.

### *3.3. Universality of participation*

Everyone were able to participate in the great deal of activities of invention and creation.

## References

[1] Altshuler G S 2013 *Explore Creativity: TRIZ Introduction* (Beijing, China: Science Press)
[2] Altshuler G S 2008 *The Innovation Algorithm: TRIZ, Systematic Innovation and Technical Creativity* (Wuhan, China: Huazhong University of Science and Technology Press)
[3] Kwatra S and Salamatov Y 2012 *Trimming, Miniaturization and Ideality via Convolution Technique of TRIZ* (Berlin,Germany: Springer Science & Business Media)
[4] Theory of inventive problem solving available from: *http://www.mazur.net/triz/*
[5] Chybowski L 2018 Use of triz su-field models in the process of improving the injector of an internal combustion engine *XV Int. Conf. Multidisciplinary Aspects of Production Engineering* (Poland: Zawiercie) **1** 257-68
[6] Tan J 2010 Development of rolling process for H section steel *Tianjin Metallurgy* 2010 17-9
[7] Feng W, Zhang X S and Kang Y L 2005 Modern pass design technique progress *Shandong*

*Metallurgy* **27** 33-5
[8] Yang Z D and Ma Y H 2010 Practice and Research on roll stability control of four roll reversing mill *China Metallurgy* **20** 32-5
[9] Wei W, Fei Y, Li J and Gou X Y 2012 Analysis and improvement of low temperature impact properties of H steel used in high cold area *China Metallurgy* **22** 18-20
[10] Liu R G, Wang F L and Xu Y 2011 Analysis of the technology of mixing pass rolling channel steel for universal continuous rolling mill *China Metallurgy* **21** 40-3
[11] Huang Z Y, Wu S F and Chen M X 2011 Study on development and application of controlled cooling process for H steel *Hot Working TECH* **40** 154-6
[12] Wang J, Wu B Q, Wu M Z, Yan J C, Xia M, Xing J and Gong Q H 2018 Analysis of crack defects in flange of S460M hot rolled H-beam *Anhui Metallurgy* **20** 8-52

IOP Conf. Series: Materials Science and Engineering **668** (2019) 012013 doi:10.1088/1757-899X/668/1/012013

# Validation of constitutive models for experimental stress-strain relationship of high-strength steel sheets under uniaxial tension

**J J Chen[1], C W Lian[2,3,4] and J P Lin[4,5]**

[1]Pan Asia Technical Automotive Center, 2199 Jufeng Road, Shanghai, 201208, China
[2]Research Institute of Baoshan Iron & Steel Co., Ltd., 655 Fujin Road, Shanghai 201900, China
[3]State Key Laboratory of Development and Application Technology of Automotive Steels (Baosteel), 655 Fujin Road, Shanghai 201900, China
[4]School of Mechanical Engineering, Tongji University, 4800 Caoan Road, Shanghai 201804, China

E-mail: jplin@tongji.edu.cn

**Abstract.** The constitutive relationship between stress and strain is one of the important issues in numerical simulation of metal forming. For steel, there are two kinds of constitutive models wildly used. One is power law like Ludwik and Swift. The other is saturation models like Voce and Hockett-Sherby. It is necessary to determine the most appropriate model and parameters for newly developed high strength steels. In order to describe the ultra-high strength dual phase and quenching and partitioning steels, uniaxial tensile tests with the aid of digital image correlation techniques were performed to analyse the strain field over the whole gauge section as well as the element close to the necking point. Through experimental data analysis and numerical fitting, the validity of different models have been confirmed. It shows that equation of Hockett-Sherby model has the best fitting accuracy and high strength steels require a model with more parameters. The extrapolated weighting coefficient of combined Swift and Hockett-Sherby models is characterized, which is proved to improve the accuracy of the numerical simulation of forming of ultra-high strength steels based on the conventional tensile tests.

## 1. Introduction

Along with the requirements of energy-saving, safety and comfort of vehicles, lightweight of car body has become an important trend in automotive industry. For the purpose of lower thickness but same strength level, application of high strength steel is one of the most economical way to reduce the weight of car body with acceptable performance. Though many high strength steels have been developed and applied to automotive industry in recent years, challenges remain for the forming simulation, which constrains the widely application of high strength steels [1,2]. Constitutive relationship, which is one of the most important parameters during metal sheet forming, describes the stress strain behavior under external force of metal sheet. An appropriate constitutive relationship could help to improve the accuracy of forming simulation, which brings much benefit to the application of high strength steels.

As early as in 1909, Ludwik put forward the concept of the strain-hardening exponent (n) [2]. Through theoretical analysis, Hollomon established the exponential equation (n) from experience in metal tensile test in 1944 [3]. Then more models of constitutive equation describing the relationship of true stress and strain were proposed in literature. For steels, models like Ludwik, Ghosh [4], Hockett-Sherby [5], Swift [6], and Voce [7] have been widely used. Combinations of models such as the Swift-Voce or Swift/Hockett-Sherby models could better optimize the modeling results [7]. The uniaxial tensile testing is the most convenient method to generate stress-strain data of homogeneous deformation stage using rectangular cross-section specimen. Kang *et al* [8] studied compatibilities of different constitutive equations for the determination of yield and ultimate tensile strengths by tensile true stress vs. true strain curves of 27 metal alloys. To identify the constitutive model of newly developed high strength steel, like dual-phase (DP) steel and quenching and partition (QP) steel, the applicability of each model needs to be scrutinized.

However, the stress vs. strain curve obtained by these methods is limited to a certain strain level due to the diffuse necking in the uniform elongation of the tensile specimens. The equivalent strain of uniform elongation during uniaxial tensile testing is normally much lower than that during metal forming. There is significant difference at larger strains by extrapolating the experimentally obtained stress-strain curve to larger strain values with assuming constitutive equations [9]. There are several experimental methods to obtain stress vs. strain curves up to a larger strain [10,11]; however, more complex experiment equipment and numerical analysis are still required. The strain hardening behavior at these large strains can be better extracted from the tensile tests with some additional non-standard measurements and associated analyses. One method is firstly to measure the current cross-section area at the narrowest section of a neck, obtaining the average axial true strain and then to compute the average axial true stress by applying a correction factor to obtain the effective true stress vs. true strain curves at larger strains [12,13]. Optical full-field measurement with digital image correlation technique is also helpful to measure local strain accurately during tensile testing [14]. Another way is an inverse method based on iteratively correcting an initially assumed post-necking strain hardening behavior by minimizing the difference between FEA and experiment results of force-displacement curve, or total works (real or virtual, external or internal) within the gauge section [15-17].

In this work, 10 typical steels from mild steel to ultra-high strength steel were selected to perform uniaxial tensile tests, extract stress vs. strain data of homogeneous deformation, and evaluate the fitting precision of different models including Ludwick, Swift, Ghosh and Hockett-Sherby. For characterizing the hardening behavior of ultra-high strength steel, digital image correlation techniques were employed to obtain accurate strain data of uniaxial tensile testing, and combinations of models were applied to fitting the experimental true stress vs. true strain data.

## 2. Experiment materials

**Table 1.** Mechanical properties of testing materials.

| Type | Steel grade | Thickness mm | Rp0.2/ReL MPa | Ae % | Rm MPa | Agt % |
|---|---|---|---|---|---|---|
| Mild Steel | DC04 | 0.8 | 141 | 0.269 | 294 | 26.0 |
| HSS IF | HC260Y | 1.0 | 303 | 0.345 | 449 | 19.3 |
| BH | HC220B | 0.7 | 225 | 0.307 | 354 | 22.0 |
| HSLA | HC420LA | 1.5 | 476 | 4.394 | 525 | 15.4 |
| DP | HC340/590DP | 1.5 | 388 | 0.383 | 620 | 18.5 |
| | HC500/780DP | 2.0 | 587 | 0.476 | 825 | 6.30 |
| | HC550/980DP | 1.2 | 667 | 0.530 | 1005 | 8.61 |
| | HC700/980DP | 1.2 | 796 | 0.584 | 1104 | 6.60 |
| QP | HC600/980QP | 1.2 | 672 | 0.525 | 1052 | 16.9 |
| | HC820/1180QP | 1.0 | 881 | 0.618 | 1192 | 12.7 |

Cold-rolled mild steel, high strength interstitial-free steel, bake-hardening steel, high strength low alloy steel, DP and QP steel are widely used in car body manufacturing. 10 grades of steel sheets were selected from various categories, with nominal thicknesses ranging from 0.7 mm to 2.0 mm. All materials were produced by Baosteel for the application in automotive industry. The basic mechanical properties of these materials are listed in table 1. The steel grade is named according to Baosteel enterprise standards(Q/BQB 408 — 2018, Q/BQB 418 — 2018 and Q/BQB 419 — 2018). Taking HC340/590DP as an example, "HC"stands for high strength steel with cold rolling, and 340/590 is the minimum yield strength and minimum tensile strength respectively.

## 3. Methodology

Uniaxial tensile testing specimens were processed into the dimensions recommended by ISO 6892-1-2009, as illustrated in figure 1. Wire cut electrical discharge machining was applied to cut the specimen for improving the edge quality. The axial direction of all specimens is along the rolling direction. Sample surface of machining area was polished before testing. The tensile test was operated on a material testing system and the crosshead speed was set to 3 mm/min. Strain was measured using video extensometer by calculating the length change of gauge area. In order to measure full-field strain of DP and QP steel during tensile process, digital image correlation method was used by spraying random speckles on one surface of each specimen. Average axial strain within gauge area (i.e. with a length of 50 mm and a width of 12.5 mm) and local true strains at different segments were calculated respectively.

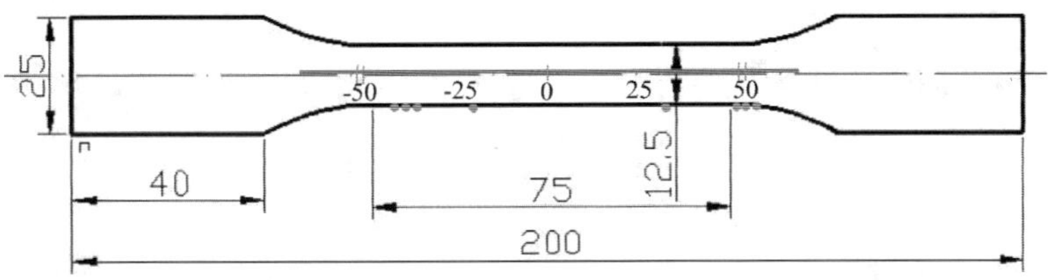


**Figure 1.** Dimensions of the uniaxial tensile testing specimens (unit in mm).

Before the onset of diffuse necking, average true stress ($\sigma$) and true strain ($\varepsilon_T$) during homogeneous deformation can be calculated by equations (1)-(3):

$$\sigma = \sigma_E(1 + \varepsilon_E) \tag{1}$$

$$\varepsilon_T = \ln(1 + \varepsilon_E) \tag{2}$$

$$\sigma_E = \frac{F}{A_0} \tag{3}$$

where $\sigma_E$ is the engineering stress, F is the tensile force recorded by the tensile testing machine, $A_0$ is the original cross section area of the parallel segment, and εE is the average engineering strain of the gauge segment. By subtracting elastic strain, true plastic strain ε is achieved by equation (4), where E the elastic modulus.

$$\varepsilon = \varepsilon_T - \frac{\sigma}{E} \tag{4}$$

Then true stress vs. true strain data of homogeneous deformation was used to fit different models using the non-linear least squares method with an algorithm of trust-region. Table 2 summarizes corresponding model equations with σ true stress, ε true plastic strain, and α, C1-C4 modeling parameters.

R-square value is describes the goodness of fit, i.e. the ratio of the sum of squares of the regression and the total sum of squares. The value closer to 1 indicates that a more suitable model for the investigating data.

IOP Conf. Series: Materials Science and Engineering **668** (2019) 012013 doi:10.1088/1757-899X/668/1/012013

**Table 2.** Equations of constitutive models.

| Model | Equation | |
|---|---|---|
| **Ludwik** | $\sigma = C_1 + C_2\varepsilon^{C_3}$ | (5) |
| **Swift** | $\sigma = C_1(C_2 + \varepsilon)^{C_3}$ | (6) |
| **Ghosh** | $\sigma = C_1(C_2 + \varepsilon)^{C_3} - C_4$ | (7) |
| **Hockett-Sherby** | $\sigma = C_2 - (C_2 - C_1)e^{-C_3\varepsilon^{C_4}}$ | (8) |
| **Swift/ Hockett-Sherby** | $\sigma = (1-\alpha)\sigma_{Swift} + \alpha\sigma_{Hockett\text{-}Sherby}$ | (9) |

## 4. Results and discussion

True stress vs. true plastic strain data before the onset of diffuse necking from tensile testing of 10 steels were extracted to evaluate the fitting precision of different models including Ludwick, Swift, Ghosh and Hockett-Sherby. For further characterization of the hardening behavior of ultra-high strength steels, accurate strain data of HC550/980DP and HC600/980QP was analyzed. By calculating the true strain of the segment close to necking points, more stress vs. strain data beyond the conventional stain calculation method was achieved. Combination of Swift and Hockett-Sherby model was applied to fitting the true stress vs. true plastic strain data and identifying the weighting parameter in the model.

Figure 2 shows the true stress vs. true plastic strain curves of the tested 10 steels and fitting curves for 4 models. Figure 3 shows the analyzed result of goodness of fit of these models for different steels by considering tensile strength and strain hardening exponent.

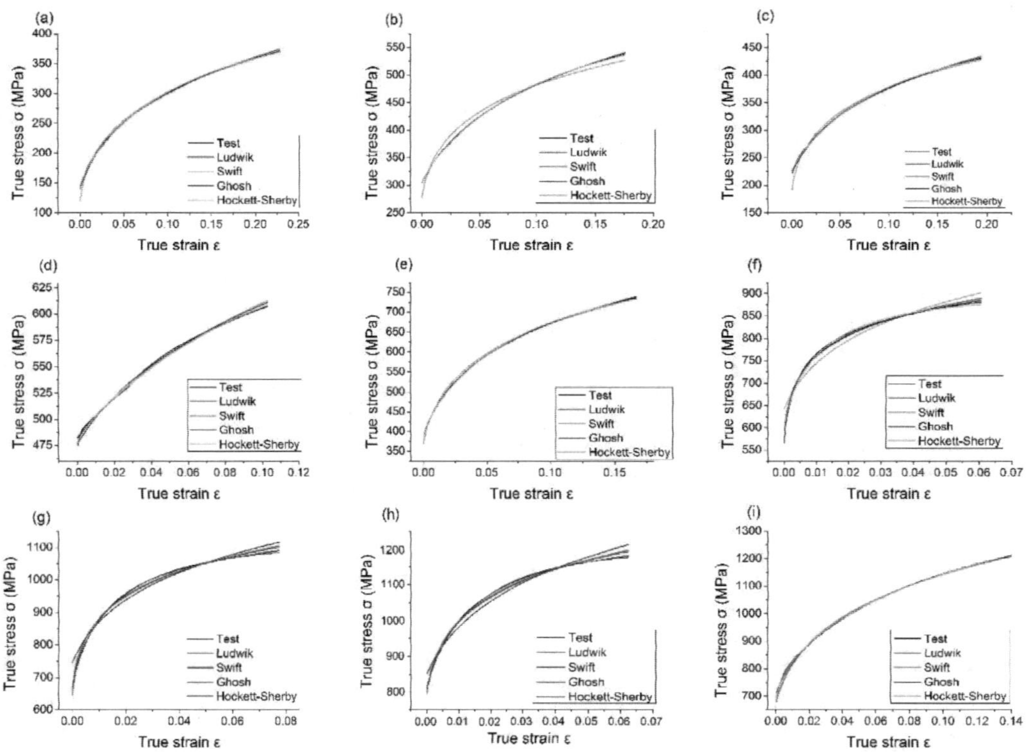

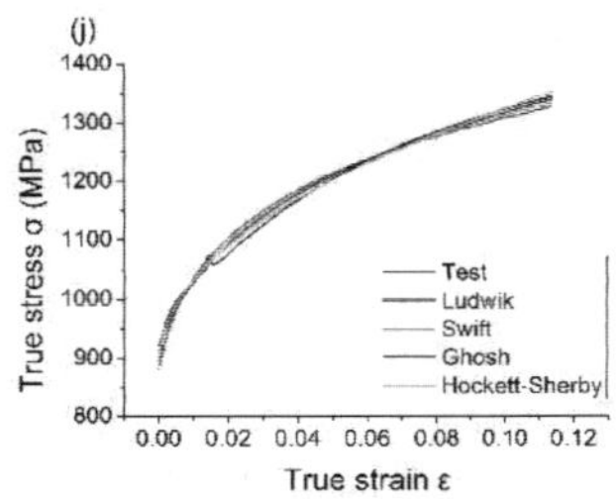


**Figure 2.** Compare of testing and models fitting results. (a) DC04; (b) HC260Y; (c) HC220B; (d) HC420LA; (e) HC340/590DP; (f) HC500/780DP; (g) HC550/980DP; (h) HC700/980DP; (i) HC600/980QP; (j) HC820/1180QP.

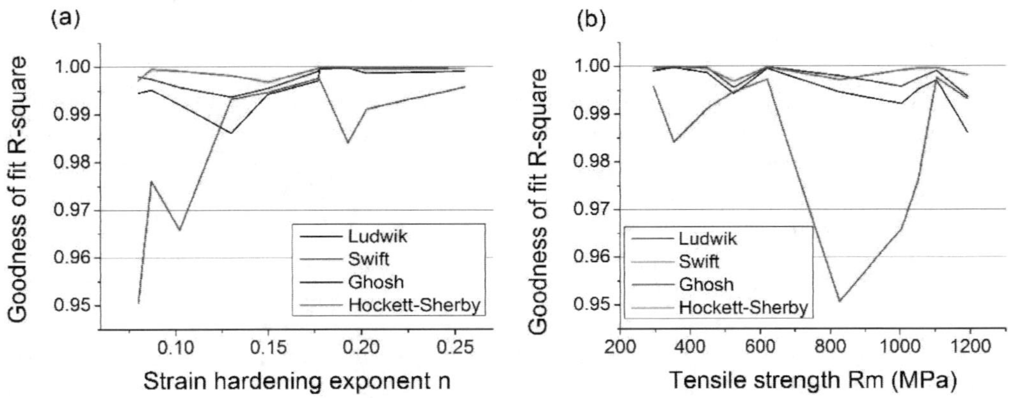


**Figure 3.** (a) Goodness of fit versus strain hardening exponent; (b) Goodness of fit versus tensile strength.

The results show that all 4 models are capable to describe the constitutive relationship of stress and strain according to data from uniform deformation stage of uniaxial tensile test. Among these models, Hockett-Sherby is the best one, followed by Ghosh and Ludwik. Equation of Hockett-Sherby model with 4 parameters has more degrees of freedom to fit the hardening behavior of steel. By examining the effect of strain hardening exponent on goodness of fit, figure 3 indicates that steel with a larger n value resulting a better fitting result. In comparison with low strength steels and high strength steels (with a tensile strength lower than 780 MPa), showing in figure 3(b), ultra-high strength steel (with a tensile strength higher than 780 MPa), are more complex and require a model with more parameters.

*4.1. Model validation at larger strains of DP and QP steels*

It has been pointed out that there is a significant difference at larger strains by extrapolating the experimentally obtained stress vs. strain curve to larger strains by assuming constitutive equations. Optical full-field measurements with digital image correlation techniques provide a feasible solution to measure local strain accurately during tensile testing. As illustrated in figure 1, 101 points of the axial section where the maximum longitudinal strain locates were selected, where '0' was the middle point, '-50' and '50' were points at both ends. The distance of adjacent points was about 0.78 mm. The strain history of each pair of points from initial deformation to fracture was obtained and the strain of all points versus time of HC600/980QP is shown in figure 4(b). By analyzing the true strain of the segment close to the final localized necking point within diffuse necking stage that is from the onset of diffuse necking "t3" to the onset of localized necking "t6" in figure 4(b), more stress vs. strain data can be acquired than conventional method by measuring the average strain over the whole gauge. Figure 4(a) shows true stress vs. true plastic strain curve of point near final localized necking point, comparing with data of whole gauge section.

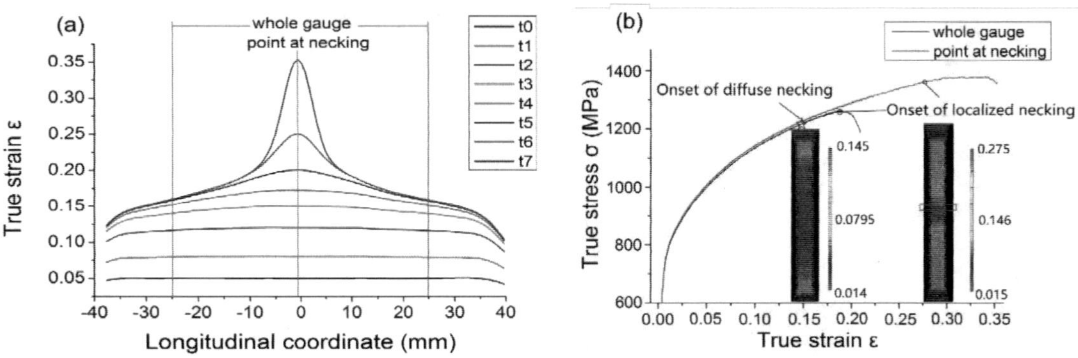


**Figure 4.** Tensile data of HC600/980QP. (a) Strain distribution at different time; (b) True stress-strain curve using different gauge.

The advantage of the digital image correlation method is that the strain at any location in the gauge section is effective until the fracture of specimen when the digital image correlation method is applied, while the strain measured by extensometer after the necking is not accurate anymore. The onset of localized necking is determined by Hill localized necking theory [18]. Without consideration of stress triaxiality variation within the diffuse necking stage, the data from the onset of diffuse necking to localized necking is able to validate the constitutive model and determine the weighting parameter of the combined model. Figure 5 shows the validation result of HC550/980DP and HC600/980QP, where the parameter of Swift and Hockett-Sherby model was calculated by fitting of data from homogeneous stage. According to equation (9), the weighting parameter $\alpha$ of combination of Swift and Hockett-Sherby model is estimated by the least square method, which is 0.66 and 0.8 for HC550/980DP and HC600/980QP, respectively. The extrapolation of combination model obtained by digital image correlation method can further improve the accuracy of forming simulation for ultra-high strength steels.

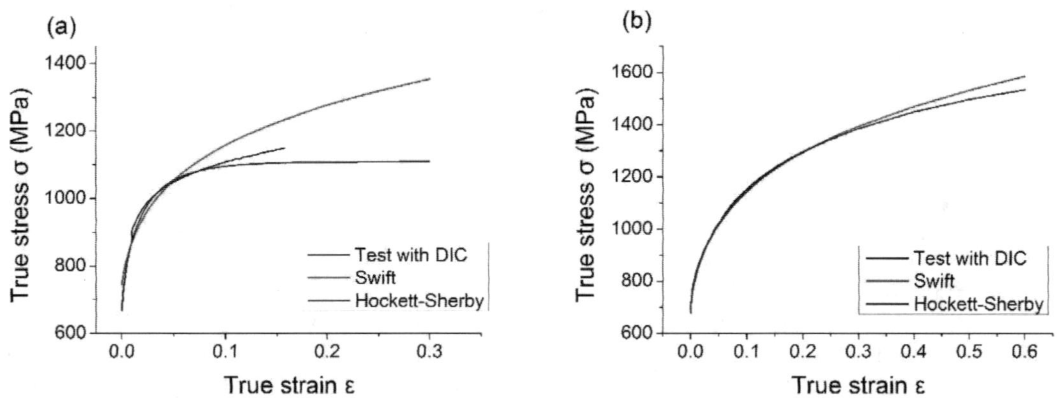


**Figure 5.** Model validation result at larger strains. (a) HC550/980DP; (b) HC600/980QP.

## 5. Conclusions

- The selected models are capable to describe the constitutive relationship of stress and strain during the homogeneous deformation in uniaxial tensile testing. Hockett-Sherby model is the best, followed by Ghosh and Ludwik models.
- In comparison with the traditional steels, high strength steels and ultra-high strength steels are more complex and require a model with more parameters.
- By analyzing the full strain field with digital image correlation method, the validity of model extrapolating to large strain can be examined, and more accurate model are achieved to

describe the hardening behavior of ultra-high strength steels.

**References**

[1] Lian C W, Chen X P and Ye Y 2017 Formability of the 3rd generation high strength steel - QP steel based on part application *J. Netshape Forming Eng.* **9** 79-84
[2] Ludwik P 1909 *Elemente der Technologischen Mechanik* (Berlin Heidelberg: Springer)
[3] Hollomon J H and Member J 1945 Tensile Deformation *Metal Technol.* **12** 268-90
[4] Ghosh A K 1977 Tensile instabiity and necking in materials with strain hardening and strain-rate hardening *Acta Metall.* **25** 1413-24
[5] Hockett J E and Sherby O D 1975 Large strain deformation of polycrystalline metals at low homologous temperatures *J. Mech. Phy. Solid.* **23** 87-98
[6] Swift H W 1952 Plastic instability under plane stress *J. Mech. Phy. Solid.* **1** 1-18
[7] Voce E 1948 The relationship between stress and strain for homogeneous deformation *J. Instit. Metal* **74** 537-62
[8] Kang S, Kim Y, Kim K, Kwon D and Kim J 2014 Constitutive equations optimized for determining strengths of metallic alloys *Mech. Mater.* **73** 51-7
[9] Ling Y 1996 Uniaxial true stress-strain after necking *AMP J. Technol.* **5** 37-48
[10] Kessler L and Gerlach J 2006 The impact of materials testing strategies on the determination and calibration of different FEM material models *Conf. Proc. IDDRG* pp 113-20
[11] Capilla G, Hamasaki H and Yoshida F 2017 Determination of uniaxial large-strain workhardening of high-strength steel sheets from in-plane stretch-bending testing *J. Mater. Processing Technol.* **243** 152-69
[12] Zhang Z L, Deg Rd J and Vik O P S 2001 Determining true stress–strain curve for isotropic and anisotropic materials with rectangular tensile bars: method and verifications *Comput. Mater. Sci.* **20** 77-85
[13] Wang L and Tong W 2015 Identification of post-necking strain hardening behavior of thin sheet metals from image-based surface strain data in uniaxial tension tests *Int. J. Solid Struct.* **75-76** 12-31
[14] Zhu F, Bai P, Zhang J, Lei D and He X 2015 Measurement of true stress–strain curves and evolution of plastic zone of low carbon steel under uniaxial tension using digital image correlation *Optics and Lasers in Engineering* **65** 81-8
[15] Zhao K, Wang L, Chang Y and Yan J 2016 Identification of post-necking stress–strain curve for sheet metals by inverse method *Mech. Mater.* **92** 107-18
[16] Coppieters S, Kim J H, Denys K, Cooreman S and Debruyne D 2017 On complete solutions for the problem of diffuse necking in sheet metal *Procedia Eng.g* **207** 2012-7
[17] Tardif N and Kyriakides S 2012 Determination of anisotropy and material hardening for aluminum sheet metal *Int. J. Solid Struct.* **49** 3496-506
[18] Ding L, Lin J, Min J, Pang Z and Ye Y 2013 Necking of Q & P steel during uniaxial tensile test with the aid of DIC technique *Chin. J. Mech. Eng.* **26** 448-53

# A modified mathematical model for end-point carbon prediction of BOF based on off-gas analysis

**W H Lin[1], J K Sun[1], K X Zhou[1], Q Liu[1,4], S Q Jiao[1,4], J S Zhang[1], X Su[2], M Liu[2], Y C Huang[2] and N Li[3]**

[1]State Key Laboratory of Advanced Metallurgy, University of Science and Technology Beijing, Beijing, China
[2]Xinyu Iron and Steel Group Co., Ltd., Jiangxi, China
[3]Western Superconducting Technologies Co., Ltd., Xi'an, China

E-mails: qliu@ustb.edu.cn (Q Liu)/ sjiao@ustb.edu.cn (S Q Jiao)

**Abstract.** Several models for end-point carbon prediction of BOF(Basic Oxygen Furnace) based on off-gas analysis were studied in this paper. The advantages and disadvantages of the integral model, the exponential decay model and the cubic fitting model were analyzed respectively. Based on analysis of the characteristics of the decarburization rate curve, a new exponential model was established by the introduction of a correction algorithm. The principle of the proposed model involves applying the decarburization rate curve and the descending gradient of the historical heats to obtain the average decarburization curve and reference decarburization efficiency coefficient using the regression fitting method. According to the deviation between the actual and the predicted decarburization curves, the decarburization efficiency coefficient was corrected to improve the prediction accuracy. Plant trials were carried out in a 210 t converter to compare the performance of the mentioned models. The results showed that the new model exhibited better adaptability and higher accuracy than the other ones. The hit ratio of the new model reached more than 90% for the prediction of end-point carbon content within a tolerance of ±0.02%.

## 1. Introduction

Off-gas analysis technique is one of the most important methods for end-point control of converter steelmaking. Its core principle is using the information of CO and $CO_2$ in the off-gas to estimate the decarburization reaction and the carbon content in the bath, so as to realize the dynamic prediction of the blowing process and the control of the blowing end-point. In recent years, Dofasco [1,2] in Canada, ILVA Taranto [3,4] in Italy, Tangshan ISCO [5] in China and many other steel plants have achieved satisfied results with the application of off-gas analysis technology.

IRSID® developed a technique for continuously measuring the CO and $CO_2$ contents of the off-gas from the basic oxygen furnace and using these data to compute the decarburization rate of the bath [6]. Based on this technique, MacFarlane *et al* [7] and Meyer *et al* [8] built a carbon integral model to calculate the variation of carbon content in the bath. However, this model usually presents large deviations because of the poor detection precision of the off-gas flow and composition as well as the unstable accuracy of the original information of the material, and the deviations are difficult to be eliminated. In order to improve the accuracy of BOF end-point carbon prediction, the optimization of algorithms also drew increasing attention. Meyer and Glasgow *et al* [9] proposed an exponential correspondence between specific decarburization rate and the carbon content of the bath, and established an exponential decay model. On this basis, Liu *et al* [10] corrected the exponential model parameters by means of isometric multi-point continuous correction. Li *et al* [11] established a modified exponential model by the introduction of an index of bath mixing degree, and the hit ratio of the model was improved. Zhang *et al* [12] proposed a cubic fitting model by comparing the off-gas data and the sub-lance detection data at the end of the blowing, and fit them according to different oxygen lance heights. As the understanding of decarburization mechanism is still inadequate and the blowing process could be dramatically different, the documented models fail to be applied in practices due to the poor hit rate and unstable prediction results. As to be applied to control the actual operations of industrial steelmaking process, the accuracy and stability of the off-gas model needs to be further improved.

The basic principles, main advantages and disadvantages of the above mentioned algorithms for carbon prediction based on off-gas analysis are systematically studied in this paper. A new exponential model correction algorithm is proposed by analysing the characteristics of the actual off-gas curve in a steelmaking process and the prediction accuracy is improved.

## 2. Characteristics of off-gas during BOF blowing process

In BOF blowing process, hot metal in the furnace and high-speed oxygen jet ejecting from the top lance undergo intense reactions of decarburization and combustion. Then, molten steel, slag and off-gas are obtained as the products of these reactions. There are three main reaction zones in the furnace, which are name as the oxygen jet impact reaction zone, the slag-metal interface reaction zone and the emulsion phase reaction zone [13], as shown in figure 1.

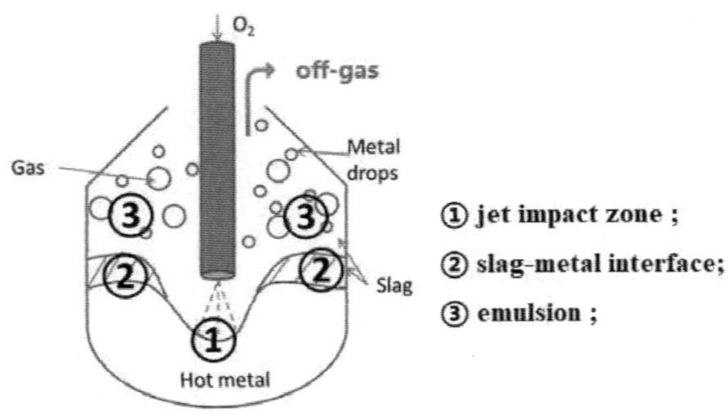


**Figure 1.** Three zone of oxidation reactions in BOF steelmaking [13].

The main chemical reactions occurring in each reaction zone are as follows:
① jet impact zone

$$[C] + {}^{1}/_{2}\,O_2 = CO \tag{1}$$

$$CO + {}^{1}/_{2}\,O_2 = CO_2 \tag{2}$$

$$[Si] + O_2 = SiO_2 \tag{3}$$

$$[Mn] + {}^{1}/_{2}\,O_2 = MnO \tag{4}$$

$$[Fe] + {}^{1}/_{2}\,O_2 = FeO \tag{5}$$

② slag-metal interface and ③ emulsion phase reaction zone

$$[C] + (FeO) = CO + [Fe] \tag{6}$$

$$[Si] + 2(FeO) = (SiO_2) + 2[Fe] \quad (7)$$

$$[Mn] + (FeO) = (MnO) + [Fe] \quad (8)$$

$$2[P] + 5(FeO) + 4(CaO) = (4CaO \cdot P_2O_5) + 5[Fe] \quad (9)$$

Among them, the CO and $CO_2$ generated by the carbon-oxygen reaction become the main component of the converter off-gas. Most of the oxidation products obtained by other oxidation reactions enter the slag phase, and the remaining become dust. A typical variation curve of CO and $CO_2$ in the off-gas is shown in figure 2.

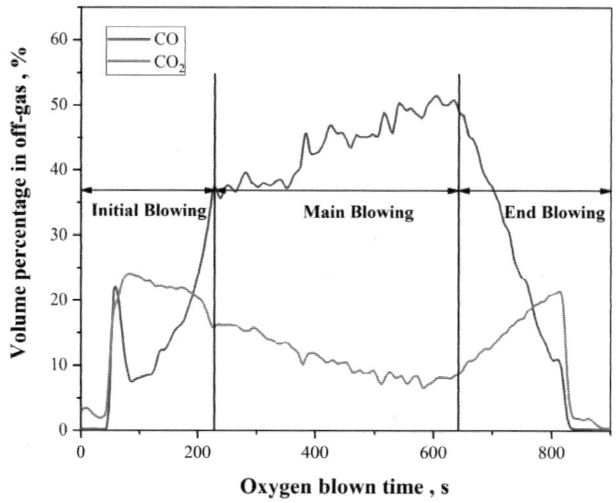


**Figure 2.** Typical variation curve of CO and $CO_2$ in the off-gas.

In the initial blowing stage of BOF (initial 240s of oxygen blown time in figure 2), only a few part of the supplied oxygen is used for the oxidation reaction of carbon, due to the preferential oxidation of [Si] and [Mn] in the melt, and the content of CO and $CO_2$ in the off-gas is low and tends to increase. With the decrease of [Si] and [Mn] contents and the increase of bath temperature, the carbon-oxygen reaction occurs intensely, and maintains a relatively stable and high decarburization rate in the main blowing stage (oxygen blown time 240~650 s). The oxygen supplied during the main blowing stage is mainly used for the decarburization. Once the carbon reaches a threshold carbon concentration in the bath, the carbon-oxygen reaction rate decreases sharply, and the CO and $CO_2$ content in the off-gas also decreases sharply. It can be seen from figure 2 that the variation trend of the BOF off-gas composition curve throughout the blowing process is basically consistent with the description of the three-stage decarburization theory [14].

## 3. Mechanism of carbon prediction by off-gas analysis

According to the information of composition and flow rate of off-gas and oxygen blown rate, the decarburization rate and specific decarburization rate of the metal bath can be obtained using the following equations.

$$v_C = Q_{\text{off-gas}} \times \varphi(CO + CO_2) \times \frac{12}{22.4} \quad (10)$$

$$R = \frac{v_C}{Q_{O_2}} \quad (11)$$

Where the decarburization rate $v_C$ is defined as the decarburization quantity of the metal bath per unit time [6], and the unit is kg/s, which is greatly affected by the oxygen blown rate $Q_{O_2}$ with the unit Nm$^3$/s; and the specific decarburization rate $R$ is defined as the quantity of carbon removed per volume of oxygen blown [9], and its unit is kg/m$^3$, which is less affected by the oxygen blown rate; and $Q_{\text{off-gas}}$ is the flow rate of off-gas, the unit is Nm$^3$/s; $\varphi(CO + CO_2)$ is the volume percentage of CO and $CO_2$ in the off-gas.

Since the composition and flow rate of the off-gas are continuously detected during the blowing process, $v_C$ and R can be obtained, which allows a real-time prediction of the carbon content from the metal bath. Thus, making full use of these two parameters is the key to establish an effective end-point carbon predict model based on off-gas analysis.

## 4. Advantages and disadvantages of existing off-gas models

The integral model, the exponential decay model, and the cubic fitting model are the main models used for end-point carbon prediction of BOF based on off-gas analysis.

### *4.1. The integral model*

During the blowing of oxygen into the converter bath, the carbon in the bath is oxidized to CO and $CO_2$. Under the premise of knowing the initial carbon content of the metal bath, the real-time carbon content of the bath can be predicted by integrating the cumulative decarburization during the process. The integral model expression is shown in the following formula:

$$C_t = C_{HM} - \frac{\int_0^t v_C d_t}{M_{(t)}} \tag{12}$$

where $C_t$ is the carbon content at time *t*,%; $C_{HM}$ is the initial carbon content of the bath,%; $v_C$ is decarburization rate, kg/s; $M_{(t)}$ is the total weight of liquid metal, kg.

Substituting the formula (10) into the formula (12), the carbon content of the bath at any time *t* can be obtained by integration during the blowing process. In theory, the carbon integral model can be used to calculate the real-time carbon content of the bath from the start to the end of the process. However, Meyer *et al* [8] and Cheng *et al* [15] found that the calculation error of the integral model caused by detection error of the initial carbon content, off-gas flow rate and component is much larger than the requirement of the end point carbon prediction.

Therefore, the carbon integral model is only applicable if the raw material data and instrumentation detection data are very accurate. Since the integral carbon model usually has large deviations that are difficult to eliminate, it is rarely used in industrial practices.

### *4.2. The exponential decay model*

According to previous studies, the oxidation reaction of carbon in main blowing stage is mainly controlled by oxygen blown rate. The decarburization rate is almost constant, and the decarburization efficiency reaches the maximum. In the end blowing stage, carbon content continues to decline and once the carbon reaches a threshold carbon concentration in the bath, the decarburization rate is limited by mass transport of carbon in the melt [16].

Meyer and Glasgow *et al* [9] studied typical paths of decarburization versus the carbon concentration which are shown in figure 3, and found that a family of curves of the type shown can be represented by a function in the following form:

$$R = A + Be^{KC} \tag{13}$$

The parameters $A$, $B$ and $K$ have unique values for any particular curve. $C$ is the carbon concentration corresponding to the specific decarburization rate. And that $A$, the asymptotic value of the curve, can be determined during actual refining. The value of the variable $K$ can then be determined from the ratio of the first derivative to the second derivative of the equation of the rate curve.

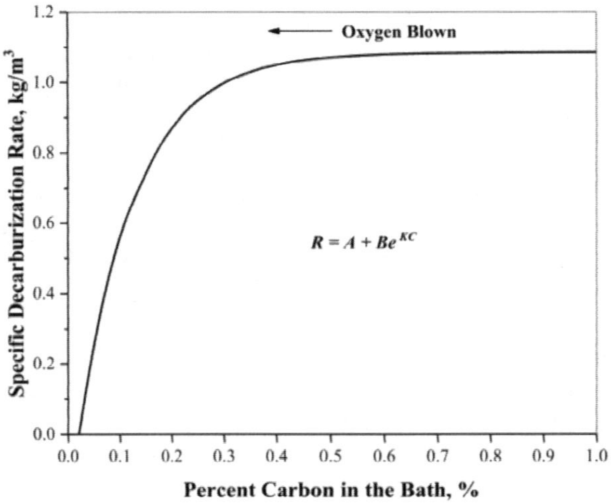


**Figure 3.** Typical end-point decarburization curves[9].

An exponential decay model was proposed based on equation (13). The main method of Meyer model is to average the decarburization curves of the historical heats to obtain a "historical average curve". Taking the "historical average curve" as a reference curve, the decarburization curve parameters of the current heat can be calculated with a single point correction method.

As shown in figure 4, when the maximum decarburization efficiency of the current heat differs greatly from the reference curve, the Meyer model calculation result will have a large calculation error. And due to the single point correction method, small fluctuations near the correction point can also cause large changes in the shape of the calculated curve.

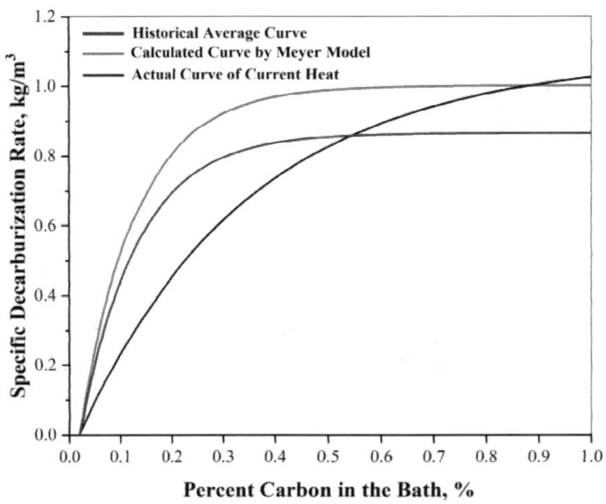


**Figure 4.** Comparison of Meyer curve and actual curve.

Based on the Meyer model, Liu *et al* [10] proposed a modified algorithm to perform isometric multi-point continuous correction on the calculation results by using the deviation between the actual decarburization curve and the calculated curve. This algorithm can eliminate the instability of single point correction method to a certain extent, but its main disadvantage is that some characteristics of the historical decarburization curves are not fully utilized. Therefore, the accuracy of this model is still not very high.

Tu *et al* [17] considered that there is a certain relationship between the decarburization rate and the carbon content in the end blowing stage of BOF process. The decarburization rate is used as an independent variable of the exponential function to predict the endpoint carbon content in their model. However, the stability of the model is not satisfactory, because the effect of the actual oxygen blowing flow rate on the decarburization rate is not considered. Li *et al* [11] considered the influence of operating parameters such as lance height, oxygen blowing rate and bottom blowing rate, and used the bath mixing degree [18] to modify the exponential model. Although the calculated hit ratio has been improved, the model fails to take the actual decarburization curve characteristics of current heat into consideration.

### *4.3. The cubic model*

Zhang *et al* [12] collected 100 heats of the off-gas and sub-lance measurement data at the end-point. The relationship between carbon content and decarburization rate of these heats were fitted by the exponential model, inverse scale model, quadratic model and cubic model, respectively. The cubic model was found to be the best, and the function is as follows:

$$w_{[C]} = C_0 + a_1 \cdot v_C + a_2 \cdot (v_C)^2 + a_3 \cdot (v_C)^3 \tag{14}$$

where $w_{[C]}$ is carbon concentration of the molten steel, %; $v_C$ is the decarburization rate, kg/s; $a_1$, $a_2$, $a_3$ are fitting coefficients of the model; $C_0$ indicates the ultimate carbon content when the decarburization rate is zero during oxygen blowing, normally about 0.02%.

This model does not consider the effect of the lance height and oxygen blowing rate on the decarburization rate. However, even if there is the same carbon content in the bath in the actual blowing process, the decarburization rate is different due to the difference in lance height and oxygen blowing rate. Generally, when using the cubic model, it is necessary to obtain different fitting coefficients for the decarburization data under different lance height to ensure the calculated hit ratio of the model. However, it has been found that the model coefficients fitted at different lance height are sometimes very different, even different in magnitude.

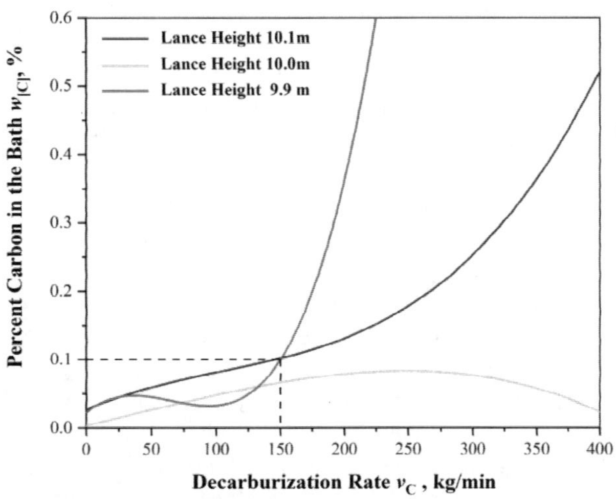


**Figure 5.** Fitting results at different lance heights using the Cubic model.

The fitting results of literature [12] are shown in figure 5, it can be seen that there is no uniform law to indicate the relationship between $w_{[C]}$ and $v_C$ according to the three fitting curves, and the cubic model fails to get reliable results when the carbon content is higher than 0.10%. So it can be considered that the cubic model does not actually have universal applicability, although it may achieve good fitting effects in a few specific conditions.

## 5. Correction algorithm for exponential decay model

### *5.1. Development of the correction algorithm*

In this section, the experiments based on above off-gas models are performed by introducing actual production data from a 210-ton BOF converter. The experimental results indicate that the exponential model in [11] reveals the highest hit ratio in comparison to other models. In summary, the exponential model is also the better candidate in terms of the overall performance.

On the basis of the morphology of the fitting curves shown in figure 6, better fitting degree of the exponential model is also found compared with cubic fitting model. Hence, a further research will be carried out based on the exponential model. To obtain a better hit ratio, a correction algorithm is proposed.

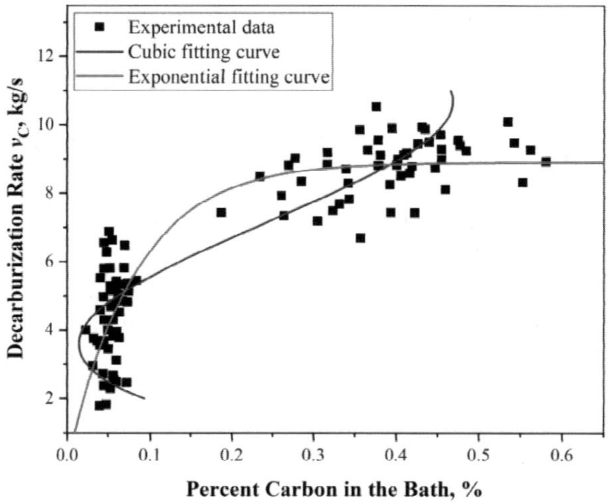


**Figure 6.** Comparison of fitting curves and experimental data.

In the literature, the study of the exponential decay model commonly includes two parts of modes: a) Decarburization rate is regarded as the relevant variable; b) Decarburization oxygen efficiency (also called specific decarburization rate by Meyer and Glasgow *et al* [9]) is regarded as the relevant variable. The details are displayed in equations (15) and (16).

$$v_C = k_1(1 - e^{-k_2(C-C_0)}) \tag{15}$$

$$R = \alpha + \beta e^{\gamma \cdot C} \tag{16}$$

where $C_0$ indicates the ultimate carbon content when the decarburization rate is zero during oxygen blowing; $k_1$, $k_2$, $\alpha$, $\beta$, $\gamma$ represent the fitting coefficients related to the decarburization rate or decarburization oxygen efficiency, in which the physical significances of $k_1$, $\alpha$ are the maximum decarburization rate , and the maximum decarburization oxygen efficiency respectively.

In previous methods, only the decarburization rate and decarburization oxygen efficiency of historical heats are used to obtain the model coefficients. In this paper, the indexes of descending gradient for the decarburization rate and decarburization oxygen efficiency are also considered, and the detailed expressions are shown as follows.

$$\nabla_{v_C} = k_1 \cdot k_2 \cdot e^{-k_2(C-C_0)} \tag{17}$$

$$\nabla_R = \beta \cdot \gamma \cdot e^{\gamma \cdot C} \tag{18}$$

Apart from the thermodynamic factors, the dynamic conditions, such as combined blowing of converter, also dramatically influence the limit decarburization of converter. Thus, the ultimate carbon content, $C_0$, is also taken as one of the regression parameters.

Then, the values of $k_1$, $k_2$, $C_0$, $\beta$, $\gamma$ could be obtained according to the decarburization rate curve and descending gradient data of historical heats.

$$\alpha = -\beta e^{\gamma \cdot C_0} \tag{19}$$

Using the values of $k_1$, $k_2$, $C_0$, $\alpha, \beta, \gamma$, the average decarburization curve of historical heats can be achieved following equations (15) and (16). Subsequently, the isometric multi-point continuous correction is performed according to the deviation degree between the decarburization results of current heat and the ones of historical heats, as shown in figure 7.

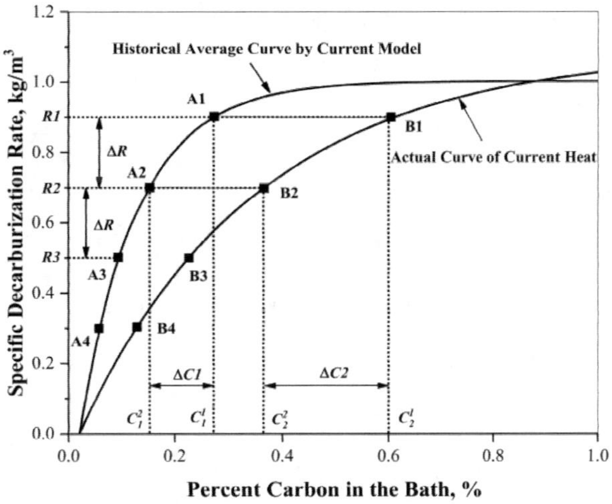


**Figure 7.** Carbon prediction use isometric multi-point continuous correction method.

The values of $\alpha$ in the two curves are the same through the normalization treatment, and meanwhile two curves both go through point ($C_0$,0).

$$R_A = \alpha + \beta_1 e^{\gamma_1 \cdot C} \tag{20}$$

$$R_B = \alpha + \beta_2 e^{\gamma_2 \cdot C} \tag{21}$$

$$\beta_2 e^{\gamma_2 \cdot C_0} = \beta_1 e^{\gamma_1 \cdot C_0} \tag{22}$$

When the coordinates of $A_1, A_2, B_1, B_2$ are set as $(C_1^1, R_1)$, $(C_1^2, R_2)$, $(C_2^1, R_1)$, $(C_2^2, R_2)$, these values can be substituted into equations (20) and (21).

$$R_1 = \alpha + \beta_1 e^{\gamma_1 \cdot C_1^1} \tag{23}$$

$$R_2 = \alpha + \beta_1 e^{\gamma_1 \cdot C_1^2} \tag{24}$$

$$R_1 = \alpha + \beta_2 e^{\gamma_2 \cdot C_2^1} \tag{25}$$

$$R_2 = \alpha + \beta_2 e^{\gamma_2 \cdot C_2^2} \tag{26}$$

The following equation can be achieved by combing equations form equation (23) to equation (26):

$$\gamma_2 \cdot (C_2^1 - C_2^2) = \gamma_1 \cdot (C_1^1 - C_1^2) \tag{27}$$

Then:

$$\gamma_2 = \gamma_1 \cdot \frac{\Delta C_1}{\Delta C_2} \tag{28}$$

where, $\Delta C_1$ is the decarburization amount of the reference curve, which can be obtained from known parameters; $\Delta C_2$ is the decarburization amount of actual curve, which can be calculated according to the off-gas composition and flow.

After obtaining, $\gamma_2$, put it into equation (22) to calculate $\beta_2$, and then put $\gamma_2$, $\beta_2$ and $\alpha$ into equation (26) to obtain the X-axis value of point $B_2$, as in follow.

$$C_2^2 = ln\left(\frac{R_2}{\beta_2} + e^{\gamma_2 \cdot C_0}\right)/\gamma_2 \tag{29}$$

The same $\Delta R$ was used to analogize $B_3$, $B_4$,..., $B_n$, until the target carbon content was achieved.

$$\Delta R = R_1 - R_2 = R_2 - R_3 = R_3 - R_4 = \cdots = R_{n-1} - R_n \tag{30}$$

### *5.2. Validation of the model*

Compared to previous algorithms, the main characteristics of the proposed correction algorithm involves three aspects: (1) Both the decarburization rate falling gradient and decarburization rate itself are used to calculate the model parameters by the method of regression fitting; (2) The ultimate carbon content of $C_0$ is also regarded as one of the regression parameters, rather than just to set a fixed value of 0.02% or other fixed values; (3) Since the method of isometric multi-point continuous correction is adopted, the information referring to historical heats and current heat could be utilized efficiently. Owing to these advantages of the proposed algorithm, a higher calculation accuracy and hit ratio can be realized.

The mentioned models were validated using actual production data from a steel plant of 210-ton converter, and the results are as follows. As shown in figure 8, the deviation between actual heats data and the historical average curve obtained by current method is smaller than the other curve obtained by previous method. And the hit ratio of the current model is the highest of the four mentioned models, as shown in table 1.

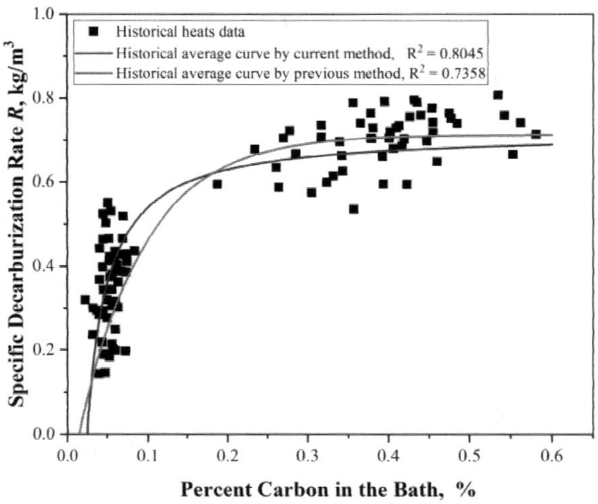


**Figure 8.** Comparison of the two historical average curves and actual heats data.

**Table 1.** Obtained hit ratios using the four models in the current study.

| Model | Hit Ratio |
|---|---|
| Integral Model | 72.30% |
| Cubic Model by Zhang *et al* [12] | 85.90% |
| Modified Exponential Model by Li *et al* [11] | 88.20% |
| Current Model | 90.70% |

## 6. Conclusions

- By the analysis on the characteristics of the carbon integral model, and cubic model, the exponential model is a better choice to reveal the changing rule of carbon content in the end blowing stage of BOF process, and the experimental results indicate improved accuracy on predicting the actual production data.
- This exponential model with the correction algorithm can obtain a better historical average decarburization curve by introducing the decarburization rate falling gradient and decarburization rate as the reference for regression fitting of model parameters, as well as selecting $C_0$ as the regression parameter.
- Due to the application of isometric multi-point continuous correction, the reference value of historical heats is fully excavated along with the sufficient utilization of the curve feather of current heat. The hit ratio of the new model reached more than 90% for the prediction of end-point carbon prediction within a tolerance of ±0.02%.

**Acknowledgments**
The financial support from Jiangxi Provincial Department of Science and Technology (20171ACE50020) is highly acknowledged.

**References**
[1] Liao D S, Sun S, Waterfall S, Boylan K, Pyke N and Holdridge D 2015 *ICS2015 Proc.* (Beijing) pp 6-12
[2] Sun S, Liao D S, Pyke N, Boylan K and Wallace G 2008 *Iron Steel Tech.* **5** 36-42
[3] Ceriani A and Aprile G 2010 *AISTech 2010 Proc.* (Pittsburg) pp 997-1003
[4] Gillgrass S, Scipolo V, Goldberg B, Maiolo J, Hole P, Spence J and Roudabush R 2009 *AISTech. 2009 Proc.*( St. Louis) pp 791-9
[5] Christian B, Hans R, Klaus W and Junan X 2012 *Asia Steel 2012 Proc.* (Beijing) p 130
[6] Dumont-Fillon J, Vayssiere P and Trentini B 1964 *JOM-US* **16** 508-11
[7] MacFarlane D 1964 *JOM-US* **16** 715-7
[8] Meyer H W, Dukelow D A and Fischer M M 1964 *JOM-US* **16** 501-7
[9] Meyer H W and Glasgow J A 1966 *Blast Furn Steel Plant* **54** 595-601
[10] Liu K, Liu L, He P and Liu W 2009 *Steelmaking* **25** 33-7
[11] Li N, Lin W H, Cao L L, Liu Q, Sun L F and Liao S S 2018 *Chin. J. Eng.* **40** 1244-50
[12] Zhang G Y, Wan X F, Lin D, Peng F and Zou Z S 2007 *Chin. J. Mater. Met.* **6** 3-6
[13] Brooks G, Coley K, Dogan N, Rhamdhani M A, Rout B, Li Z and Van Boggelen J W K 2016 *AISTech 2016 Proc.* (Pittsburgh) pp 1047-53
[14] Blanco C and Diaz M 1993 *ISIJ Int* **33** 757-63
[15] Cheng R, Zhang J M, Zhang L J and Ma H T 2019 *10th Int Symp High-Temperature Metallurgical Processing* (Springer) pp 579-90
[16] Rout B K, Brooks G, Akbar Rhamdhani M, Li Z, Schrama F N H and Overbosch A 2018 *Metall. Mater. Trans. B* **49** 1022-33
[17] Tu H, Hong X and Xie S Y 2001 *Chin. J. Baotou. Univer. Iron. Steel Tech.* **20** 289-92
[18] Li G H, Wang B, Liu Q, Tian X Z, Zhu R, Hu L N and Cheng G G 2010 *Int. J. Min. Met. Mater.* **17** 715-22

# Probability based global sensitivity analysis of fatigue reliability of steel structures

**Z Kala**[1,2]

[1]Department of Structural Mechanics, Brno University of Technology, Faculty of Civil Engineering, Czech Republic

E-mail: kala.z@fce.vutbr.cz

**Abstract.** This article focuses on reliability-oriented global sensitivity analysis of the fatigue limit state of a steel member stressed by many times repeated loading. The fatigue limit state is associated with fatigue failure, which is caused by brittle fracture due to propagation of a fatigue crack from initial to critical size. The fatigue crack propagation is analysed using linear fracture mechanics. The fundamental question in terms of structural reliability is how significant is the effect of input random quantities on the probability of failure. A new type of global sensitivity analysis subordinated to a contrast identified the equivalent stress range and initial edge crack length as random quantities that most significantly influence the failure probability. The new findings obtained using the contrast-based global sensitivity analyses show interaction effects that are unusually strong in comparison with some results of Sobol's sensitivity analysis aimed at the reliability of structures. The sensitivity indices are estimated using double-nested-loop simulation of the Latin Hypercube Sampling method.

## 1. Introduction

Numerous bridges worldwide are exposed to an increasingly stronger influence of traffic, which causes damage and reduces their service life. In particular, the fatigue phenomenon is one of the most common mechanisms of damage of steel bridges that lead to failure. Society continuously seeks measures to extend the functionality and operability of old bridges that are both economical and efficient, providing a basic framework for planning interventions [1]. The issue of fatigue analysis and evaluation of old steel bridges is, therefore, still the subject of research of reliability-based models in many countries [2].

Currently, a number of researches are focused on the experimental investigation of fatigue characteristics of steel of old bridges [3,4] as well as on the development of probabilistic approaches for assessing fatigue of these bridges [5,6]. The study of the propagation of fatigue damage of steel load- carrying structures is an essential part of identification of crucial input random quantities [7]. The results of experiments allow the verification of numerical models of bridges used in the fatigue assessment of load-carrying bridge structures [8].

Great interest has been devoted to fatigue cracks in steel girders for many years [9,10]. A procedure based on linear elastic fracture mechanics was used to evaluate the growth of a fatigue crack in [11]. The introduction, application and development of fracture mechanics combined with stochastic models has been a subject of interest in technical sciences [12,13]. A computational model based on fracture mechanics can be part of the stochastic evaluation of the fatigue characteristics of materials of steel girders, see for e.g. [5,8]. The stochastic model can be applied to study the sensitivity of the resulting reliability to the variability of input random quantities.

The presented paper deals with global sensitivity analysis investigating the effects of input quantities on the failure probability of a steel bridge member subjected to many times repeated bending loading. The sensitivity analysis of reliability can be evaluated in a number of ways, see e.g. [14-16]. The sensitivity measure of failure probability evaluating common structural reliability can be computed using unconditional and conditional probabilities, see for e.g. [17,18]. An advanced kind of global sensitivity analysis of failure probability [19], which identifies all the main and interaction effects of input random quantities on time-dependent reliability of a steel bridge member with a propagating fatigue crack [20], is applied in the paper.

A recently published innovative sensitivity measurement [19] applied to failure probability, which identifies all important effects of input random quantities on the time-dependent reliability of a steel bridge member with a propagating fatigue crack [20], is applied in the paper. The novelty of the results of the case study is the consistent analysis of the influence of five random variable factors on the fatigue reliability of a bridge member using the decomposition of the failure probability. The results of the work can be primarily used in probability studies of reliability of steel bridge girders. It is shown that the effects of input factors on the fatigue reliability differ over time. However, the weight of passing vehicles represents a permanently dominant factor clearly influencing the reliability of both new and old bridge members.

## 2. Fatigue reliability assessment of steel structures

The fatigue reliability is investigated using linear elastic fracture mechanics, which uses mathematical equations to describe the propagation of the (considered edge) crack. Fatigue crack growth is generally described using Paris' law (also known as the Paris-Erdogan law), which was suggested by Paris and Erdogan [21]. Paris' law can be mathematically expressed as a linear dependence provided that log vs log plot of the range of the stress intensity factor during the fatigue cycle $\Delta K$ vs the crack growth rate $da/dN$ is introduced.

$$\log\left(\frac{\mathrm{d}a}{\mathrm{d}N}\right) = m \cdot \log(\Delta K) + \log C \tag{1}$$

where $a$ is the length of the crack, $N$ is the integral number of loading cycles and $C$, $m$ are experimentally determined Paris' constants that depend on the characteristics of material, environment and stress ratio. Equation (1) can be simplified as equation (2).

$$\frac{\mathrm{d}a}{\mathrm{d}N} = C \cdot \Delta K^m \tag{2}$$

Parameter $C$ is determined as

$$\log(C) = c_1 + m \cdot c_2 \tag{3}$$

where $c_1$, $c_2$ can be considered for common structural steel as $c_1$=-11.141, $c_2$=-0.507 [8]. The value of $\Delta K$ can be determined by Broek [22]:

$$\Delta K = \Delta\sigma_E \cdot \sqrt{\pi a} \cdot F\left(\frac{a}{W}\right) \tag{4}$$

Where $\Delta\sigma_E$ is the equivalent stress range [23]. The stress range is obtained as the difference between maximal and minimal stress values. $F(a/W)$ is the calibration function, which is obtained from experimental research [3,4]. In the case study presented here, results obtained from experimental research [3,4] on fatigue-loaded rectangular steel bodies, which cannot be applied to other geometries, are used for the introduction of equation(5). Equation (5) is sometimes interpreted and written as the course of crack growth. The calibration function for uniform bending (5) is applied in the presented case study.

$$F\left(\frac{a}{W}\right) = 1.114\left(1 - 0.806\left(\frac{a}{W}\right) + 2.4704\left(\frac{a}{W}\right)^2 + 1.01643\left(\frac{a}{W}\right)^3\right), \quad \frac{a}{W} \in [0.01, 0.5] \tag{5}$$

Equation (2) can be integrated with the aim of obtaining a mathematical equation that links the crack growth from size $a_0$ to $a_{cr}$ with the integral number of cycles from $N_0$ to $N$.

$$\int_{a_0}^{a_{cr}} \frac{1}{\left[F\left(\frac{a}{W}\right)\cdot\sqrt{\pi \cdot a}\right]^m} \mathrm{d}a = \int_{N_0}^{N} C \cdot \Delta\sigma_E^m \mathrm{d}N \tag{6}$$

The crack increases its length in nonlinear dependence from $a_0$ to $a_{cr}$ during change in the integral number of cycles from $N_0$ to $N$, where $a_0$ is the initial length of the crack and $a_{cr}$ is the critical length of the crack. If we assume that the crack propagates from the first cycle, then we can consider $N_0$=0 and the right side of the equation can be written as

$$A = C \cdot N \cdot \Delta\sigma_E^m \tag{7}$$

The left side of equation (6) represents the resistance $R_{cr}$, which can be expressed as equation (8).

$$R_{cr} = \int_{a_0}^{a_{cr}} \frac{\mathrm{d}a}{\left[F\left(\frac{a}{W}\right)\cdot\sqrt{\pi \cdot a}\right]^m} \tag{8}$$

Then the reliability function can be expressed as equation (9).

$$G = R_{cr} - A \tag{9}$$

Equation (9) represents a general mathematical formulation of the reliability function with five random quantities $\Delta\sigma_E$, $a_0$, $m$, $W$, $N$, whose probability density function can be considered according to [5-8]. Probabilistic analysis of reliability is based on the probabilistic approach to the reliability condition equation (9):

$$P_f = P(G \le 0) \tag{10}$$

Equation (10) expresses the probability of failure due to brittle fracture. This is the probability with which the crack propagated from $a_0$ to $a_{cr}$ after $N$ load cycles. Equation (10) is a general mathematical notation of the calculation of probability that has five basic random quantities listed in table 1 in this article.

**Table 1.** Input random quantities.

| | Name of quantity | Density | Arith. mean | Stand. dev. |
|---|---|---|---|---|
| $\Delta\sigma_E$ | Equivalent stress range | Gauss | 31 MPa | 3 MPa |
| $a_0$ | Initial crack size | log-normal | 0.2 mm | 0.06 mm |
| $m$ | Parameter $m$ | Gauss | 3 | 0.03 |
| $N$ | Stress peaks per year | Gauss | 1E6 | 1E5 |
| $W$ | Specimen width | Gauss | 326 mm | 15 mm |

All input quantities listed in table 1 have zero correlation between each other. The failure probability $P_f$ is computed using a time step of one tenth of a year; see full line in figure 1. $P_f$ was evaluated at each time step using two million samples of the Monte Carlo method.

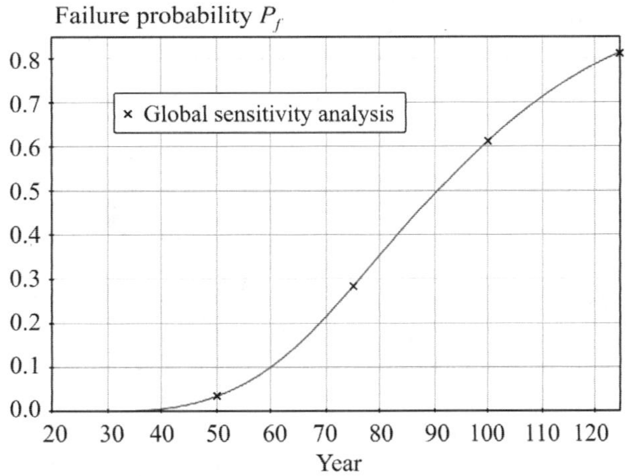


**Figure 1.** The failure probability analysis based on inputs in table 1.

## 3. Global sensitivity analysis of failure probability subordinated to contrasts

The failure probability $P_f$ is a subject of interest in the reliability analysis of steel structures. Let us consider a computational model in the form $Y = f(X_1, X_2, \ldots, X_M)$, with $Y$ a scalar. The input quantities $(X_1, X_2, \ldots, X_M)$ are random quantities described using probability distributions reflecting the uncertainty of information on material parameters, load actions, etc. Failure ensues if $Y<0$. The contrast function $\psi$ for the probabilistic appraisement ($P_f$) of the reliability can be expressed using real parameter $\theta$ as

$$\psi(\theta) = E(\psi(Y,\theta)) = E(1_{Y\leq 0} - \theta)^2 \tag{11}$$

The sensitivity measurement is based on the assumption that $\psi(\theta^*)$ is a contrast associated with parameter $\theta^*$, where

$$\theta^* = \underset{\theta}{\mathrm{Argmin}}\, \psi(\theta) \tag{12}$$

Using [19], the main effect (first order sensitivity index) based on the contrast is defined as

$$P_i = \frac{\min\limits_{\theta} \psi(\theta) - E\left(\min\limits_{\theta} E(\psi(Y,\theta)|X_i)\right)}{\min\limits_{\theta} \psi(\theta)} \tag{13}$$

The parameter $\theta^*$ has the significance of $P_f$ and can be estimated from equation (10) using the Latin Hypercube Sampling (LHS) method [24,25]. For the unconditional failure probability, the minimal value of equation (11) can be estimated using $K$ samples of the LHS procedure as

$$\min_{\theta} \psi(\theta) \approx \frac{1}{K} \sum_{k=1}^{K} \left(1_{Y<0} - \theta^*\right)^2 \tag{14}$$

In the numerator of equation (13), the second member can be computed similarly using double-nested-loop algorithm of LHS, where the conditional failure probability $\theta^*$ is calculated for $K$ runs of $Y$ using random quantities $X_{\sim i}$ (nested loop), where $X_i$ is a fixed (constant) run from $L$ runs of LHS (outer loop). The second order contrast index $P_{ij}$ can be expressed as

$$P_{ij} = \frac{\min\limits_{\theta} \psi(\theta) - E\left(\min\limits_{\theta} E(\psi(Y,\theta)|X_i, X_j)\right)}{\min\limits_{\theta} \psi(\theta)} - P_i - P_j \tag{15}$$

and estimated using LHS method analogously. The other probability contrast indices can be expressed similarly. After computing all sensitivity indices, their sum must be 1.

$$\sum_i P_i + \sum_i \sum_{j>i} P_{ij} + \sum_i \sum_{j>i} \sum_{k>j} P_{ijk} + \ldots + P_{123..M} = 1 \tag{16}$$

Global sensitivity analysis based on the contrast measures the effect of the variability of five input random quantities $\Delta\sigma_E$, $a_0$, $m$, $N$, $W$ (table 1) on the failure probability $P_f$ (10). The global sensitivity analysis is applied to the fatigue limit state of a steel member with a small initial edge crack.

## 4. Sensitivity analysis results

Sensitivity analysis based on the contrast measures the global effect of the variability of five input random quantities: $\Delta\sigma_E$, $a_0$, $m$, $N$, $W$ (table 1) on the failure probability $P_f$ (10). The numbers of LHS samplings are $K$=2000000, $L$=2000.

The sensitivity analysis results are shown in figures 2 and 3 for the 50th, 75th, 100th and 120th year of the lifetime of the bridge member. It is evident that $P_f$ is strongly influenced by $\Delta\sigma_E$ in each reference period. It is not only influenced by the main effect of $\Delta\sigma_E$, but also by the higher-order interactions of $\Delta\sigma_E$ with other quantities. The second dominant quantity is $a_0$ in the main and interaction effects. The third influential quantity is Paris parameter, which depends on the material, environment and stress ratio.

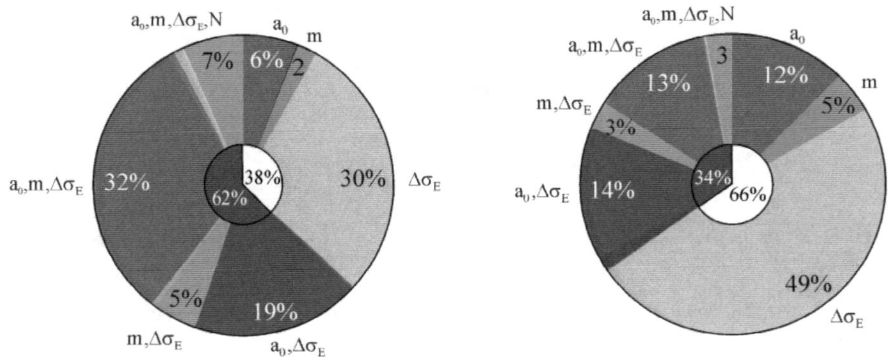


**Figure 2.** Sensitivity analysis results for year 50 and 75.

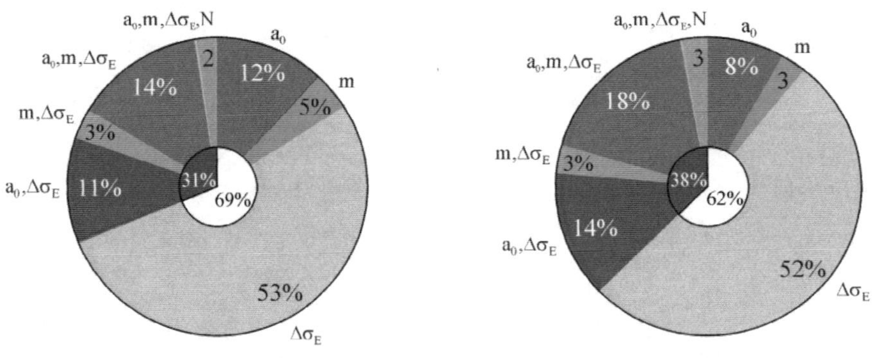


**Figure 3.** Sensitivity analysis results for year 100 and 120.

The interaction effects are relatively strong at all monitored times of bridge operation, which is in contrast with the results of Sobol's sensitivity analysis in structural reliability studies [26,27]. The relatively strongest interactions are observed at 50 years of bridge operation, see figure 2. Conversely, the weakest interactions are observed at 100 years of bridge operation, see figure 3.

Similarly strong interactions were observed in the results of quantile-based sensitivity analysis of the design value of resistance, where the design (lower quantile) resistance is computed as 0.1 percentile [28,29].

The variability of quantities *N*, *W* does not have a great influence on $P_f$, neither in the main effects nor in the interaction effects. These quantities can be fixed at any value of their domain without any significant effect on $P_f$. In practical terms, it is not mandatory to identify the types of probability density functions and statistical characteristics of quantities *N*, *W* and as precisely as the dominant quantities $\Delta\sigma_E$, $a_0$. This knowledge can simplify and streamline stochastic computational models and contribute to a deeper understanding of the limit states of steel bridges exposed to the effects of fatigue.

It can be noted that the presented concept of global sensitivity analysis of the reliability (failure probability) can be applied to other limit states of steel structures. For instance, applications studying the interaction effects between fatigue behaviour and material degradation due to corrosion may present one of the possible goals of further fundamental or applied research, which will further elaborate the results of the presented study [30].

## 5. Conclusion

The probabilistic reliability assessment of a steel bridge member subjected to fatigue phenomena is presented in the case study in which time-dependent effects of input random quantities are taken into account. The probability of failure $P_f$ was investigated using an advanced kind of contrast-based global sensitivity analysis. The outcomes identify the effects of input quantities on $P_f$ in the 50$^{th}$, 75$^{th}$, 100$^{th}$ and 120$^{th}$ year of the lifetime of the bridge member. The input random quantities are divided into groups: influential and non-influential, according to their effects on the fatigue limit state.

The equivalent stress range influences $P_f$ most significantly at all monitored times. The reliability of the steel members under bending can be improved by restricting very heavy vehicles, which can prolong the lifetime of an aging bridge in cases when repairs are no longer economical. Such justified limitation of extreme vehicle weights is common in old, very old and historic bridges, where the number of lanes is intentionally reduced or shuttle services are introduced if no other alternatives are economical or possible. Nevertheless, overloading does not benefit the lifetime of middle-aged bridge members. This article not only contributes to understanding the reasons for weight limitations, but also shows mathematically quantified sensitivities between the influence of five factors including their combinations on the fatigue limit state.

The initial edge crack length is the second random quantity with a strong influence on $P_f$. Both random quantities contribute strongly to the interaction effects between each other and with other quantities. The strongest interaction effects were found during fifty years of bridge operation, when the value of $P_f$ is the lowest of all four times in which the sensitivity analysis was evaluated.

The sensitivity of $P_f$ to the width of the specimen and load cycles in the considered year of solution is relatively very low to negligible. Fixing one, the other or both quantities at any value of their domain will have very little influence on $P_f$. The uncertainty of such minor to stochastically negligible quantities can be identified approximately (expert judgement) or not at all. The introduction of these quantities as deterministic can be considered during optimization of stochastic computational models.

The presented global sensitivity analysis methodology can be used in other case studies, for example welded joints or corrosion. However, such studies will require solving numerous issues, such as how to define random parameters such as the coating effect of the rate of corrosion during modelling. The results of sensitivity analysis can be used to make decisions on inspections, maintenance, repair and the loading capacity of bridges at different periods of its lifetime.

**Acknowledgment**

Financial support provided by the Czech Science Foundation (grant No. 17-01589S) and by No. LO1408 “AdMAs UP” are greatly appreciated.

## References

[1] Leander J, Honfi D, Ivanov O L and Björnsson Í 2018 *Eng. Fail. Anal.* **91** 306-14
[2] Leander J 2018 *J. Constr. Steel Res.* **141** 1-8
[3] Seitl S, Miarka P and Kala Z 2018 *Transactions of the VSB – Technical University of Ostrava, Civil Engineering Series* **18** 44-9
[4] Seitl S, Miarka P, Malíková L and Krejsa M 2017 *Key Eng. Mater.* **754** 353-6
[5] Krejsa M, Koubova L, Flodr J, Protivinsky J and Nguyen Q T 2017 *Frattura ed Integrità Strutturale* **39** 143-59
[6] Krejsa M, Kala Z and Seitl S 2016 *Procedia Eng.* **142** 146-53
[7] Kala Z, Omishore A, Seitl S, Krejsa M and Kala J 2019 *Int. J. Mech.* **13** 69-78
[8] Kala Z 2018 *Int. J. Mech.* **12** 121-30
[9] Marques F, Correia J A F O, de Jesus A M P, Cunha A, Caetano E and Fernandes A A 2018 *Eng. Fail Anal.* **94** 121-44
[10] Alencar G, de Jesus A, da Silva J G S and Calçada R 2019 *Eng. Fail Anal.* **104** 154-76
[11] Omishore A 2017 *IOP Conf. Ser. Mater. Sci. Eng.* **245** 1-6
[12] Omishore A 2019 *IOP Conf. Ser. Mater. Sci. Eng.* **471** 1-10
[13] Leander J and Al-Emrani M 2016 *Int. J. Fatigue* **93** 82-91
[14] Wang Y, Xiao S and Lu Z 2018 *Aerosp. Sci. Technol.* **79** 364-72
[15] Xu L, Lu Z and Xiao S 2019 *Appl. Math. Model.* **66** 592-610
[16] Zhou C, Zhang Z, Liu F and Wang W 2019 *Chinese J. Aeronaut.* **32** 948-53
[17] Wang Y, Xiao S and Lu Z 2019 *Mech. Syst. Signal Pr.* **115** 607-20
[18] Yun W, Lu Z and Jiang X 2019 *Reliab. Eng. Syst. Saf.* **187** 174-82
[19] Fort J C, Klein T and Rachdi N 2016 *Commun Stat–Theor M* **45** 4349-64
[20] Kala Z 2019 *Eng. Struct.* **194** 36-45
[21] Paris P C and Erdogan F 2963 *J. Basic. Eng.* **85** 528-34
[22] Broek D 1986 *Elementary Engineering Fracture Mechanics* (The Hague: Springer)
[23] Kwon K and Frangopol D M 1010 *Int. J. Fatigue* **32** 1221-32
[24] Iman R L, Johnson M E and Watson Jr. C C 2005 *Risk Anal.* **25** 1277-97
[25] McKey M D, Conover W J and Beckman R J 1979 *Technometrics* **1** 239-45
[26] Kala Z and Valeš J 2017 *Eng. Struct.* **134** 37-47
[27] Kala Z and Valeš J 2018 *Arch. Civ. Mech. Eng.* **18** 1207-18
[28] Kala Z 2019 *J. Civ. Eng. Manag.* **25** 297-305
[29] Kala Z and Valeš J 2017 *J. Constr. Steel Res.* **139** 110-22
[30] Peng D, Jones R, Cairns K, Baker J and McMillan A 2018 *Theor. Appl. Fract. Mech.* **97** 385-99

# Formation and decomposition of Widmanstätten austenite in GOES belt-casted strips

**V Vodárek[1,4], C-P Reip[2], B Smetana[3] and A Volodarskaja[1]**

[1]Department of Materials Engineering, Faculty of Materials Science and Technology, VSB – Technical University of Ostrava, 708 33 Ostrava, Czech Republic
[2]SMS group, GmbH, Eduard-Schloemann-Str. 4, 40237 Dűsseldorf, Germany
[3]Department of Physical Chemistry and Theory of Technological Processes, Faculty of Materials Science and Technology, VSB – Technical University of Ostrava, 708 33 Ostrava, Czech Republic

E-mail: vlastimil.vodarek@vsb.cz

**Abstract.** This contribution deals with the formation and decomposition of Widmanstätten austenite during solidification of the thin belt-casted strips made of a grain oriented electrical steel (GOES). Solidification of liquid steel started with the formation of δ-ferrite. Cooling in the δ + γ phase field resulted in the formation of a small fraction of Widmanstätten austenite by displacive mechanism accompanied by carbon partition. Widmanstätten austenite laths formed flat low energy interface facets along ferrite grain boundaries. In order to minimize the interfacial energy, ferrite grain boundaries in the vicinity of flat austenite/ferrite facets migrated. It resulted in the formation of either straight or zig-zag ferrite boundaries. Intensive precipitation of sulphides along ferrite/austenite interfaces made it possible to study the early stages of austenite decomposition at the end of the δ + γ phase field. Complex sulphides containing chromium, manganese, iron and copper were identified as the $Cr_2CuS_4$ phase. Decomposition of austenite started with the formation of epitaxial ferrite. This was accompanied by further partitioning of carbon into remaining austenite. The growth of epitaxial ferrite into the flat ferrite/austenite interface facets along ferrite grain boundaries resulted in a wavy shape of these boundaries. Finally, remaining carbon rich austenite transformed either to pearlite or plate martensite.

## 1. Introduction

Strip casting is an efficient and economical technology for production of thin steel sheets [1]. It reduces the total hot rolling reduction during subsequent processing of as–cast strips. Recent progress in strip casting makes it possible to apply this technology for the production of a grain oriented electrical steel (GOES).

Recent results have demonstrated that the microstructure of belt-casted GOES strips can consist of ferrite, pearlite and a mixture of plate martensite and residual austenite [2-4]. GOES strips firstly solidify as δ-ferrite, and during subsequent cooling in the δ + γ field a small fraction of austenite and several minor phases precipitate in the parent phase. Austenite is a harder phase than ferrite and δ→γ transformation is accompanied by a volume shrinkage [3]. Austenite in GOES strips usually forms Widmanstätten-like patterns and its dominant morphology can be defined as laths [3]. At the end of the δ + γ field austenite decomposes. Products of austenite decomposition depend on the cooling rate, chemical composition and size of Widmanstätten austenite laths. Minor phases play an important role in the production of the final Goss texture in GOES sheets [5]. Intensive precipitation of minor phases in the δ-ferrite makes it possible to study the beginning of austenite decomposition [4].

Formation mechanisms, morphology and crystallography of Widmanstätten ferrite in steels have been studied extensively but less is known about Widmanstätten austenite [6,7]. This paper deals with studies on the formation and decomposition of Widmanstätten austenite in the GOES belt-casted strip. The basic goal is a better understanding of the microstructure evolution in GOES strips produced by advanced strip casting technology.

## 2. Material and experimental procedures

The thin strip was manufactured at TU Claustahl using the horizontal belt casting process [1]. In this case, liquid steel flows from a ladle into a tundish system and is dispensed onto a water–cooled belt. The top surface of the solidifying strip is protected by argon. The cooling rate varies across the thickness of the strip. The calculated cooling rate on the top surface of the strip in the temperature range from 1573 K to 973 K was about 2 $K \cdot s^{-1}$. The width and thickness of the belt-casted strip was 0.30 m and 0.014 m, respectively. Chemical composition of the strip is shown in table 1.

**Table 1.** Chemical composition of the cast investigated, mass %.

| C | Si | Mn | S | Cr | Cu | Al |
|---|---|---|---|---|---|---|
| 0.034 | 2.81 | 0.06 | 0.024 | 0.20 | 0.15 | 0.002 |

Microstructure investigations were carried out on longitudinal sections in the middle width, across the whole thickness of the strip. Evolution of the microstructure during cooling of the solidified strip was studied using a combination of light microscopy (LM), X-ray diffraction (XRD), scanning electron microscopy (SEM), electron backscattered diffraction (EBSD), X-ray microanalysis (EDX), transmission electron microscopy (TEM) and differential scanning calorimetry (DSC) techniques. Specimens for electron backscattered diffraction (EBSD) in SEM (Quanta 450 FEG) were prepared by mechanical grinding and polishing, the final polishing was done using colloidal silica. The OIM Analysis™ software was used for acquisition and indexing of Kikuchi diffraction patterns and for evaluation of the EBSD orientation data. Minor phases were studied in TEM (JEM 2100) on carbon extraction replicas using both selected area electron diffraction and EDX. DSC analyses were carried out on the Setaram MHTC 96 Line equipped with a 3D heat flux DSC-B type sensor. Measurements were performed using specimens with dimensions of $\phi$5 x 8 mm in the temperature interval between 900 and 1650 K in an inert gas atmosphere (He).

## 3. Results and discussion

Microstructure across the thickness of the belt-casted strip consisted of coarse ferrite grains and decomposed Widmanstätten austenite laths [4]. Coarse ferrite grains were fragmented into a number of subgrains with misorientation less than 2°. The fraction of decomposed Widmanstätten austenite, as determined by image analysis, was about 2% [4]. Particles of Widmanstätten austenite preferentially nucleated at ferrite grain boundaries where they formed low energy flat austenite/ferrite interface facets. The formation of Widmanstätten austenite only on one side of a ferrite grain boundary resulted in its straightening, figure 1. Such straight grain boundaries are actually planar in three dimensions [8]. If Widmanstätten austenite laths nucleated on both sides of a ferrite grain boundary, two sets of flat austenite/ferrite interface facets arose [3]. The planar interface facets for both sets of Widmanstätten austenite laths were not parallel due to misorientation between the ferrite grains. In order to minimize the interfacial energy, segments of ferrite grain boundaries were forced to migrate. This was probably favoured by the high formation temperature of Widmanstätten austenite [3]. It resulted in zig-zag shape of ferrite grain boundaries composed of alternating segments of straight boundaries, figure 2. Furthermore, fine austenite particles also nucleated intragranularly.

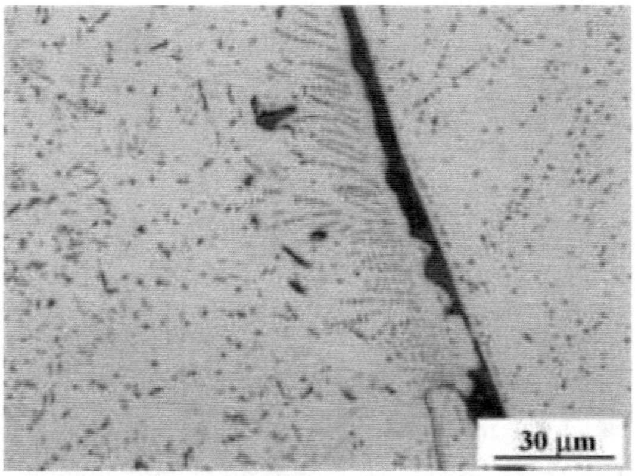


**Figure 1.** The straight ferrite grain boundary decorated by decomposed Widmanstätten austenite "smithiomorphs" [6] growing into one ferrite grain, LM.

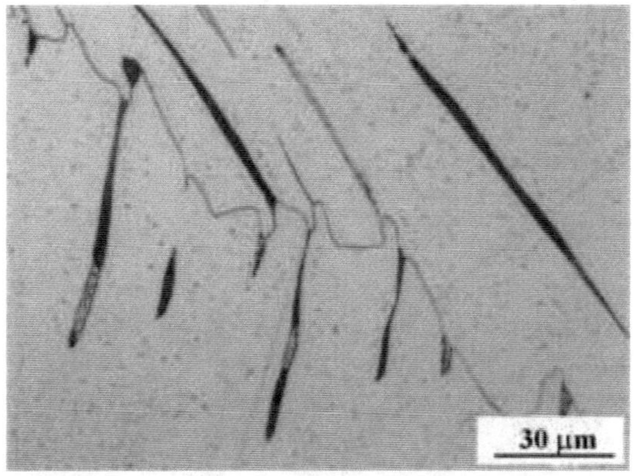


**Figure 2.** Widmanstätten austenite laths growing into two adjacent ferrite grains, migration of the ferrite grain boundary resulted in its zig-zag shape, LM.

Figures 1 and 2 also document heavy precipitation along grain/subgrain ferrite boundaries, ferrite/austenite interfaces and in ferrite grains. In the area of final solidification of the belt-casted strip, i.e. at about ¼ depth under the top surface of the strip, relatively coarse sulphides decorated ferrite grain boundaries, figure 3. EDX analyses revealed that these sulphides were rich in iron, table 2. Needle-like intragranular particles in the δ-ferrite were identified as cementite, figure 4. Most fine globular precipitates were complex sulphides. Chemical composition of fine sulphides was variable. Majority of these sulphides were rich in chromium, manganese, iron and copper. The average chemical composition of fine sulphides is shown in table 3. Electron diffraction studies proved that the crystal structure of fine sulphides was consistent with the $Cr_2CuS_4$ phase (cubic, Fd3m, a = 0.961 nm), figure 5.

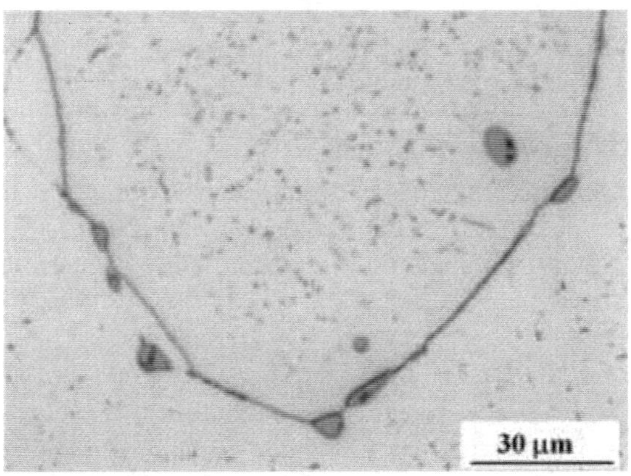


**Figure 3.** Coarse iron sulphides along ferrite grain boundaries and fine intragranular complex sulphides, LM.

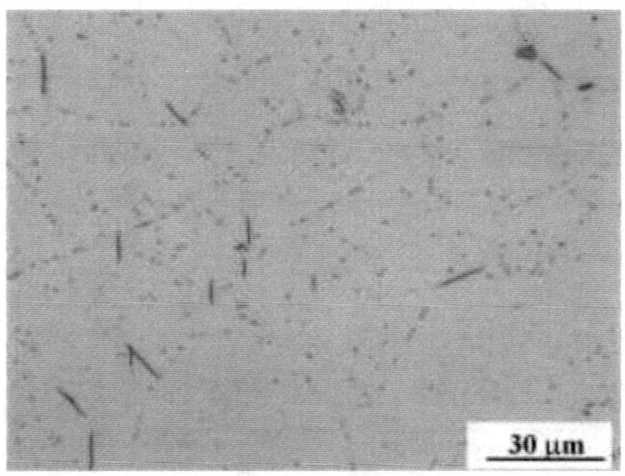


**Figure 4.** Needle-like particles of cementite, fine sulphides along subgrain boundaries and in ferrite grains, LM.

**Table 2.** Results of semiquantitative EDX analyses of coarse sulphides, mass %.

| No. | Si | S | Ti | Cr | Mn | Fe | Cu |
|---|---|---|---|---|---|---|---|
| 1 | 0.3 | 33.6 | 0.9 | 8.9 | 9.5 | 46.0 | 0.8 |
| 2 | 0.5 | 35.5 | 1.9 | 8.5 | 2.0 | 50.5 | 1.0 |
| 3 | 1.1 | 25.8 | 0.2 | 2.0 | 19.3 | 51.0 | 0.7 |

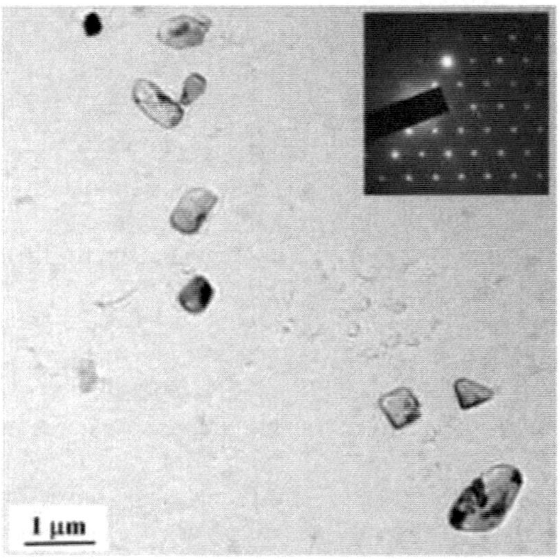


**Figure 5.** Intragranular $Cr_2CuS_4$ sulphides, insert: zone axis $[111]_{Cr2CuS4}$, TEM.

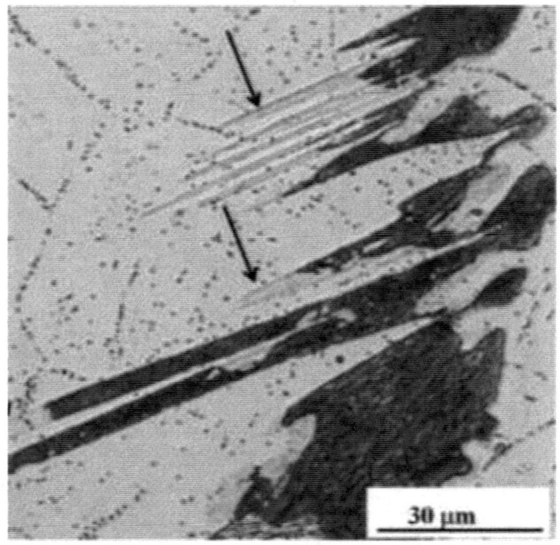


**Figure 6.** Decomposition of Widmanstätten austenite to epitaxial ferrite and pearlite, arrows mark epitaxial ferrite, LM.

At the end of the $\delta + \gamma$ field Widmanstätten austenite decomposed either to pearlite or plate martensite. In some laths of decomposed Widmanstätten austenite both products coexisted. Discontinuous networks of sulphides at ferrite/austenite interfaces made it possible to study the early stages of austenite decomposition. The decomposition of austenite started with the formation of epitaxial ferrite with an identical orientation as the surrounding ferrite. Arrows in figure 6 mark the parts of Widmanstätten austenite laths which transformed to epitaxial ferrite. This transformation was preferentially observed in thin Widmanstätten austenite laths and especially close to the tips of these laths. It was accompanied by a further partition of carbon into remaining austenite. The remaining austenite in this area transformed to pearlite, figure 6.

**Table 3.** Results of semiquantitative EDX analyses of fine complex sulphides, mass %.

| No. | S | Cr | Mn | Fe | Cu |
|---|---|---|---|---|---|
| 1 | 43.3 | 33.3 | 9.3 | 7.5 | 6.6 |
| 2 | 43.6 | 32.4 | 13.6 | 4.5 | 5.9 |
| 3 | 42.1 | 34.0 | 13.5 | 3.4 | 7.1 |
| AVE[a] | 43.0 | 33.2 | 12.1 | 5.1 | 6.5 |
| STD[a] | 0.8 | 0.5 | 3.0 | 2.9 | 0.4 |

[a] AVE is the average value and STD is the standard deviation.

Arrows in figure 7 mark the original positions of austenite/ferrite interfaces decorated by fine sulphides. Decomposition of austenite started with the formation of epitaxial ferrite and the remaining austenite in this case transformed to a mixture of pearlite and plate martensite. The formation of plate martensite suggested, that the Widmanstätten austenite lath was significantly enriched in carbon. Plate martensite usually forms in Fe – C alloys containing more than 0.6 wt.% C [9]. Plate martensite always coexists with some retained austenite, figure 8.

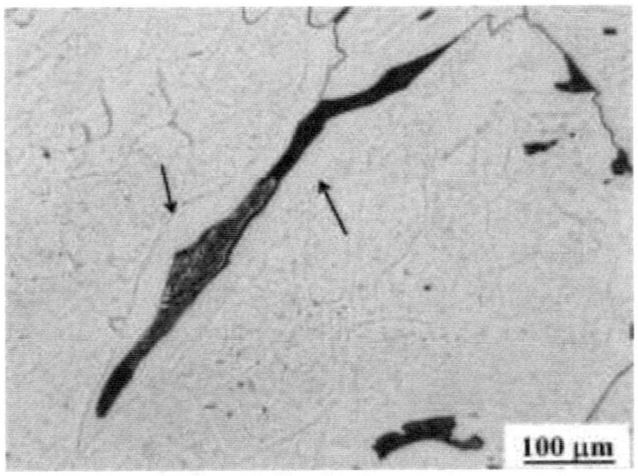


**Figure 7.** Widmanstätten austenite lath decomposed to a mixture of pearlite and plate martensite. Arrows mark the original positions of the ferrite/austenite interface, decomposition of Widmanstätten austenite started with the formation of epitaxial ferrite, LM.

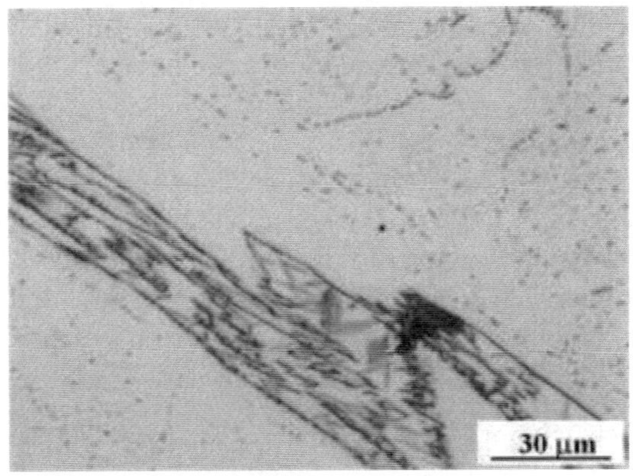


**Figure 8.** Decomposition of Widmanstätten austenite laths in a mixture of plate martensite and residual austenite, LM.

Except for the straight and zig-zag ferrite grain boundaries, some ferrite grain boundaries exhibited in the vicinity of decomposed Widmanstätten austenite laths a wavy character. The IPF orientation map in figure 9 shows two parallel decomposed Widmanstätten austenite laths growing from the ferrite grain boundary into the grain A.

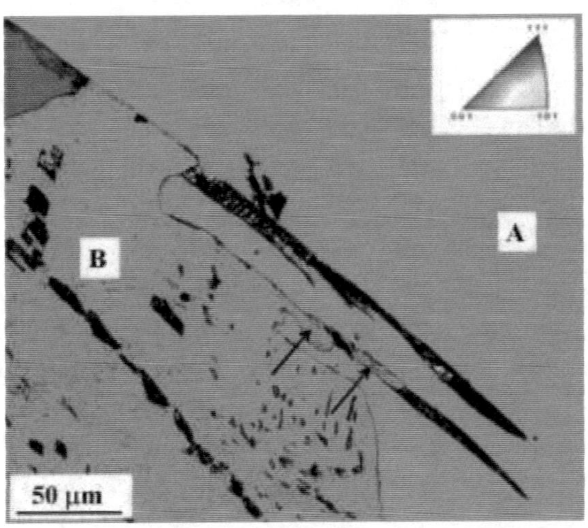


**Figure 9.** IPF orientation map, partial decomposition of Widmanstätten austenite to epitaxial ferrite - arrows mark the growth of epitaxial ferrite from the grain B into the originally flat austenite/ferrite interface facet.

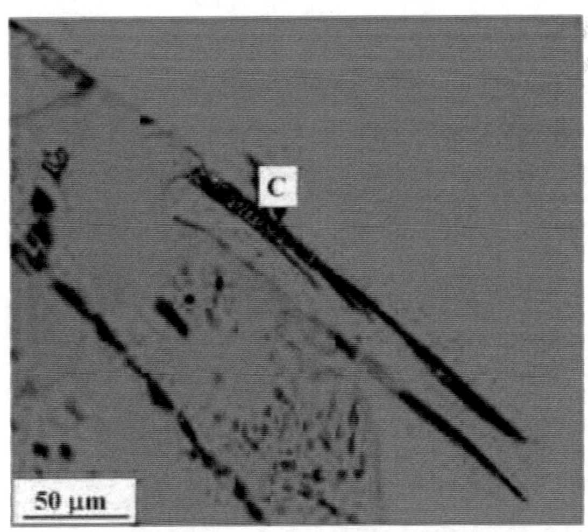


**Figure 10.** Phase map of ferrite (red) and retained austenite (green) + high angle grain boundaries ($\alpha$ >15°). Retained austenite forms islands between martensite plates (area C) and fine intragranular particles.

IOP Conf. Series: Materials Science and Engineering **668** (2019) 012016 doi:10.1088/1757-899X/668/1/012016

Products of Widmanstätten austenite laths decomposition are plate martensite + retained austenite and pearlite. Arrows at the bottom lath show the growth of epitaxial ferrite with the orientation identical to the ferrite grain B across the originally flat interface facet into the Widmanstätten austenite lath. It resulted in a wavy character of the ferrite grain boundary. The phase map in figure 10 documents ferrite and retained austenite distribution in this area. Islands of retained austenite coexist with plate martensite in the decomposed Widmanstätten austenite lath – see area C in figure 10. Furthermore, fine intragranular particles of austenite were resistant to decomposition during cooling of the strip.

Figure 11 shows thermal effects recorded during the DSC experiment in the temperature range of 900 – 1650 K. Heating and subsequent cooling of the GOES specimen with the starting microstructure consisting of the δ-ferrite, pearlite and sulphides was carried out at the rates of 15 K·min$^{-1}$ and 40 K·min$^{-1}$, respectively. The heating curve shows the peak related to the change of magnetic properties of the δ-ferrite (Curie temperature, 1027 K) and two minor peaks. The thermal effects at 1093 – 1125 K can be related to the formation of austenite. The broad peak at 1442 – 1590 K probably corresponds to dissolution of sulphides and austenite.

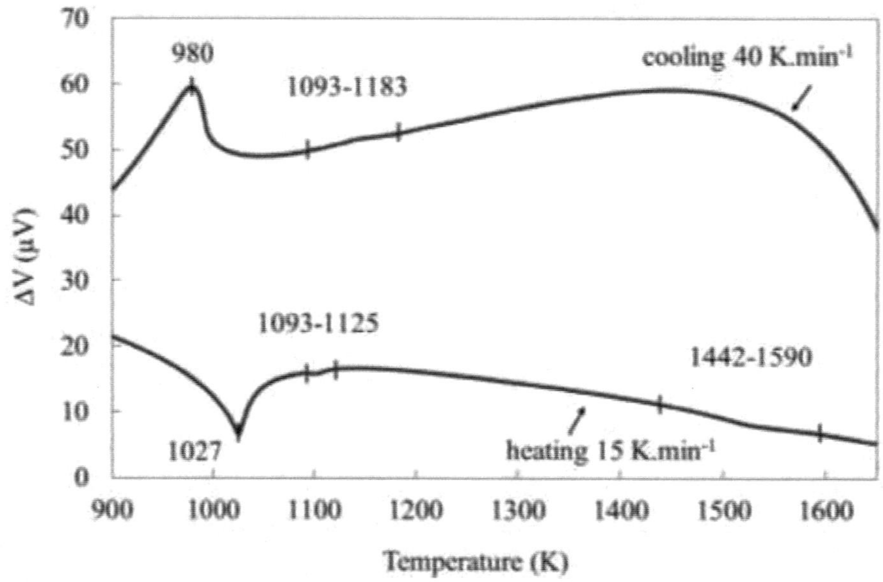


**Figure 11.** DSC curve showing peaks during heating and cooling of the GOES specimen between 900 and 1673 K.

Nevertheless, the volume fraction of sulphides in the microstructure was too low to cause distinct thermal effects on the DSC curve during their dissolution. Thermal stability of complex sulphides in GOES steels increases with decreasing copper content [5]. On the cooling curve the peak at 1183 – 1093 K was observed. These thermal effects can be linked with the austenite decomposition to epitaxial ferrite and pearlite. The peak at 980 K corresponds to the change in magnetic properties of the parent phase.

## 4. Conclusions

Solidification of the belt-casted strip started with the formation of the δ-ferrite. During cooling of the strip a small fraction of Widmanstätten austenite precipitated in the δ-ferrite by displacive mechanism accompanied by partitioning of carbon. Widmanstätten austenite laths preferentially nucleated on ferrite grain boundaries and formed the flat low energy austenite/ferrite interface facets along the ferrite grain boundaries, which were forced to migrate. If Widmanstätten austenite laths nucleated on both sides of the ferrite grain boundary, zig-zag boundaries arose. Heavy precipitation of complex sulphides developed in the δ-ferrite, grain/subgrain ferrite boundaries and ferrite/austenite interfaces. Majority of fine sulphides corresponded to the $Cr_2CuS_4$ phase. Decomposition of Widmanstätten austenite laths started with the formation of epitaxial ferrite. The growth of epitaxial ferrite into the flat ferrite/austenite interface facets along ferrite grain boundaries resulted in a wavy shape of these boundaries. The formation of epitaxial ferrite resulted in further enrichment of remaining austenite islands in carbon. The final step of Widmanstätten austenite decomposition resulted in the formation of pearlite or plate martensite. The decomposition mechanism was affected by the cooling rate, chemical composition and dimensions of Widmanstätten austenite. The fraction of retained austenite in the final microstructure of the belt-casted strip was very low (about 0.5 %).

**Acknowledgment**
The authors gratefully acknowledge the support by the project "Damage Prediction of Structural Materials", No. CZ.02.1.01/0.0/0.0/17_048/0007373 within the Operational Programme Research, Development and Education financed by the European Union and by the state budget of the Czech Republic and the GAČR project No. 17-18668S.

**References**
[1] Ge S, Isac M, Gutrie R I L 2013 Progress in Strip Casting Technologies for Steel, Technical Developments *ISIJ Int.* **53** 729-42
[2] Song H Y, Lu H H, Liu H T, Li H Z , Geng D Q, Misra R D K, Liu Z Y and Wang G D 2014 Microstructure and Texture of Strip Cast Grain-oriented Silicon Steel after Symmetrical Asymmetrical Hot Rolling *Steel Res. Int.* **85** 1477-82
[3] Song H Y, Liu H T, Wang G D and Jonas J J 2017 Formation of Widmanstätten Austenite in Strip Cast Grain Oriented Silicon Steel *Metall. Mater. Trans. A* **48A** 1959-68
[4] Vodárek V, Reip C P and Volodarskaja A 2019 Microstructure Evolution in Belt-Casted Strip of Grain Oriented Electrical Steel *Key Eng. Mater.* **810** 82-8
[5] Vodárek V, Volodarskaja A, Miklušová Š, Holešinský J, and Žáček O 2016 Precipitation Reactions in a Copper-Bearing GOES *Procedia Mater. Sci.* **12** 77-82
[6] Zin J, Hillert M and Borgenstam A 2017 Morphology of Proeutectoid Ferrite *Metall. Mater. Trans. A* **48A** 1425-43
[7] *Phase Transformations in Steels. Vol 1: Fundamentals and Diffusion Controlled Transformations* 2012 ed E Pereloma and D V Edmonds (Oxford: Woodhead Publishing Limited) chapter 6 pp 187-216
[8] Ohmori Y, Nakai K, Ohtsubo H and Isshiki Y 1995 Mechanism of Widmanstätten Austenite Formation in δ/γ Duplex Steels *ISIJ Int.* **35** 969-75
[9] Bhadeshia H K D H and Honeycombe R W K 2017 *Steels: Microstructure and Properties* 4th edition (Oxford: Butterworths – Heinemann) chapter 5 pp 98-127

# Analysis of the mechanical properties of Q345R steel in deep-regulating units by the spherical indentation method

**M C Sun[1,3], C Yang[1], S Y Xiao[1], W Tan[2] and G P Zhou[2]**

[1]Electric Power Research Institute of State Grid Liaoning Electric Power Co. Ltd, Shenyang 110006, China
[2]State Grid Liaoning Electric Power Supply Co. Ltd, Shenyang 110004, China

E-mail: 36813992@qq.com

**Abstract.** The principles of testing the mechanical properties of metallic materials through the use of the spherical indentation test have been analysed and studied in this paper. The error between the representative indentation stress from spherical indentation and the theoretical stress of the power exponential equation was taken as the convergence condition using the optimization function in Matlab. The yield strength, strain hardening index and elastic modulus were optimized between the stress-strain curve from spherical indentation and the theoretical stress-strain curve of the power exponential equation. The smallest error that could be achieved between them was obtained and from this the mechanical properties of the material could then be calculated and analysed. Q345R steel, which is used for pressure vessels, was taken as an example and the mechanical properties were tested using the spherical indentation test, and the results were compared with the conventional tensile test. Comparative analysis of the mechanical properties of the steel between the results from the spherical indentation test and the conventional tensile test showed that the difference was small. The accuracy of the spherical indentation method when testing the mechanical properties of metallic materials is high. The results have shown that the spherical indentation test can be used as a replacement for the conventional tensile test and can be applied in the field of testing the mechanical properties of Q345R steel for pressure vessels.

## 1. Introduction

The safety of industrial equipment during its operation is very important, but there has always been a bottleneck in ascertaining the material properties that affect the safety assessment of industrial equipment. Usually, it is only after pressure vessel equipment is in service that it can be regularly checked for defects. Samples of the material are required for conventional testing of the mechanical properties and it is difficult to test its mechanical properties by conventional methods while in situ.

Instrumented indentation testing technology is a new technology to test the mechanical properties of materials, which was developed on the basis of the traditional Brinell hardness test. The indentation method obtains the mechanical properties of materials, such as the elastic modulus, the yield strength, the tensile strength and the Brinell hardness, by simultaneously measuring and recording the load and displacement curves during the indentation process. This technology is known as a material's mechanical properties probe [1].

The indentation method has many advantages, such as utilizing compact and portable equipment, which can be used in the field to directly test a material's mechanical properties; it can be used to test in-service equipment; the test site is small, which can be used to determine the local characteristics of the materials; the indentation method determines the mechanical properties of materials using a load-depth curve, and does not require the use of optical microscopy to measure the indentation.

Tabor first proposed the formulas for calculating stress and strain based on the indentation method in 1951, and also used the indentation method to test the tensile properties of metals [2]. In the 1980s, Haggag *et al* established a method of calculating the tensile properties of a material without the need to measure the indentation diameter, based on automatic collection of the load and displacement. Many studies have been conducted to calculate the tensile properties of materials by automatically collecting data from displacement and load sensors [1-3]; to this end, various formulas can be used to calculate stress and strain. In this paper, automatic spherical indentation tests were carried out on Q345R steel and then the results were compared with the conventional tensile test; the method of obtaining the mechanical properties of Q345R steel using spherical indentation was also studied.

## 2. Experimental principles of the spherical indentation method

### *2.1. The spherical indentation test*

The spherical indentation tester is driven by a motor which is loaded onto the spherical indentation head and then pressed vertically into the surface of the material. Load and displacement sensors are used to measure the load and displacement throughout the entire process, and from this the continuous indentation curve (load-displacement curve) for the indentation test process can be obtained. The load-displacement curve is then transformed into the stress-strain curve of the material, and the mechanical properties of the material, including the yield strength, tensile strength, elastic modulus, strain hardening index and strength coefficient can be obtained.

### *2.2. Load-displacement curve for spherical indentation*

The deformation of the material that is caused during the indentation process includes both elastic deformation and plastic deformation. The extent of the elastic deformation can be calculated using the Oliver-Pharr method [4]:

$$h_d = \omega \frac{F_{max}}{S} \tag{1}$$

$$h_d = \omega(h_{max} - h_r) \tag{2}$$

$$h_c^* = h_{max} - h_d \tag{3}$$

$$a^{*2} = 2Rh_c^* - h_c^{*2} \tag{4}$$

where. $F_{max}$ is the maximum load; $h_d$ is the depth of the elastic deformation; $\omega$ is the shape coefficient of the indenter; $h_c^*$ is the depth when the pile-up or sink-in effect are not considered; $h_{max}$ is the maximum depth; $h_r$ is the intersection point of the tangent line of the unloading curve and the depth axis; $h_p$ is the permanent indentation depth after the load has been removed; $R$ is the radius of the spherical indenter.

Figure 1 has shown an illustration of a spherical indenter being pressed into a metal surface. The plastic deformation at the indentation is characterized by the pile-up or sink-in that occurs around the indentation. The extent of the pile-up or sink-in depends on the strain hardening index. The influence of the material's pile-up or sink-in near the indenter should also be considered when calculating the contact radius, and various correction methods have been proposed that take the material pile-up or sink-in into consideration. Hill introduced dimensionless parameters in the literature in order to consider the effect of the material's pile-up or sink-in [5], which can be written as follows:

$$c^2 = \frac{a^2}{a^{*2}} = \frac{5}{2}\frac{2-n}{4+n} \tag{5}$$

$$a^2 = \frac{5}{2}\frac{2-n}{4+n}(2Rh_c^* - h_c^{*2}) \tag{6}$$

where, $c$ is a dimensionless parameter; $a^*$ is the contact radius without taking the pile-up or sink-in effect into consideration; $a$ is the contact radius after taking the pile-up or sink-in effect into consideration; $n$ is the strain hardening index.

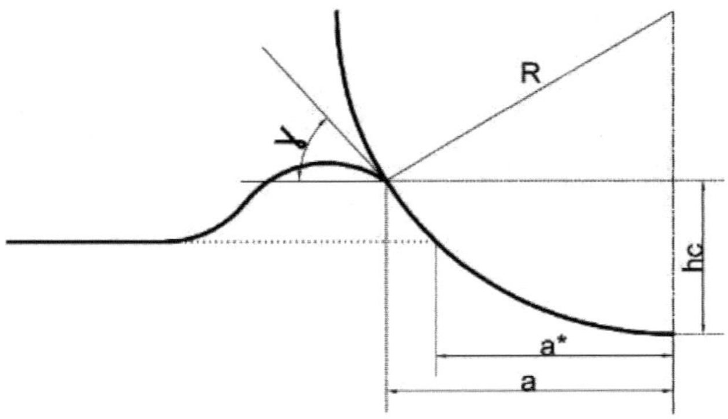


**Figure 1.** Schematic representation of a ball indentation into a metal surface.

*2.3. Stress in the spherical indentation method*

In the case of complete plastic deformation, Tabor proposed that the equivalent stress of the indentation can be replaced by the average stress $\sigma_r$, expressed by the load $F$ and the contact area $A$ [2], which can be written as:

$$\sigma_r = \frac{1}{\Psi}\frac{F}{A} = \frac{1}{\Psi}\frac{F}{\pi a^2} \tag{7}$$

where, $\Psi$ is the plastic constraint factor, which is related to the extension of the plastic zone, i.e. the yield strain and strain hardening index of the materials. From the results of many studies, the plastic constraint factor was recommended to be $\Psi$=3 in the Dutch indentation guide in the literature [6].

*2.4. Strain in the spherical indentation method*

The earliest representative strain $\varepsilon_r$ was obtained by Tabor using traditional optical techniques based on experience, as shown in equation (8), where the contact angle between the indenter and the specimen was $\gamma$ and $K = 0.2$. The maximum strain value obtained by the strain expression is 0.2:

$$\varepsilon_r = K\frac{a}{R} = K\sin\gamma \tag{8}$$

Ahn and Kwon proposed the concept of shear strain at the contact edge of the indenter in the literature [7], which expanded the scope of the representative strain; the representative strain $\varepsilon_r$ is defined as:

$$\varepsilon_r = \alpha \frac{1}{\sqrt{1-(a/R)^2}} \frac{a}{R} = \alpha \tan \gamma \tag{9}$$

where, $\alpha$ is a material constant. The expression of the shear strain at the contact edge of the indenter has been demonstrated to be in good agreement with the experimental results.

*2.5. Elastic modulus of the spherical indentation method*

According to Sneddon's early research into the elastic contact system, the relationship between the elastic modulus and the elastic contact stiffness of metallic materials can be written as follows [8]:

$$S = \frac{dF}{dh} = \frac{2}{\sqrt{\pi}} E_r \sqrt{A} \tag{10}$$

$$\frac{1}{E_r} = \frac{1-\nu^2}{E} + \frac{1-\nu_i^2}{E_i} \tag{11}$$

where, $S$ is the elastic contact stiffness; $E_r$ is the reduced elastic modulus; $A$ is the contact area of the indenter; $E$ and $\nu$ are the elastic modulus and the Poisson's ratio of the materials being tested; $E_i$ and $\nu_i$ are the elastic modulus and the Poisson's ratio of the indenters.

*2.6. Stress-strain curve of the power exponential equation*

In the uniform deformation stage, the plastic behavior of metals can be approximately described by the power exponential law. At present, a two-parameter power exponential equation is commonly used to describe the elastic-plastic properties of materials when testing them using the spherical indentation method. The yield strength of the materials $\sigma_y$ is obtained from the point where the line of 0.2% strain intersects with the true stress-strain curve.

Dao *et al* used a three-parameter modified elastic-plastic model to describe the true stress-strain curve [9], the specific formula of which can be written as follows:

$$\sigma_r = \sigma_y (1 + \frac{E}{\sigma_y} \varepsilon_r)^n \tag{12}$$

This equation is a three-parameter elastic-plastic equation for use with the spherical indentation method.

*2.7. Tensile strength*

The tensile strength $\sigma_u$ is the maximum value that is reached in a stress-strain curve. Therefore, when the value of the true strain $\varepsilon = n$, the stress that corresponds to the true stress is the tensile strength [10], which can be written as:

$$\sigma_u = K \left( \frac{n}{e} \right)^n \tag{13}$$

## 3. Computational method and flow chart

*3.1. Analytical method*

In this paper, the indentation test was carried out fifteen times and the cyclic curves of the stress and strain from the tests were obtained, including fifteen unloading points. Fifteen stress and strain points from the spherical indentation were obtained, and then these points were fitted as stress-strain curves for the spherical indentation process. The theoretical stress-strain curve could be obtained using the theoretical power exponential equation.

The elastic modulus, the yield strength and the strain hardening index of the materials are unknown variables that must be solved. The trial and error method were used to optimize the selection, that is, to select a group of different parameters, and then compare the errors between the stress-strain curve of the spherical indentation test and the stress-strain curve of the theoretical power exponential equation [11]; theoretically the two curves should coincide. The combination of the parameters of the material that gives the smallest error is the actual performance parameters of the materials that need to be determined.

### *3.2. Optimization conditions*

The three material parameters of the improved elastic-plastic model, namely the elastic modulus, the yield strength and the strain hardening index, are taken as unknown variables and can be solved using the optimization method [12]. The optimum condition that was adopted is where the sum of the square of the error between the stress from the indentation test and the stress from the theoretical power exponential equation has the smallest value, that is:

$$error = \min \sum_i [\sigma_{ri} - \sigma_y (1 + \frac{E}{\sigma_y} \varepsilon_{ri})^n]^2 \tag{14}$$

## 4. Experimental results

### *4.1. Indentation curve test results*

The load-displacement curve from the indentation test was acquired using an instrument indentation instrument. The maximum load of the instrument is 2000 N and the displacement precision is 0.2 μm. The indenter was a tungsten carbide ball with a diameter of 1mm, and the test material was Q345R steel.

Figure 2 has shown the multi-cycle load-displacement curve obtained from the test; there were 15 loading and unloading cycles.

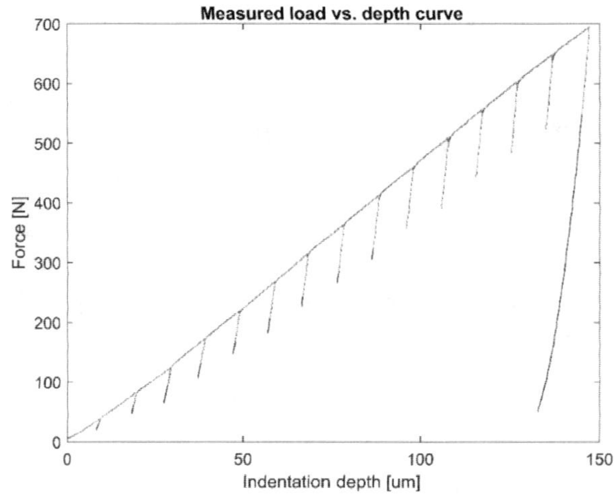


**Figure 2.** Multiple load-depth curve from the ball indentation test and the calculated curve.

The optimized results of the stress-strain curve and the theoretical power exponential stress-strain curve for the spherical indentation have been shown in figure 3. The red line denotes the stress-strain curve for the spherical indentation test and the blue line denotes the stress-strain curve of the theoretical power exponential law. The error between the spherical indentation stress and the theoretical stress was the smallest when the yield strength of the material was 381 MPa, the strain hardening index was 0.1421 and the elastic modulus was 210120 MPa which had been optimized using Matlab.

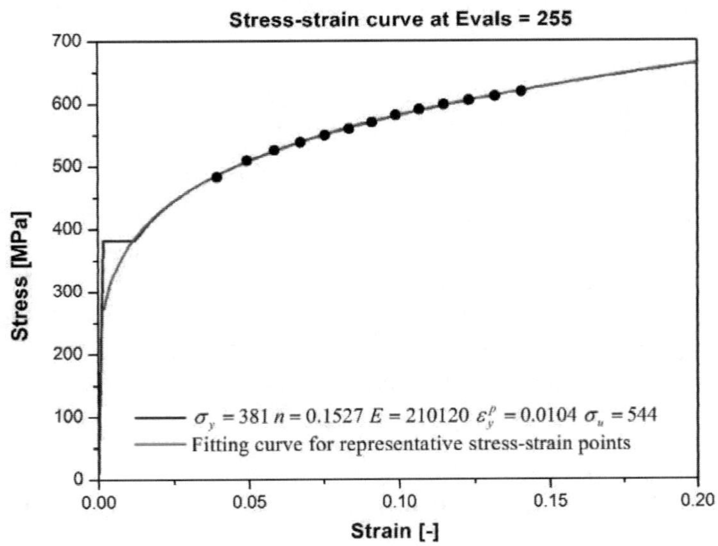


**Figure 3.** Representative stress-strain curve and the fitted true stress-strain curve.

*4.2. Comparison of the two methods*

According to standard GB/T 228.1-2010 Tensile Tests of Metal Materials Part 1: Room Temperature Testing Method; uniaxial tensile tests were carried out using an electronic universal testing machine CMT5205. The material tested was Q345R steel for pressure vessels, which had been processed into standard specimens with dimensions of 10 mm × 50 mm.

Table 1 has shown the comparison of the results of the stress-strain method for spherical indentation and the uniaxial tensile test. The errors of the yield strength, the strain hardening index, the elastic modulus and the tensile strength between the two methods were small, and the accuracy of the measurement of the mechanical properties of the metallic materials using the spherical indentation method was high.

**Table 1.** Comparison of the results from the indentation test and the uniaxial tensile test.

| | Uniaxial tensile test | Indentation test | Error |
|---|---|---|---|
| **Yield strength /MPa** | 380 | 381 | +0.26% |
| **Tensile strength /MPa** | 549 | 520 | -5.28% |
| **Strain hardening index** | 0.1335 | 0.1421 | +6.44% |
| **Elastic modulus /MPa** | 206420 | 210120 | +1.79% |

## 5. Conclusion

- In this paper, three parameters, the elastic modulus, the yield strength and the strain hardening index, were automatically optimized using the optimization function in Matlab. The error between the stress from the spherical indentation test and the stress from the theoretical power exponential law was taken as the convergence condition. The mechanical properties of Q345R steel, such as the yield strength $\sigma_y$ 381 MPa, the strain hardening index $n$ 0.1421, the elastic modulus $E$ 210120 MPa and the tensile strength $\sigma_u$ 520 MPa, were then obtained.

- The errors between the results from the spherical indentation method and the uniaxial tensile test were small and they were all less than 10% for all of the values investigated. The results showed that the accuracy of the spherical indentation method is high and meets the requirements of the field. Therefore, the spherical indentation method can be used to analyse the mechanical properties of Q345R steel for field testing in engineering.

**Acknowledgments**

This work is supported by the National Key Research and Development Program of China (No. 2017YFB0902100).

**References**

[1] Cao Y J and Lu J 2004 *Acta Mater.* **52** 4023-32
[2] Tabor D 1951 *Hardness of Metals* (Oxford: Clarendon Press) 35-50
[3] Yan W Y and Pun C L 2011 *Mater. Res. Innov.* **15** 41-9
[4] Oliver W C and Pharr G M 1992 *J Mater Res.* **7** 1564-83
[5] Hill R, Storakers B and Zdunek A B 1989 *P. Roy. Soc.* **423** 301-30
[6] ISO 2008 Metallic materials-Measurement of mechanical properties by an instrumented indentation test-indentation tensile properties *NPR-ISO/TR 29381* 1-29
[7] Ahn J H and Kwon D 2001 *J Mater Res.* **16** 3170-8
[8] Sneddon I N 1965 *Int. J. Eng. Sci.* **3** 47-57
[9] Dao M, Chollacoop N, Vliet K J. V, Venkatesh T A and Suresh S 2001 *Acta Mater.* **49** 3899-918
[10] Kang J J, Becker A A and Sun W 2012 *Int. J. Mech. Sci.* **62** 34-46
[11] Cao Y P, Qian X Q and Huber N 2007 *Mater. Sci. and Engin. A* **454** 1-13
[12] Li Y Z, Stevens P, Sun M C, Zhang C Q and Wang W 2016 *Int. J. Mech. Sci.* **117** 182-96

# Study on the microstructure of 2124 aluminum alloy subjected to stress assisted aging

**L W Quan[1,2] and C C Wang[3,4]**

[1]Department of Resources and Materials Science, Northeastern University at Qinhuangdao, Qinhuangdao, 066004, China
[2]Advanced Materials Testing and Analysis Center, Hebei University of Technology, Tianjin, 300401, China
[3]Tianjin Sino-German University of Applied Sciences, Tianjin, 300350, China

E-mail: wangchuancai@tsguas.edu.cn

**Abstract.** The effect of a load on the microstructure of 2124 aluminum alloy was investigated by employing transmission electron microscopy. When the alloy was aged at 190°C under a load of 6.05 kg, the deformation was increased with the aging time prolonging. Meanwhile the length of $Al_2CuMg$(S) phase was shorter than the length of S phase in the alloy without loading. At the same time, there was no preferential orientation effect on S phase or Genier Preston Bagaryatsky (GPB) zones.

## 1.Introduction

The wing of airplane is commonly formed by bend forming with a load applying on aluminum alloy [1]. The bend forming has been applied for the manufacture of B-1B long range combat aircraft, A330/340 and A380 [2]. The load is in the range of elastic stress and applied during the aging process. It is necessary to understand the effect of load on the microstructure and properties. The effect of stress on the microstructure has been investigated by some researchers [3-9]. The precipitates in Al-Cu alloy and Al-Cu-Mg-Ag alloy show preferential orientation under an elastic stress during aging treatment. The mechanism of the orientation effect of $\theta'$ phase and $\Omega$ phase was explained by Eshelby elastic inclusion theory [5,10,11]. However the bend forming of 2124 aluminum alloy has not been studied. 2124 aluminum alloy is widely applied for the manufacture of the wing and fuselage of airplane. It is generally regarded that the main strengthening second phases in 2124 alloy are GPB zones and S phase. GPB zones are short range ordered phase with several nanometers. S phase is lath with crystal lattice parameters a=0.400 nm, b=0.923 nm, c=0.714 nm [12]. And there are 12 variants of S phases. The orientation effect of GPB and S phase is not clear up to now. It is essential to investigate the effect of load on the microstructure of 2124 aluminum alloys. Therefore, in this paper the microstructure and formability of 2124 aluminum alloy under loading condition was investigated.

## 2.Experimental methods

Table 1 shows the chemical composition of commercial 2124 alloy used in this paper. The dimension of the specimens is 260 mm×25 mm×2 mm. Bend forming experiment was carried out in a self-made furnace. The specimens were solution treated at 495°C for 1h and quenched into water at room-temperature. And then the specimens were immediately transferred into the furnace and aged at 190°C with a load of 6.05 kg. Vickers hardness was tested with 450SVDTM with a load of 5 kg and dwell time of 20 s. Thin foils for transmission electron microscopy (TEM) were prepared by cutting, grinding and punching into 3 mm discs. Electro-polishing was employed in a twin-jet Tenupol by a 33% nitric acid solution in methanol operated at -25°C and 13.8 V. FEI CM20 operated at 200kV was carried out to observe the microstructures.

**Table 1.** Chemical composition of commercial 2124 alloy (wt%).

| Si | Fe | Cu | Mn | Mg | Cr | Zn | Ti | Al |
|---|---|---|---|---|---|---|---|---|
| 0.08 | 0.16 | 4.31 | 0.65 | 1.51 | 0.01 | 0.02 | 0.01 | balance |

## 3.Results and discussion

### *3.1. The deformation mechanism*

3-point bending was adopted to study the bending formation of 2124 aluminum alloy. The middle part of the specimen showed the largest deformation of the specimen. It can be calculated that the maximum stress of the specimen was 231 MPa when the load is 6.05 kg. This is lower than the yield strength 280 MPa of this alloy.

Figure 1 illustrates the relationship of the deformation and the aging time. It can be clearly seen that the deformation increases with the time prolonging. At the initial stage, the deformation gradually increases. But it increases rapidly at higher temperature. This can be explained by the creep mechanism with Arrhenius equation as shown in equation (1).

$$\varepsilon = A\sigma^{n} \exp(-Q / RT) \tag{1}$$

Q is the activation energy; R is the gas constant; T is the absolute temperature. Higher temperature leads to larger deformation according to the equation (1). Therefore the deformation mechanism can be explained by creep mechanism.

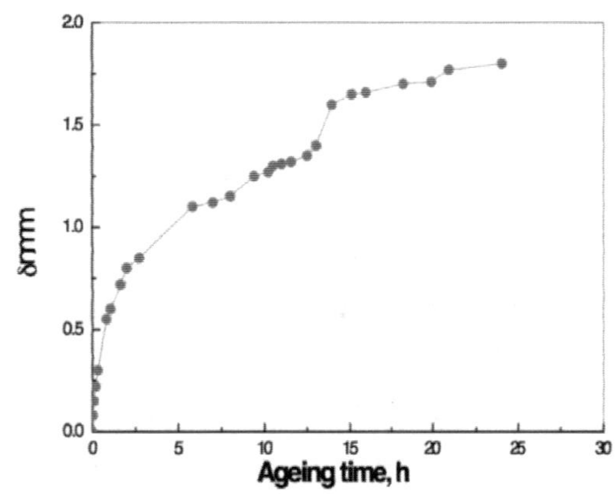


**Figure 1.** The change of deformation curves of 2124 alloy in bend-form.

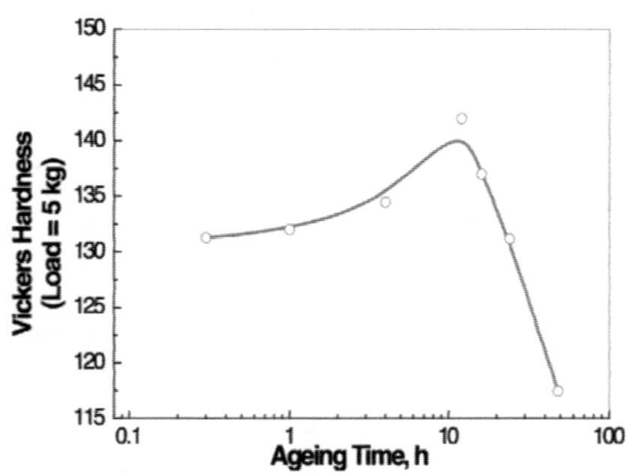


**Figure 2.** Ageing hardness curves of 2124 alloy aged at 190°C.

### *3.2. The mechanical properties of the alloy*

The aging curve of 2124 alloy aged at 190°C is shown in figure 2. The peak aging time is 12h for the bend formed 2124 alloy. And the peak hardness is 142VHN, which is higher than 134VHN of the alloy that without a load.

### *3.3. The microstructure of the bend formed alloy*

As discussed above, the largest deformation part is the half length of the specimen. It suffers a tensile stress downside of the specimen and a compressive stress on the top of the specimen. The microstructure of the alloy aged to peak temper with and without a load is shown in figure 3. S phases and GPB zones are the predominant phases in the peak-aged 2124 aluminum alloy. When the alloy is aged without a load seen in figure 3(a), the length of S phase is 445 nm, which is longer compared to the length of 405 nm in the alloy with a load of 6.05 kg as shown in figure 4. This means the average length of S phases is shorter under the role of load, which can be explained by the vacancies. Quantities of vacancies generate when the alloy is solution treated and quenched [13]. As a load is applied, on one hand, dislocations are generated which can increase the nucleation sites for the second phases. On the other hand, lots of vacancies will migrate to the dislocations which lead to the concentration of vacancies decreasing. The growth of second phase will be retarded due to the decreasing of the vacancies. Therefore the applying load can affect the growth of S phase and decrease the length of S phase.

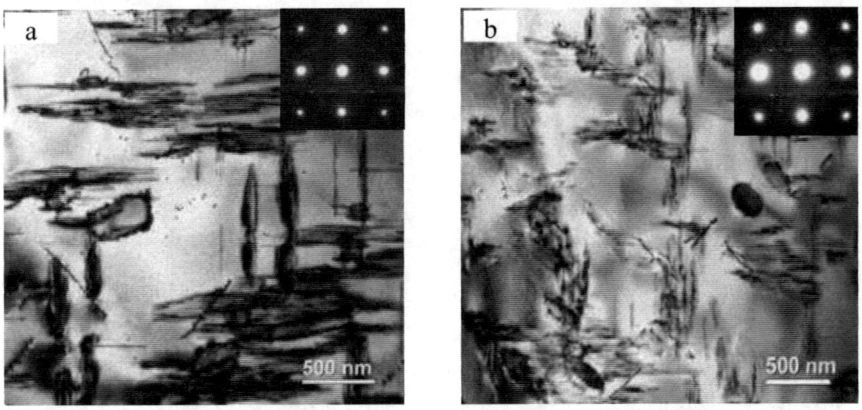


**Figure 3.** Bright field images and corresponding SAD pattern of S phases in 2124 Al alloy aged at 190°C for 12h. B=[001] (a) without load (b) with load of 6.05 kg.

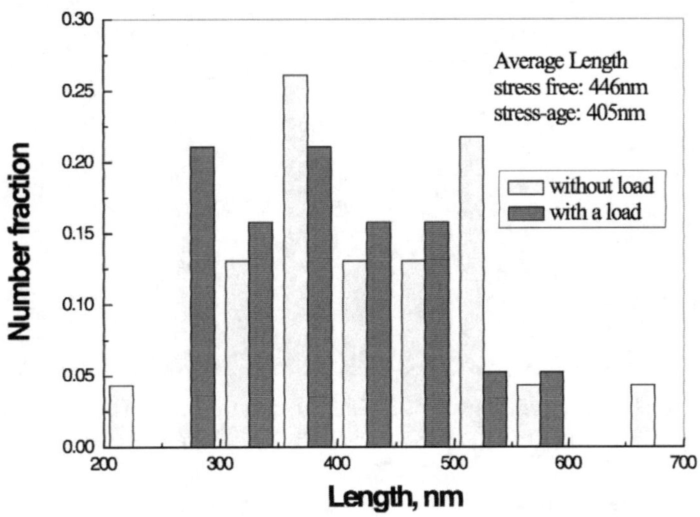


**Figure 4.** The number fraction and the dimension distribution of S phases.

It can be seen that the S phases are in large dimensions and distribute equivalently on all {210} planes in the aged alloy without a load. Meanwhile, S phases are also growing equivalently on different directions when the alloy subjected a load of 6.05 kg as shown in figure 3(b). The load has no obvious effect on the distribution of S phases. This can be also proved by selected area diffraction patterns (SAD). The intensity of the reflection spots of S phases is the same in $[200]_\alpha$, $[020]_\alpha$ directions, which means that the quantity of S phases along $[200]_\alpha$, $[020]_\alpha$ are the same. This indicates that load has little role on the orientation of S phases.

GPB zones play a dominant role on strengthening the alloy [14,15]. Many tiny needle GPB zones distribute homogeneously in the alloy as seen in figure 5. The dot pattern phases in the figure 5 are the cross-section of GPB zones. It indicates that the number density of GPB zones is almost the same with or without stress. In the magnified images (figure 6), it can be also clearly seen that dimension of the cross-section of GPB zones has no significant difference. Additionally, the SAD pattern agrees with this finding. And the strength and morphology of the streaks along $[200]_\alpha$, $[020]_\alpha$ directions are almost the same with or without a load. As a result the load has no effect on GPB zones.

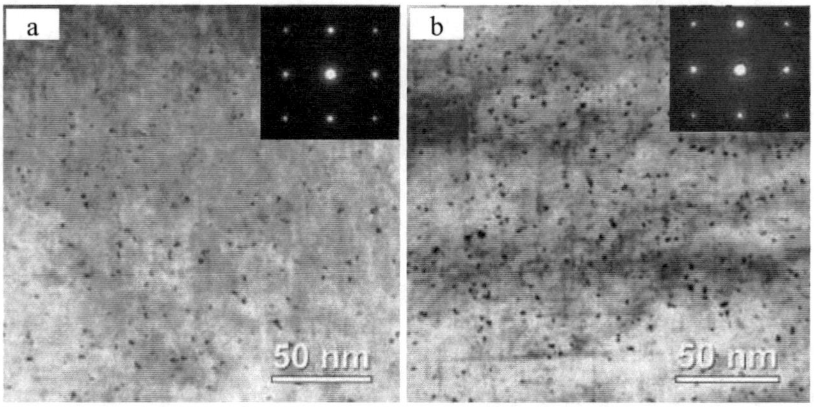


**Figure 5.** Bright field images and corresponding SAD pattern of GPB zones in 2124 Al alloy aged at 190°C for 12h. B=[001] (a) without load (b) with load of 6.05 kg.

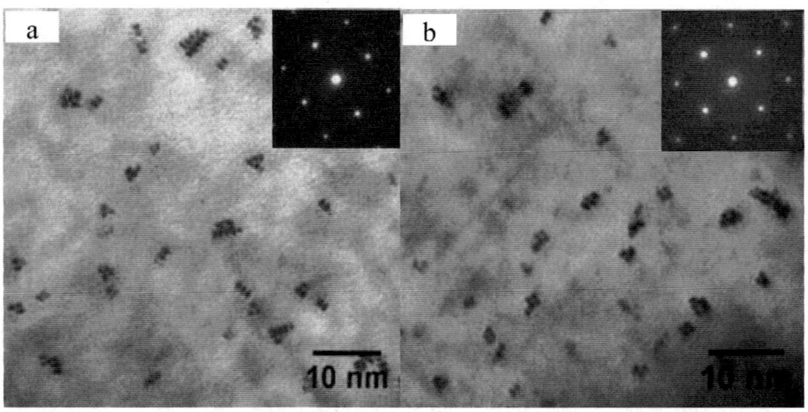


**Figure 6.** High magnification of GPB zones in 2124 Al alloy aged at 190°C for 12h (a) without a load (b) with a load of 6.05 kg. B=[001].

In the previous studies, several models of GPB zone by other researchers are proposed [16-18], among which GPB proposed by Kovarik is adopted in this paper [18]. The orientation relationship of the GPBII zone with the matrix is $[100]_{Al}//[100]_{GPBII}$ and $[010]_{Al}//[010]_{GPBII}$. It is known that GPB zones possess an orthorhombic lattice and space group Cmmm. There are three kinds of variants of GPB zones. One of the variants is parallel to the incident beam, while the other two are perpendicular to it. So the two variants are observed to study the effect of external load. When applying a load, the forming energy can be expressed by phase transformation strain and the uniform strain derived from external load. Due to the structure of GPB zones are symmetrical, the uniform strain can be considered as a constant. Then the transformation strain is essential for the total energy. It can be expressed as equation (2). Variant1 and variant 3 can be simplified as equation (3).

$$\varepsilon_{ij}=\begin{pmatrix}\varepsilon_{11} & \varepsilon_{12} & \varepsilon_{13}\\ \varepsilon_{21} & \varepsilon_{22} & \varepsilon_{23}\\ \varepsilon_{31} & \varepsilon_{32} & \varepsilon_{33}\end{pmatrix}=\begin{pmatrix}\varepsilon_{11} & \varepsilon_{12}\\ \varepsilon_{21} & \varepsilon_{22}\end{pmatrix}=\begin{pmatrix}\varepsilon_{xx} & \varepsilon_{xy}\\ \varepsilon_{yx} & \varepsilon_{yy}\end{pmatrix} \tag{2}$$

$$\varepsilon^0(1)=\begin{pmatrix}\varepsilon_3 & 0\\ 0 & \varepsilon_1\end{pmatrix}\quad \varepsilon^0(3)=\begin{pmatrix}\varepsilon_1 & 0\\ 0 & \varepsilon_3\end{pmatrix} \tag{3}$$

Where $\varepsilon_1=\dfrac{ma_p-na_0}{na_0}$, $\varepsilon_3=\dfrac{c_p-a_0}{a_0}$, the atomic parameter of Al $a_0$=0.405 nm. Assuming the ratio of precipitate atom and the matrix atom is 1:3. Then it can be calculated $\varepsilon_1=-0.25\%$, $\varepsilon_3=-0.25\%$, and $\varepsilon^0(1)$ is equal to $\varepsilon^0(3)$. Therefore the total energy is the same for different variants, which means the load has no obvious effect on the GPB zones.

The crystal structure of S phase is orthogonal with the crystal parameters *a*=0.400 nm, *b*=0.923 nm, *c*=0.714 nm, and the orientation relationship with the matrix is $[100]_{Al}//[100]_S$, $[02\bar{1}]_{Al}//[010]_S$, $[012]_{Al}//[100]_S$ [19,20]. The misfit can be calculated by the equation as following:

$$\delta_i=\frac{md_p-nd_{Al}}{nd_{Al}} \tag{4}$$

Where $d_p$ is the lattice parameter of precipitate, and $d_{Al}$ is the lattice parameter of aluminum. The results are as shown in table 2. The misfit of *a* axis is small, so that the interface energy to form the plane perpendicular to *a* axis is very small. Moreover the number of matching lattices is 1. S phase is coherent with Al matrix along *a* axis, which means S phases are easy to grow along *a* axis. However, when the number of Al planes is up to 5 and 4 for *b* axis and *c* axis respectively, the misfit can reach to the minimum value. In this case, due to the number of Al planes is large, it is difficult to grow fast along *b* and *c* axis. Therefore, the morphology of S phase is lath. When applying a load on the alloy, S phases suffer strain energy and interface energy. Strain energy is quite small compared to interface energy. Consequently, as for part-coherent and incoherent S phase, the external load has no effect on S phases.

**Table 2.** The misfit between S phase and Al matrix.

| Planes | Al plane space (nm) | Number of Al planes, n | S phase space (nm) | δ (%) |
|---|---|---|---|---|
| $(100)_{Al}//(100)_S$ | $d_{100}$=0.405 | 1 | 0.400 | +1.23 |
| $(012)_{Al}//(010)_S$ | $d_{012}$=0.1811 | 5 | 0.923 | −1.93 |
| $(02\bar{1})_{Al}//(001)_S$ | $d_{02\bar{1}}$=0.1811 | 4 | 0.714 | +1.44 |

## 4.Conclusions

In this work, the effect of a load on the 2124 aluminum alloy during aging process was studied. And mechanical properties and microstructure of 2124 alloy were investigated. When the alloy was aged at

190°C under a load of 6.05 kg, the deformation was increased with the aging time prolonging. Due to the increased nucleation sites and decreased vacancies by introduction of a load, the length of the S phases was shorter in the bended alloy compared with the reference alloy without loading. However the load has no obvious effect on the preferential orientation of S phase or GPB zones because the total energy of different variants of S phases or GPB zones is almost the same.

**Acknowledgments**
The authors are greatly thankful to the financial support from the "Fundamental Research Funds for the Central Universities" (N172304047) and the Natural Science Foundation of Hebei Province in China (E2018501088 and E2018501030).

**References**

[1] Holman M C 1989 *J. Mech. Work. Technol* **20** 477-88
[2] Eto T, Sato A and Mori T 1978 *Acta Meter* **26** 499-508
[3] Adachi T, Kimura S, Nagayama T, Takehisa H and Shimanuki M 2004 *Materials Forum* **28** 202-7
[4] Hosford W F and Agrawal S P 1975 *Metallurgical Transactions A* **6** 487-91
[5] Li D Y and Chen L Q 1998 *Acta Mater* **46** 2573-85
[6] Zhu A W and Starke E A 2001 *Acta Mater* **49** 3063-9
[7] Zhu A W and Starke E A 2001 *J Mater Process Tech* **117** 354-8
[8] Zhu A W, Chen J and Starke E A 2000 *Acta Mater* **48** 2239-46
[9] Skrotzki B, Shiflet G J and Starke E A 1996 *Metall Mater Trans A* **27** 3431-44
[10] Bakavos D, Prangnell P B and Dif R 2004 *Materials* **28** 124-31
[11] Zheng Z Q, Chen D Q and Li S C 2004 *Materials Forum* **28** 439-44
[12] Perlitz H and Westgren A 1943 *Arkiv Kemi Mineral Geol B* **16** 1-5
[13] Zheng Z Q, Chen D Q, Li S C, Chen Z G and Liu Z Y 2004 *Acta Metall Sin* **40** 799-804
[14] Hull D and Bacon D J 2001 *Introduction to Dislocations* 4th ed (Oxford Butterworth)
[15] Ringer S P, Hono K, Polmear I J and Sakurap T 1996 *Acta Mater* **44** 1883-98
[16] Ringer S P, Sakurai T and Polmear I J 1997*Acta Mater* **45** 3731-44
[17] Silcock J M 1961 *Journal of the Institute of Metals* **89** 203-10
[18] Wolverton C 2001 *Acta Mater* **49** 3129-42
[19] Kovarik L, Gouma P I, Kisielowski C, Court S A and Mills M J 2004 *Acta Mater* **56** 2509-20
[20] Radmilovic V, Kilaas R, Dahmen U and Shiflet G J 1999 *Acta Mater* **47** 3987-97

IOP Conf. Series: Materials Science and Engineering **668** (2019) 012019 doi:10.1088/1757-899X/668/1/012019

# Preparation and properties of GO-$SiO_2$ anticorrosive coatings

**X L Bai[1,2], B Wu[1], K Jiang[1], C J Liu[1], H Zhang[1], F Zhang[1], X Huang[1], M J Cai[1] and P F Wang[1]**

[1]PLA Army Academy of Artillery and Air Defense, Hefei 230031, China

E-mail: xlbai@ustc.edu

**Abstract.** Epoxy resin is often used as raw material of anticorrosion, but in the curing process of resin, a large number of micropores are prepared in the solvent evaporation method, which reduces the performance of anticorrosion. In order to reduce internal defects and improve corrosion resistance, GO was compounded with $SiO_2$ modified by KH550 to prepare for GO-$SiO_2$. By means of mechanical stirring and ultrasonic dispersion, GO-$SiO_2$ was added to epoxy resin to prepare for the anticorrosive coating. SEM was used to observe the coating morphology, and the physical and electrochemical properties of the coatings with different mass fractions of GO-$SiO_2$ are tested for corrosion resistance. The results show that 3 wt% GO-$SiO_2$ coating shows the best performance, the coating hardness reaches H, the adhesion grade is II.

## 1. Introduction

Currently, corrosion costs are increasing, and preventive strategies have become an important industrial demand. Anticorrosive coatings on metal surface can play the role of shielding, passivation and electrochemical protection, which is of great significance to the research of metal corrosion and protection. Epoxy resin (EP) is a common anticorrosive coating with good chemical resistance, high temperature resistance, toughness and bonding ability. However, the outdoor weather resistance is poor, the brittleness of the coating film is high, and the volatilization during the curing process will cause many micro-holes inside the coating, which will reduce the corrosion resistance of the coating.

Graphene oxide (GO), a product of chemical stripping of graphite, has a two-dimensional lamellar structure and similar unique properties to graphene [1,2]. At the edge of GO, there are a large number of oxygen-containing functional groups such as epoxy groups, hydroxyl groups, ketones and carboxyl groups. These functional groups provide many active sites for the functionalization of GO, enabling it to be functionalized with organic and inorganic materials through chemical crosslinking to form stable chemical bonds [3,4]. Krishnamoorthy *et al* were prepared to have good antibacterial properties and prevent fouling heat resistance of graphene oxide modified alkyd resin coatings. The results show that the mechanical properties and corrosion resistance of graphene oxide coatings are improved, and the most important and anticorrosive properties are improved [5]. Jiang *et al* studied the influence of modified GO by active amino silane precursor on the corrosion resistance of epoxy coating and found that the modified GO by adding silane coupling agent could effectively promote the interface adhesion strength of coating and improve the corrosion resistance of coating [6].

Ramezanzadeh developed a sol-gel film on the basis of silane film, which is filled with GO nanosheets to improve corrosion resistance of EP coating and cathodic stripping performance on steel substrate [7].

Compared with GO, the coating with GO-$SiO_2$ has better mechanical properties, which may be due to the fact that GO-$SiO_2$ can increase the crosslinking density between coatings, make the arrangement between particles in the coating more orderly, effectively improve the permeability resistance of the coating, and further improve the corrosion resistance of the coating [8-10]. Ma *et al* found that by adding nano $SiO_2$/graphene oxide compounds, epoxy resin coating anticorrosion performance compared to the pure epoxy resin coating as well as adding $SiO_2$ separately and GO nano filler of epoxy resin coatings, the anticorrosion performance improved significantly [11] Wang *et al* studied the self-healing and corrosion resistance of GO-mesoporous silicon layer nanostructure coating, and believed that there was a synergistic enhancement effect [12]. Compared with the pure EP coating, the hardness, adhesion and corrosion resistance of nano $SiO_2$-GO/EP are obviously enhanced [13]. The mutual barrier between GO and $SiO_2$ is used to inhibit the agglomeration of each other, and meanwhile, the synergistic modification effect of the two on EP is given full play, which is also reflected in other fields [14,15]. At present, many scientific mechanisms related to GO-$SiO_2$ anticorrosive coatings, such as dispersion problems, synergistic enhancement effects and interface problems have not been clarified.

GO-$SiO_2$ composite material is an ideal resin filler with excellent properties. $SiO_2$ particles are evenly distributed on graphene sheets, the two-dimensional structure of graphene is increased to three-dimensional structure. With such a special three-dimensional structure, it is not easy to overlap between graphene sheets and form a "stacked" structure coating. This method overcomes the disadvantage that GO and $SiO_2$ are easy to agglomerate and improves the dispersion performance significantly. The epoxy resin can be strengthened and toughened by adding the composite material to the resin matrix, and the corrosion resistance of the coating can be improved.

## 2. Experiment

### *2.1. Materials*
Nano $SiO_2$ is provides by Shanghai Aladdin Biochemical Technology Co., LTD. Silane coupling agent, gamma-aminopropyl triethoxy silane (KH550), concentrated $H_2SO_4$, $KMnO_4$, $NaNO_3$, HCL, anhydrous ethanol, N, n-dimethyl formamide (DMF), acetone, NaCl are obtained from Sinopharm Chemical Reagent Co., LTD. All of the above are pure analysis. GO(>99 %) is provided by SuZhou TanFeng Graphene Technology Co., LTD..

### *2.2. Preparation of GO–$SiO_2$*
$SiO_2$ was dried for 24 h in an electric thermostatic air drying oven at 100°C.10 g $SiO_2$ was added to 100ml anhydrous ethanol and deionized water mixture with a volume ratio of 3:1 for 30 min. Add 5 g KH550 to 50 ml anhydrous ethanol, adjust pH to 3 with acetic acid, stir magnetically for 1h, add to the above solution and stir in constant temperature water bath at 75°C for 3h. Then wash repeatedly with anhydrous ethanol to remove the physically adsorbed silane coupling agent and reaction by-products on the surface, and dry in vacuum drying box at 60°C for 12 h to obtain modified $SiO_2$.

0.1 g of modified nano $SiO_2$ was added to 200 ml of n, n-dimethyl formamide DMF, and ultrasonic dispersion was used for 30 min. 0.3 g of GO was added to the above solution, and ultrasonic dispersion was continued for 30 min. The mixed solution was put in an oil bath at 105°C for about 4 h of magnetic stirring. After the reaction, the solution was filtered, washed with anhydrous ethanol for 3 times, and then dried in a vacuum drying box at 60°C for 24 h for grinding.

### *2.3. Preparation of coatings*
The mass percentage of GO–$SiO_2$ was 0 wt%, 1 wt% and 3 wt% of coating powder, and the samples were numbered as GS-0(EP), GS-1 and GS-3, respectively. The epoxy resin and GO were mixed evenly by dispersion-machine agitation, and the rotation speed was 1000 R/M for 20~30 minutes. Appropriate defoaming agent and curing agent are added in the state of mechanical agitation. Ultrasonic dispersion of the stirred slurry was conducted for 2 hours to obtain the GO-$SiO_2$ coated slurry. Planet ball mill (Nanjing NanDa Instrument Plant) is made of zirconia. After consulting

relevant literature, set the appropriate milling parameters, zirconia ball of the ball diameter $\Phi$5, respectively, $\Phi$3, the ratio of its quality for $\Phi$5: $\Phi$3 = 3:2. Ball material ratio of 2:1, ball mill speed of 220 R/M, ball grinding time of 4 h. Steel 616 (22Si$Mn_2$TiB) is a low alloy high strength steel. The thickness of the plate used in this paper is 200 microns, and the alloy substrate is cut into 4 cm×4 cm block by the plate cutter. The glass rod is evenly coated on the treated steel sheet, and the hanging sheet is solidified for 3 days in an electric thermostatic air-drying oven at 60°C.

### *2.4. Characterization*

Scanning electron microscope (SEM, jeol-jsm-6700f) was used to observe the surface morphology and interface morphology of the coating. The corrosion resistance of salt spray coating is characterized by salt spray test. The coated steel 616 sheet was put into the salt spray test chamber, and the temperature of the salt spray test chamber was controlled at 35°C. The relative humidity of the salt spray test chamber was > 95%. 5% NaCl solution was sprayed with steam through the action of air compressor, and the pH of the atomized NaCl solution remained neutral. The steam volume flow rate is sprayed at the speed of 1.2 ~ 2.0 mL.$h^{-1}$ per 80 $cm^2$. Water resistance was tested according to GB/ T11733-1993, acid/alkali resistance according to GB/ T9274-1988. Mechanical properties test coating flexibility test according to GB/T1731-1993, pencil hardness test according to GB/T173l-2006, adhesion test according to GB/T 9286-1998.

## 3. Results and discussion

### *3.1. Morphology of GO–$SiO_2$ coating*

SEM was used to observe the surface and cross-sectional morphology of the coating, and to compare the differences of the microstructural morphology of GS-0 samples (EP). GS-1 GS-3 was used to investigate the effect of GO-$SiO_2$ on the microstructural morphology of the coating. It can be seen from figure 1 that after mixing with the ball mill, the coating materials are evenly distributed, and the mixing effect is good. No particles of large size appeared on the coating surface. Particle size was refined by ball grinding to achieve satisfactory results. No agglomeration of go was observed on the surfaces of the doped GO coatings (samples GS-1 and GS-3), suggesting that GO was well mixed with the coating slurry and dispersed in the coating slurry after mechanical agitation and ultrasonic dispersion. It can be seen from the SEM photos of the surface morphologies of samples GS-0, GS-1 and GS-3 that the coating surface is relatively flat, indicating that the coating prepared by coating method has a good surface state. As shown in figure 1(c), cracks 5 to 10 microns wide appeared on the surface of sample GS-1. The cracks on the coating surface are caused by the volume shrinkage of $SiO_2$ based coating paste during solidification and sintering, which makes it easy to produce relatively large internal stress in the coating, and the internal stress will be concentrated in local areas. When the concentrated stress is greater than the bonding strength inside the coating, the crack will occur. In figures 1(b), 1(d) and 1(f) it can be seen that pore structures of different sizes appear on the surfaces of samples GS-0, GS-1 and GS-3.Compared with sample GS-1, the aperture of pore structure on the surface of sample GS-3 is smaller, and the size and quantity of pore structure on the coating surface of sample GS-3 are smaller than that of sample GS-0.This means that the doped go expands in the coating. Other coating components attach to the surface of GO.

Compared with samples GS-0 and GS-1, the pore structure on the surface of sample GS-3 is smaller in diameter, and the pore structure size of the coating surface of sample GS-3 is smaller than that of sample GS-0, and the number is less, and the flake increases. It indicates that GO-$SiO_2$ expands in the coating, and the coating is connected into a whole, and the coating organization is more compact.

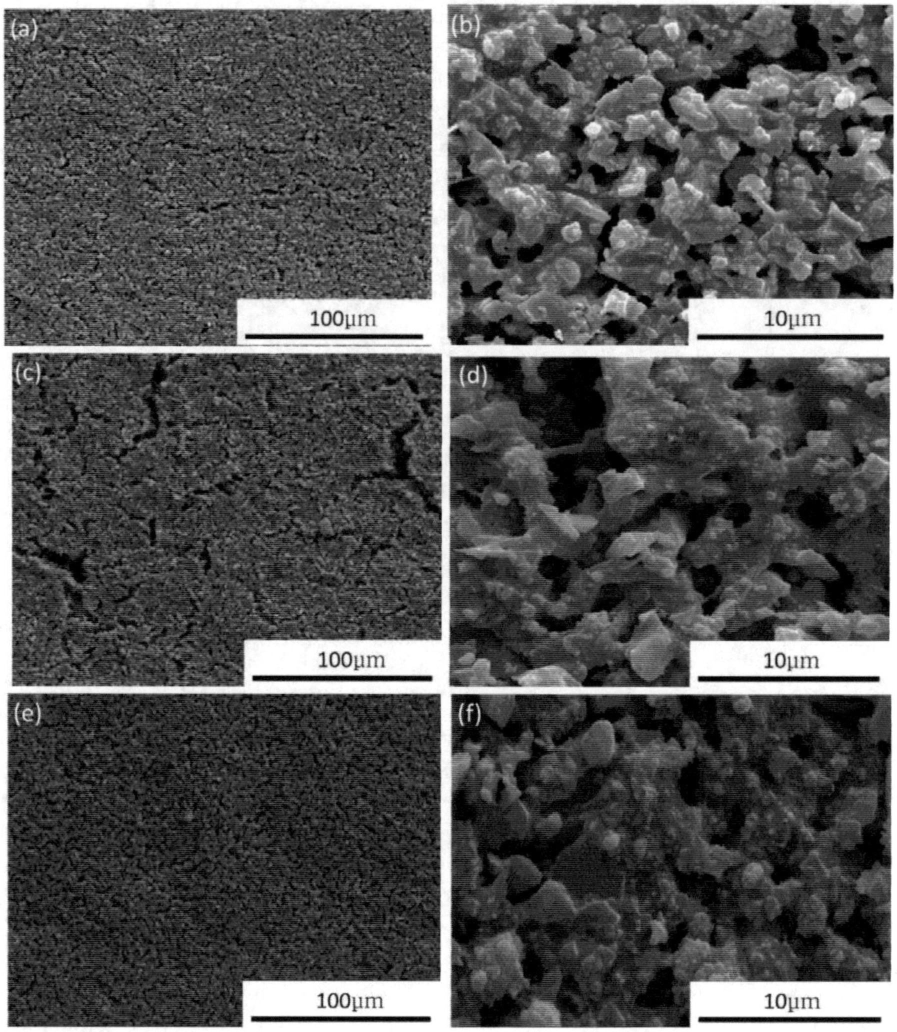


**Figure 1.** SEM photos of coating surface morphology. (a) (b) sample GS-0; (c) (d) sample GS-1; (e) (f) samples GS-3.

### *3.2. Morphology of coating*

**Figure 2.** SEM photos of cross-section morphology. (a) (b) sample GS-0; (c) (d) sample GS-1; (e) (f) samples GS-3.

Figure 2 shows SEM photos of section morphologies of the coatings doped with GO (samples GS-0,

IOP Conf. Series: Materials Science and Engineering **668** (2019) 012019 doi:10.1088/1757-899X/668/1/012019

GS-1, and GS-3). The thickness of each coating can be roughly estimated from the SEM photos in figures 2(a), 2(c) and 2(e). The thickness of sample GS-0 is 67 microns, the thickness of sample GS-1 is 43 microns and the thickness of sample GS-3 is 56 microns. It can be seen from the figure that the coating is tightly bound to the substrate without gaps, and there is a clear interface between them, indicating that the coating prepared by brush coating method can be well bound to the substrate. As shown in figures 2(d) and 2(f), the flake structure appears in the coating doped with GO, and the more doped go content in the coating, the flake structure becomes more and more obvious. However, no similar flake structure appears in the undoped go coating. The sheet structures introduced by doped GO form a large number of interfaces and barrier layers in the coating.

### *3.3. Mechanical properties and corrosion resistance*

The mechanical properties of GS-0, GS-1 and GS-3 coatings prepared were tested, and the results are shown in table 1. With the increase of the amount of GO-$SiO_2$, coating flexibility are all meet the national standard (3 mm or less), coating adhesion increased, when the dosage of 3% reach level II coating adhesion. This is because the oxygen-containing functional groups such as the hydroxyl group on the surface of go react with the epoxy resin, and the epoxy group can also chemically react with the hydroxyl group on the surface of the steel plate, thus increasing the adhesion of the film. The increase of GO also increased the hardness, flexibility and viscosity of the film.

**Table 1.** Mechanical properties of coating.

| | **GS-0** | **GS-1** | **GS-3** |
|---|---|---|---|
| Adhesive grade | III | III | II |
| Pencil hardness | H | H | H |
| Viscosity/mPa · s | 2510 | 2519 | 2531 |

Three coatings were tested for corrosion resistance, as shown in table 2. It was found that adding GS-3 coating had better corrosion resistance. This may be due to $O_2$, some corrosive medium, such as water and ions can be generated by EP surface of microchannel reach the surface of the metal substrate, corrosion reactions, with metal layers GO with special favour winding path on the surface of the electrolyte to enter the body, the surface load of the $SiO_2$ can fill the coating porosity, slowing the rate of electrolyte to metal substrate, reduce the corrosion of the metal substrate. The combination of GO and $SiO_2$ can reduce the agglomeration of $SiO_2$ and improve the compatibility between the material and EP, thus protecting the metal substrate. Figure 3 shows the photos of 20 days soaked in 20% HCL solution.

**Table 2.** Main technical indexes and test results of three coatings.

| | **GS-0(EP)** | **GS-1** | **GS-3** | |
|---|---|---|---|---|
| Water resistance | 20d, no significant change | 30d, no significant change | 30d, no significant change | 23±2°C |
| Acid resistance | 20d, scattered vesicles | 20d, no significant change; 30d, scattered vesicles | 20d, no significant change; 30d, scattered vesicles | 20%HCL |
| alkali resistance | 20d, scattered vesicles | 20d ,no significant change; 30d, scattered vesicles | 20d, no significant change; 30d, scattered vesicles | NaOH saturated solution |
| salt water resistance | 20d, scattered vesicles | 20d, scattered vesicles | 20d, no significant change; 30d, scattered vesicles | NaCl saturated solution |

IOP Conf. Series: Materials Science and Engineering **668** (2019) 012019 doi:10.1088/1757-899X/668/1/012019

**Figure 3.** Photos of 20 days soaked in 20% HCL solution (left to right: GS-0, GS-1, GS-3).

Both nano $SiO_2$ and GO have large specific surface areas. After modification by silane coupling agent surface, nano $SiO_2$ can easily interact with EP to form a dense grid structure. The lamellar GO surface contains many oxygen-containing organic functional groups, which can form excellent compatibility and combination structure with EP and enhance the mechanical properties of EP coating. Since the coating hardness is also related to the dispersion uniformity of nanomaterials in the coating, fully mixed GO-$SiO_2$ can reduce the agglomeration of $SiO_2$ particles, and evenly dispersed GO-$SiO_2$ can fill the gap between EP molecular chains, making the coating close to the metal matrix and improving the permeability.

## 4. Conclusions

GO was compounded with $SiO_2$ modified by KH550. GO-$SiO_2$ was added to epoxy resin to prepare for the anti-corrosion coating. The GO-$SiO_2$ coating was prepared on steel 616 sample. The anti-corrosion performance of GO-$SiO_2$ coating was tested, and its anti-corrosion mechanism was preliminarily discussed. The results show that the coating hardness reaches H, the adhesion grade is III, and 3 wt% GO-$SiO_2$ coating shows the best performance.In addition, lamp-structured composite materials like GO-$SiO_2$ may open up a new anti-corrosion strategy and have a positive application prospect in the field of nano-thin films of anti-corrosion coatings.

## Acknowledgment

This work is supported by the Self-Topic Fund of PLA Army Academy of Artillery and Air Defense.

## References

[1] Nanda S S, Papaefthymiou G C and Dong K Y 2015 Functionalization of graphene oxide and its biomedical applications *Rev. Solid State Sci.* **40** 291-315

[2] Jiang F W, Zhao W J, Wu Y M, Dong J D, Zhou K H, Lu G M and Pu J B 2019 Anti-corrosion behaviors of epoxy composite coatings enhanced via graphene oxide with different aspect ratios *Prog. Org. Coat.* **127** 70-9

[3] Ye Y W, Zhang D W, Li J Y, Liu T, Pu J B, Zhao H C and Wang L P 2019 One-step synthesis of superhydrophobic polyhedral oligomeric silsesquioxane-grapheneoxide and its application in anti-corrosion and anti-wear fields *Corros. Sci.* **147** 9-21

[4] Ramezanzadeh B, Niroumandrad S, Ahmadi A, *et al* 2016 Enhancement of barrier and corrosion protection performance of an epoxy coating through wet transfer of amino functionalized graphene oxide *Corros. Sci.* **103** 283-304

[5] Krishamoorthy K, Jeyasubramanian K, Preamanathan M, Subbuah G, Shin S H and Kim S J 2014 Graphene oxide nanopaint *Carbon* **72** 328-37

[6] Jiang M Y, Wu L K, Hu J M *et al* 2015 Silane-incorporated epoxy coatings on aluminum alloy (AA2024), Part 2: Mechanistic investigations *Corros. Sci.* **92** 127-35

[7] Ramezanzadeh B, Ahmadi A and Mahdavian M 2016 Enhancement of the corrosion protection performance and cathodic delamination resistance of epoxy coating through treatment of

steel substrate by a novel nanometric sol-gel based silane composite film filled with functionalized graphene oxide nanosheets *Corros. Sci.* **109** 182-205

[8] Haeri S Z, Ramezanzadeh B and Asghari M 2017 A novel fabrication of a high performance SiO2-graphene oxide nanohybrids: Characterization of thermal properties of epoxy nanocomposites filled with $SiO_2$-GO nanohybrids *J. Colloid Interf. Sci.* **493** 111-22

[9] Guo L *et al* 2018 Enhanced dispersion of graphene in epoxy-acrylic waterborne anticorrosion coating: Bifunctional ligands linking graphene to $SiO_2$ *Int. J. Electrochem. Sci.* **13** 11867-81

[10] Ramezanzadeh B, Haeri Z and Ramezanzadeh M 2016 A facile route of making silica nanoparticles-covered graphene oxide nanohybrids ($SiO_2$-GO); fabrication of $SiO_2$-GO/epoxy composite coating with superior barrier and corrosion protection performance *Chem. Eng. J.* **303** 511-28

[11] Ma Y *et al* 2016 Fabrication of silica-decorated graphene oxide nanohybrids and the properties of composite epoxy coatings research *Appl. Surf. Sci.* **360** 936-45

[12] Wang W *et al* 2019 Self-healing performance and corrosion resistance of graphene oxide-mesoporous silicon layer-nanosphere structure coating under marine alternating hydrostatic pressure *Chem. Eng. J.* **361** 792-804

[13] Sui L L *et al* 2018 Preparation and corrosion resistance of nano $SiO_2$-graphene oxide/epoxy composite coating *Acta Materiae Compositae Sinica* **35** 1716-24

[14] Hossein H, Mohsen M, Ehsan S and Amin R B 2017 Microstructure, deposition mechanism and corrosion behavior of nanostructured cerium oxide conversion coating modified with chitosan on AA2024 aluminum alloy *J. Alloy. Compd.* **725** 968-75

[15] Li Y, Wu F, Zhao W J, Li H, Jin G, Wu X D and Xue Q 2015 Tribological performance of acrylic resin coatings modified by micro/nano $SiO_2$ particles *Lubrication Eng.* **40** 11-22

IOP Conf. Series: Materials Science and Engineering **668** (2019) 012020 doi:10.1088/1757-899X/668/1/012020

# Phase composition and microstructure formation mechanism of in-situ Cu-Fe micro-composites

**K M Liu[1,3], Y Jin[2], X C Sheng[1], Z K Huang[1], J Zou[2], W Chen[2], M C Zhang[1], H M Huang[1] and D P Lu[2]**

[1]Jiangxi Key Laboratory for Precision Drive and Control, Nanchang Institute of Technology, Nanchang 330099, P R China
[2]Jiangxi Key Laboratory for Advanced Copper and Tungsten Materials, Jiangxi Academy of Sciences, Nanchang 330029, P R China

E-mail: jokeyliu@163.com

**Abstract.** The phase composition and microstructure formation mechanism of in-situ Cu-Fe micro-composites were investigated. The microstructures of longitudinal and transverse sections were analyzed by light microscopy and scanning electron microscopy. The phase analysis was executed by X-ray diffraction. The common microstructure characteristic of Cu-XFe (X = 11, 14 and 17) alloys was that the second phase α-Fe dendrites were uniformly distributed in the Cu matrix. The disorderly distributed Fe dendrites of Cu-14Fe alloy underwent initial inhomogeneous deformation and then were gradually changed into the directionally arranged Fe fibers of in-situ Cu-14Fe micro-composite in the longitudinal section, and were gradually transformed into the irregular V-shaped Fe fibers in the transverse section. The initial inhomogeneous deformation and the irregular V-shaped Fe fibers in the transverse section are closely related to the formation of <110> texture.

## 1. Introduction

In-situ Cu-based micro-composites have been developed gradually since the late 1970s. It has been found by Harvard University scholar Bevk *et al* [1] that as-cast Cu-20% Nb alloys after severe plastic deformation can form composites with Nb fibers distributed on the copper matrix, and their tensile strength and conductivity can reach about 2000 MPa and 70% IACS, respectively. This kind of composites with in-situ formation of fiber in the process of cold deformation is called in-situ micro-composites. The excellent comprehensive properties matching ultra-high strength and good conductivity are the remarkable characteristics of this kind of materials.

A large number of studies have shown that these alloys consisting of Cu with face-centered cubic (fcc) lattice and transitional metals W, V, Mo, Cr and Fe with body-centered cubic (bcc) lattice or Ag, have similar microstructural characteristics, but due to the different kinds of alloying elements and preparation methods, there are some differences in mechanical properties and conductivity of the materials [2-5]. In recent years, a series of progress has been made in high strength and high conductivity Cu-based conductive materials. Among them, binary in-situ Cu-based micro-composites such as Cu-Nb, Cu-Ta, Cu-Ag, Cu-Fe and Cu-Cr have good strength and conductivity [6-11].

Over the past two decades, in-situ Cu-Fe micro-composites have become a hot spot for researchers all over the world because of the relatively low cost of Fe [12-15]. In this work, the phase composition of Cu-Fe alloys and the microstructure evolution of in-situ Cu-Fe micro-composites were investigated. The microstructure formation mechanism during cold deformation was analyzed.

## 2. Experimental details

Ingots of Cu-XFe (X = 11, 14 and 17 wt.%) alloys were produced by a vacuum induction furnace. The casting temperature ranged from 1300 to 1400°C and the ingots were cooled by furnace-cooling. The raw materials were commercial Fe with 99.94 wt.% purity and electrolytic Cu with 99.96 wt.% purity. The diameter of cylindrical ingots was about 36 mm, and the length was about 380 mm. The ingots of Cu-XFe (X = 11, 14 and 17 wt.%) alloys were hot rolled, cold drawn, and heat treated to in-situ Cu-XFe micro-composites with different cold deformation strains. The detailed preparation process was described elsewhere [16]. The cold deformation strain $\eta$ was obtained by cold drawing and calculated by $\eta = \ln(A_0/A_f)$ due to the large cumulative cold deformation strain, where $A_0$ and $A_f$ are the transverse sectional area before and after cold deformation, respectively [2,7].

The specimens were sectioned from the Cu-XFe ingots and in-situ micro-composites with different cold deformation strains, mounted, grinded, polished and then etched in an etching solution of 20 ml HCl, 5 g $FeCl_3$ and 120 ml $H_2O$. The microstructures of longitudinal and transverse section were investigated by light microscopy (LM) and scanning electron microscopy (SEM). The phase composition was analysed by X-ray diffraction.

## 3. Results and discussion

### *3.1. Microstructure and phase composition*

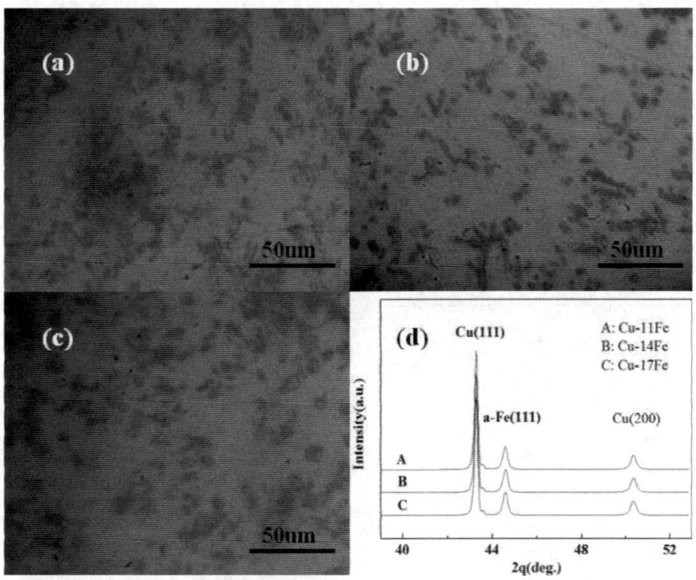


**Figure 1.** LM microstructures of (a) Cu-11Fe alloy, (b) Cu-14Fe alloy and (c) Cu-17Fe alloy; (d) X-ray diffraction curves of Cu-11Fe, Cu-14Fe and Cu-17Fe alloys.

Figures 1(a), 1(b) and 1(c) show the LM microstructures of as-cast Cu-11Fe, Cu-14Fe and Cu-17Fe alloys, respectively. It can be seen that the microstructures of the three alloys are basically similar, mainly consisting of the matrix and the second phase dendrites distributed uniformly in the matrix. The Cu matrix is lighter than the Fe dendrites because the atomic number of Cu is bigger than that of Fe.

Figure 1(d) shows the X-ray diffraction curves of the three alloys. The XRD phase analysis presents that the three alloys are all composed of copper matrix and α-Fe. Obviously, the second phase dendrites in the as-cast microstructure are α-Fe and the matrix is Cu-based solid solution. That is to say, the common phase composition characteristic of the three as-cast alloys is that the second phase α-Fe is relatively uniformly distributed in the Cu matrix.

*3.2. Longitudinal section microstructure*

According to the phase diagram of binary Cu-Fe alloys, the Fe-rich solid solution of Cu-Fe alloys with 10-20% Fe is first precipitated in liquid Cu, and the solubility of Cu in γ-Fe decreases with the decrease of temperature. The eutectoid transformation of γ-Fe occurs at 850°C to form α-Fe. When the temperature further drops below 600°C, the two solvents are almost insoluble with each other. Figure 2 shows the microstructures of longitudinal sections of Cu-14Fe alloy after different cold deformation strains. The bright area is the Cu matrix and the dark area is the Fe phase. Figure 2(a) is the SEM microstructure of as-cast Cu-14Fe alloy after hot rolling at 850°C before cold deformation. The distribution of Fe dendrites has not changed significantly. The secondary dendrite arm elongates slightly along the drawing direction, but the size change is not obvious. Figures 2(b)-2(d) presents the formation of Fe fibers in the drawing direction of in-situ Cu-14Fe micro-composite during cold deformation. It can be seen that the disorderly distribution of Fe dendrites on the longitudinal section of as-cast Cu-14Fe alloy is gradually elongated into fibers arranged along the drawing direction with the increase of cold deformation strain. At η=1, the Fe dendrite has been broken and elongated partly in the drawing direction, but the deformation of the Fe dendrite is not uniform, some fine particles are elongated, some particles are elongated only at one end, while the other end remains unchanged, and some coarse particles remain unchanged, as shown in figure 2(b). At η=4, the directional arrangement of elongated Fe grains is gradually formed, but there are still uneven tadpole-like Cr grains, as shown in figure 2(c). With the further increase of cold deformation strain, the Fe fibers are basically formed, and the average size and spacing of the fibers decrease gradually, as shown in figure 2(d).

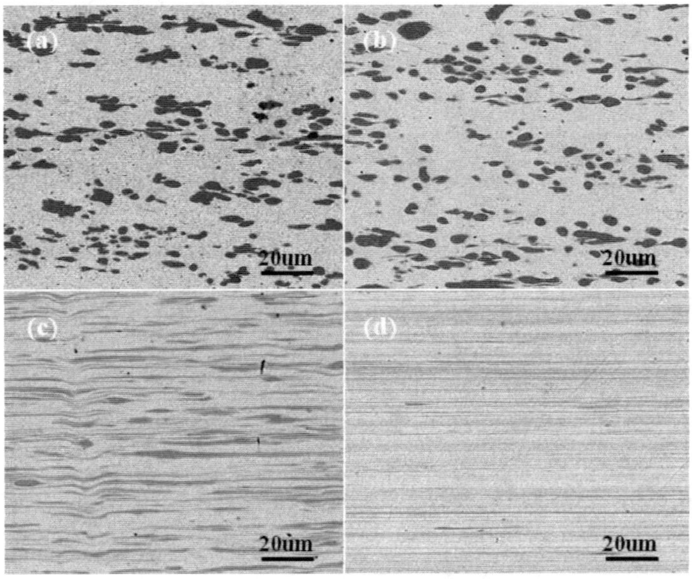


**Figure 2.** SEM microstructures of longitudinal section of Cu-14Fe alloy after different cold deformation strains: (a) η=0; (b) η=1; (c) η=4; (d) η=7.8.

### *3.3. Transverse section microstructure*

Figure 3 shows the SEM microstructures of transverse section of Cu-14Fe alloy after different cold deformation strains. The bright area is the Cu matrix and the dark area is the Fe phase. Figure 3(a) is the microstructure of as-cast Cu-14Fe alloy after hot rolling at 850°C before cold deformation. The as-cast Fe dendrites were broken mainly by fracturing the primary dendrite arm. Figures 3(b)-3(d) presents the microstructure evolution of Fe grains in the transverse section of in-situ Cu-14Fe micro-composites during cold drawing. At $\eta=1$, the width of most Fe grains did not change obviously, but the thickness decreased, as shown in figure 3(b). With the increase of cold deformation strain, the thickness change of Fe grains was gradually discernible, and the end of the grains began to bend, as shown in figure 3(c). With the further increase of cold deformation strain, at $\eta=7.8$, the twist and bending of the grains were more obvious, and the deformation and distribution were basically uniform, as shown in figure 3(d). The result suggests that the width and thickness of the hot-rolled dendrites decrease with the increase of cold deformation strain, but the speed of reduction is quite different. The thickness deceleration is obviously faster, which increases the width-thickness ratio. Accordingly, the cold-drawn second phase gradually bended and broke, and the transverse section of the grains gradually transformed irregular V-shaped morphology.

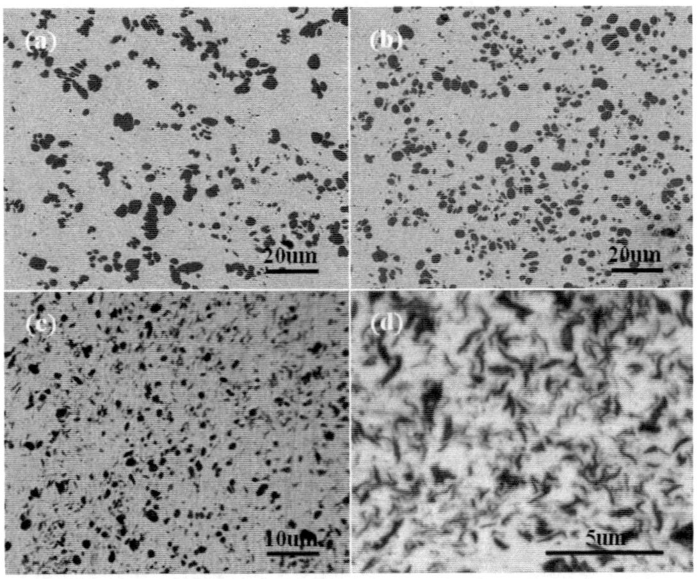


**Figure 3.** SEM microstructures of transverse section of Cu-14Fe alloy after different cold deformation strains: (a) $\eta=0$; (b) $\eta=1$; (c) $\eta=4$; (d) $\eta=7.8$.

### *3.4. Microstructure formation mechanism*

The non-uniform deformation of in-situ Cu-Fe micro-composites at the initial stage of deformation is related to the non-uniform distribution of radial shear strain during cold drawing and the rotation of the second phase along the drawing direction during deformation. In the initial stage of deformation, the critical shear stress changes under a given stress state, and the partially oriented grains begin to deform, while the partially oriented grains rotate and reorient. When the plastic deformation reaches a certain degree, the bcc Fe phase orientation tends to be the same, and the <110> texture is formed along the principal strain direction [17].

The irregular morphology of Fe fibers in transverse section of in-situ Cu-14Fe micro-composites is also closely related to the formation of <110> texture [17]. The deformation process of bcc Fe phase can be analyzed by figure 4 [18]. The slip plane and orientation of bcc Fe phase are {110} and <111>

respectively. The deformation direction in figure 4 is parallel to [$\bar{1}01$]. There are four possible <111> slip orientations of Fe phase [$\bar{1}1\bar{1}$], [$\bar{1}\bar{1}1$], [111] and [$11\bar{1}$]. The [$\bar{1}1\bar{1}$] and [$\bar{1}\bar{1}1$] orientations are within the slip plane (110), so they can produce slip during the whole deformation process. Strains perpendicular to the direction of deformation can be produced in the orientations of [111] and [$11\bar{1}$]. If these two directions are not slippable, they will be plane strain deformation. The slip systems of the Cu matrix of fcc are {111} <110> or {100} <110>. In order to coordinate the deformation with the axial uniform flow of the matrix, the (110) [111] and (110) [$11\bar{1}$] slip systems of the Fe phase of bcc must be movable [18,19]. As a result, the Fe phase is forced to bend or break around the tensile axis, forming the irregular V-shaped morphology in the transverse section.

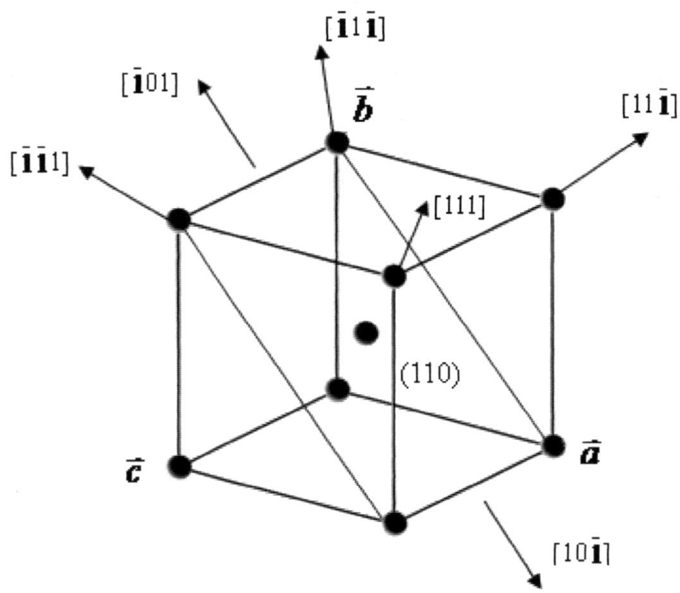


**Figure 4.** Analysis of bcc phase deformation.

## 4. Conclusions

- The common phase composition characteristic of Cu-XFe (X = 11, 14 and 17) alloys is that the second phase α-Fe is relatively uniformly distributed in the Cu matrix.
- The disorderly distributed Fe dendrites of Cu-14Fe alloy underwent initial inhomogeneous deformation and then were gradually changed into the directionally arranged Fe fibers of in-situ Cu-14Fe micro-composite in the longitudinal section.
- The randomly distributed Fe dendrites underwent initial inhomogeneous deformation and then were gradually transformed into the irregular V-shaped Fe fibers in the transverse section.
- The initial inhomogeneous deformation and the irregular V-shaped Fe fibers in the transverse section of in-situ Cu-14Fe micro-composite are closely related to the formation of <110> texture.

## Acknowledgments

This project was supported by the Key Science and Technology Program of Jiangxi Education Department (GJJ170976), the National Natural Science Foundation of China (51861025; 51561010; 51461018), and the Natural Science Foundation of Jiangxi Province (20192BAB206001; 20171BAB206040).

**References**

[1] Bevk J, Harbison J P and Bell J D 1978 *J. Appl. Phys.* **49** 6031-8
[2] Gao H Y, Wang J and Sun B D 2009 *J. Alloys Compd.* **469** 580-6
[3] Liu K M, Wang Z X, Jiang Z Y, Atrens A, Huang Z K, Guo W, Zhang X W, Yu J M, Lu Q G and Lu D P 2018 *J. Mater. Eng. Perform.* **27** 933-8
[4] Liu K M, Lu D P, Zhou H T, Yang Y L, Atrens A and Zou J 2013 *J. Mater. Eng. Perform.* **22** 3723-7
[5] Deng L P, Liu Z F, Wang B S, Han K and Xiang H L 2019 *Mater. Charact.* **150** 62-6
[6] Venugopal T, Prasad R K and Murty B S 2007 *Acta Mater.* **55** 4439-45
[7] Liu K M, Jiang Z Y, Zhao J W, Zou J, Chen Z B and Lu D P 2014 *J. Alloys Compd.* **612** 221-6
[8] Liu K M, Jiang Z Y, Zhou H T, Lu D P, Atrens A and Yang Y L 2015 *J. Mater. Eng. Perform.* **24** 4340-5
[9] Raabe D, Ohsaki S and Hono K 2009 *Acta Mater.* **57** 5254-63
[10] Liu J B, Zhang L and Meng L 2008 *Mater. Sci. Eng. A* **498** 392-6
[11] Liu K M, Lu D P, Zhou H T, Chen Z B, Atrens A and Lu L 2013 *Mater. Sci. Eng. A* **584** 114-20
[12] Liu K M, Huang Z K, Zhang X W, Lu D P, Atrens A, Zhou H T, Yin Y, Yu J M and Guo W 2016 *Mater. Sci. Eng. A* **673** 1-7
[13] Liu K M, Lu D P, Zhou H T, Atrens A, Zou J, Yang Y L and Zeng S M 2010 *Mater. Sci. Eng. A* **527** 4953-8
[14] Stepanov N D, Kuznetsov A V, Salishchev G A, Khlebova N E and Pantsyrny V I 2013 *Mater. Sci. Eng. A* **564** 264-72
[15] Sun B D, Gao H Y, Wang J and Shu D 2007 *Mater. Lett.* **61** 1002-6
[16] Liu K M, Jiang Z Y, Zhao J W, Zou J, Lu L and Lu D P 2015 *Metall. Mater. Trans. A* **46** 2255-61
[17] Gao H Y, Wang J, Shu D and Sun B D 2005 *Scipta Mater.* **53** 1105-9
[18] Gao H Y 2007 Study on high strength and high conductivity deformation processed Cu-Fe-Ag in situ composites (Shanghai, China: Shanghai Jiao Tong University) pp 51-3
[19] Snoeck E *et al* 1998 *Scipta Mater.* **38** 1643-8

IOP Conf. Series: Materials Science and Engineering **668** (2019) 012021 doi:10.1088/1757-899X/668/1/012021

# Mechano-chemical synthesis of crystalline superionic conductors $CsAg_4Br_{3-x}I_{2+x}$ (0<x<1) and their silver-conducting properties

**W B Zuo[1], V O Pelenovich[1], A B Tolstogouzov[1,2,3] and D J Fu[1,4]**

[1]Key Laboratory of Artificial Micro and Nanostructures of Ministry of Education and Hubei Key Laboratory of Nuclear Solid Physics, School of Physics and Technology, Wuhan University, Wuhan 430072, China
[2]Ryazan State Radio Engineering University, Gagarin Str. 59/1, 390005 Ryazan, Russian Federation
[3]Centre for Physics and Technological Research (CeFITec), Dept. de Física da Faculdade de Ciências e Tecnologia (FCT), Universidade Nova de Lisboa, 2829-516 Caparica, Portugal

E-mail: djfu@whu.edu.cn

**Abstract.** Silver ion conductive materials attract numerous interest of research nowadays, due to their high potentiality in the solid electrolyte ion source (SEIS) intended for the application in the ionic thruster. In the work, mechano-chemical synthesis of crystalline superionic conductors $CsAg_4Br_{3-x}I_{2+x}$ (0<x<1) was performed by means of high-energy ball milling for the first time. The structure, morphology, and elemental composition were investigated and the effects of iodine content on these properties were studied. Silver-conducting properties were characterized based on the generation of $Ag^+$ ion beams. The ion current of about 15 nA was obtained at 152°C and 15 kV.

## 1. Introduction

Superionic conductors, also called solid electrolytes (SEs), are a class of solid-state materials that have high values of ionic conductivity (>$10^{-4}$ S/cm) and low values of activation energy (~$10^{-1}$ eV) [1]. Since the discovery of crystalline $MAg_4I_5$ (M=K, Rb, and $NH_4$) in the 1960s [2], the silver-conducting SEs are widely applied in electrochemistry (solid-state batteries and supercapacitors), electric device, and ion emission device [3-9]. Among them, $RbAg_4I_5$ (silver-rubidium pentaiodide) solid electrolyte possessing one of the highest room-temperature ionic conductivity (0.26 S/cm) is inclined to degradation at this temperature, especially in the environmental conditions [10]. Apart from the silver-conducting $MAg_4I_5$, numerous superionic conductors have been found, for example, sodium-conducting β-alumina [11] and copper-conducting $Rb_4Cu_{16}I_7Cl_{13}$ compounds [12]. However, it still attracts much interests to search for novel superionic conductors.

Mechanochemical processing is a powerful technique for powder synthesis, which is often exploited in the preparations of metals and alloys, amorphous phases, and micro(nano) composites [13-15]. Among them, the application in superionic conductor synthesis just starts recently, and to the best of our knowledge, only a few studies are reported on crystalline SEs prepared by high-energy ball-milling [9]. In the present work, the crystalline $CsAg_4Br_{3-x}I_{2+x}$ (0<x<1) was prepared by the mechano-chemical synthesis and the effect of I-doping content on the structure, morphology, and ionic conducting properties was studied.

## 2. Experimental details

### *2.1. Mechano-chemical processing and characterizations*

In the work, the raw materials silver iodide β-AgI (≥99%), cesium iodide CsI (≥99.999%) and silver bromide AgBr (≥99.9%) powder (Sigma-Aldrich) were weighted according to the intended molar ratio of CsI:AgI:AgBr = 1: (1+$x$): (3-$x$), where $x$ is within the range of 0 to 1. The powders were mixed in an agate mortar for 5 minutes beforehand. Then, the well-mixed powder was put in an agate milling container with agate balls (ɗ~6 mm). High-energy ball-milling was conducted under dry argon atmosphere using a planetary ball-milling apparatus (Nanjing, QM-QX2).

The prepared samples were characterized by X-ray diffraction (XRD, Bruker D8 Advance), energy dispersive X-ray spectroscopy (EDS), and scanning electron microscope (SEM, Zeiss Sigma). Cu $k_\alpha$ radiation ($\lambda$=0.154184 nm) was used for the measurements of XRD spectra.

### *2.2. Test setup for Silver-conducting properties*

To examine the silver-conducting properties of the synthesized solid electrolytes $CsAg_4Br_{3-x}I_{2+x}$, $Ag^+$ ion emitters were fabricated using the method described in [9]. An overview of the setup for ion emission measurements is shown in figure 1. The test setup consists of an emitter, ion collector (Faraday cup), and high voltage supply. The ion emitter is made of a silver tip covered by $CsAg_4Br_{3-x}I_{2+x}$ coatings. In principle, the generation of $Ag^+$ ion beam follows three successive processes: (i) transformation of Ag atoms into mobile ions via an electrochemical reaction occurring at the interface between silver metal and electrolyte film under an influence of the applied electric field, (ii) moving/migration of $Ag^+$ ions along the fast ion-transport channels existing insides the $CsAg_4Br_{3-x}I_{2+x}$ film towards the surface, and (iii) overcoming the surface potential barrier and field-stimulated emission into vacuum.

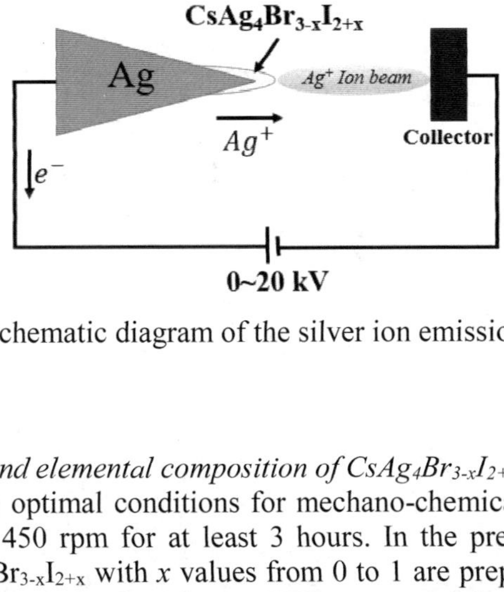


**Figure 1.** Schematic diagram of the silver ion emission test setup.

## 3. Results and discussions

### *3.1. Structure, morphology, and elemental composition of $CsAg_4Br_{3-x}I_{2+x}$*

In our previous work [9], the optimal conditions for mechano-chemical synthesis of $CsAg_4Br_{2.68}I_{2.32}$ powder were found to be at 450 rpm for at least 3 hours. In the present work, various crystalline superionic conductors $CsAg_4Br_{3-x}I_{2+x}$ with $x$ values from 0 to 1 are prepared by means of high-energy ball milling at rotation rate of 450 rpm for 3 hours. XRD spectra of the prepared powder are shown in figure 2, where the order from up to bottom corresponds to the samples of $x$=0 to 1. All diffraction peaks can be indexed to the high-conductive $CsAg_4Br_{3-x}I_{2+x}$ phase except for small contributions of impurities $CsAgBr_2$ and Ag(Br,I) in the $x$=0 sample. All $CsAg_4Br_{3-x}I_{2+x}$ ($0<x<1$) crystals belong to $MAg_4I_5$ families having a simple cubic lattice. The (*hkl*) indices of all diffraction peaks are indicated. The strongest characteristic peaks are assigned to be (311) plane and present a left shift in the XRD spectra with an increasing of *x* values, indicating an expanding of lattice. The lattice constant *a* is related to the inter-plane spacing *d* and the corresponding Miller index (*hkl*):

$$d = \frac{a}{(h^2 + k^2 + l^2)^{1/2}} \tag{1}$$

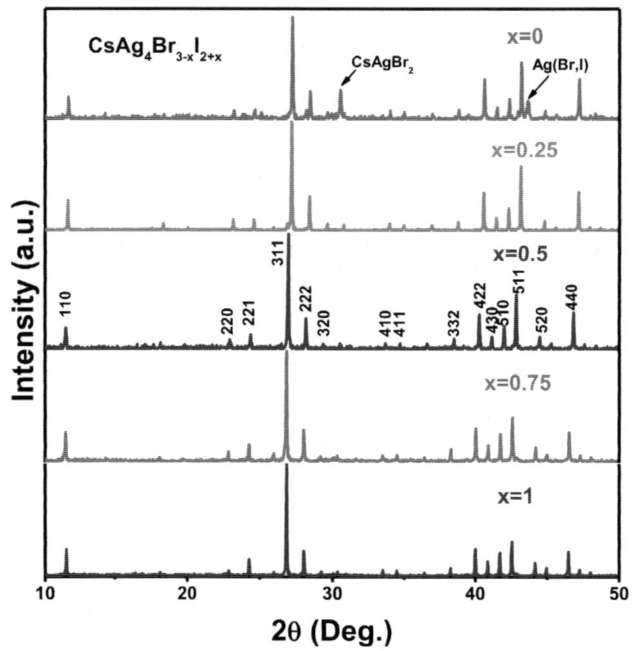


**Figure 2.** XRD spectra of the ball-milled samples with *x* value of 0 to 1 from up to bottom.

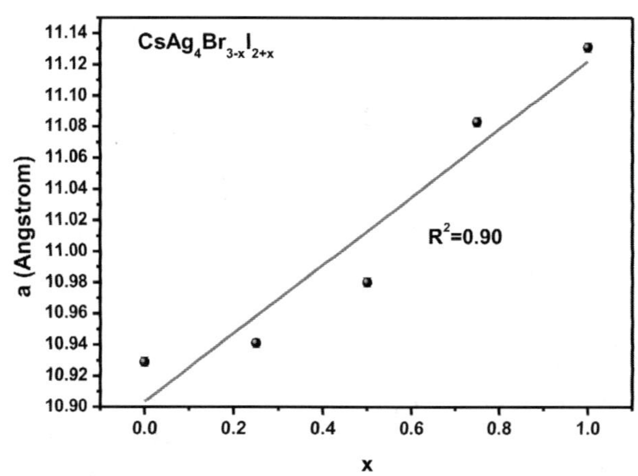


**Figure 3.** Lattice constant *a* as a function of *x* value.

The calculated values of lattice constant are plotted as a function of *x*-value in figure 3. It is observed that the lattice constant *a* shows a roughly linear increase with the *x*-value due to the bigger atomic radius of iodine than that of bromine. Hence, it is reasonable to conclude that the $CsAg_4Br_{3-x}I_{2+x}$ lattice expands with the increase of iodine content. The grain size can be estimated from the Scherrer equation:

$$D = \frac{0.89\lambda}{\beta cos\theta} \qquad (2)$$

Where $\lambda$ is the wavelength of X-ray, $\beta$ and $\theta$ indicate the FWHM and diffraction angle of the corresponding *(hkl)* peaks. The estimated values of grain size *D* for various crystalline $CsAg_4Br_{3-x}I_{2+x}$ are listed in table 1.

Figure 4 presents the SEM images of the $CsAg_4Br_{3-x}I_{2+x}$ powder, where the grain structure with the size of 0.2-1 μm is clearly observed. The average grain size is estimated for each sample using the ImageJ software, as listed in table 1. It is investigated that the grain grows with the increase of the iodine content resulting from the expanding of lattice, which is consistent with the analysis of XRD spectra. The EDS spectrum of $CsAg_4Br_3I_2$ powder is shown in figure 4(f), where the peaks at about 1.5 keV, 3.0 keV, 3.9 keV, and 4.25 keV are ascribed to Br, Ag, I, and Cs elements, respectively. The EDS spectra of the other samples are similar to figure 4(f) and therefore, not involved in the paper. The elemental compositions of $CsAg_4Br_{3-x}I_{2+x}$ crystals obtained from the analysis of EDS spectra are listed in table 1. The atomic ratio of Cs, Ag, and (Br+I) elements in all samples is rather close to the intended value of 1:4:5, revealing a successful synthesis of superionic conductors $CsAg_4Br_{3-x}I_{2+x}$ ($0<x<1$) by the method of mechano-chemical processing.

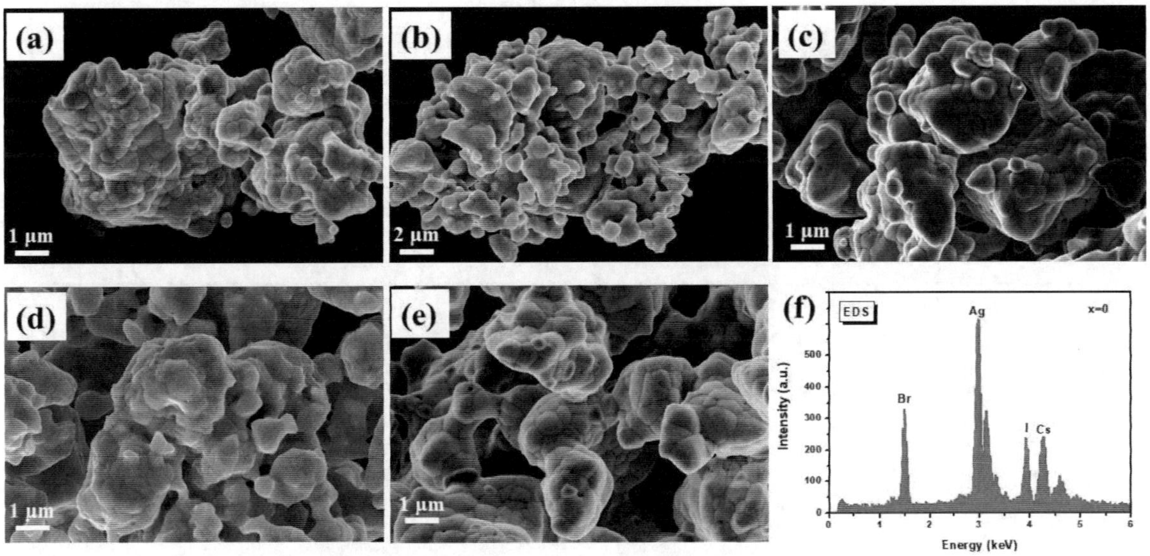


**Figure 4.** (a)-(e): SEM images of crystalline $CsAg_4Br_{3-x}I_{2+x}$ powder with *x* value of 0 to 1, respectively. (f) EDS spectrum of the $CsAg_4Br_3I_2$ ($x$=0) sample.

**Table 1.** Elemental composition and grain size for $CsAg_4Br_{3-x}I_{2+x}$ powder.

| Sample | Composition (at%) | | | | Grain size (nm) | |
|---|---|---|---|---|---|---|
| | Cs | Ag | Br | I | SEM [a] | XRD [b] |
| x=0 | 10.81 | 39.31 | 33.99 | 15.89 | 360 | 90 |
| x=0.25 | 8.3 | 40.53 | 25.93 | 25.24 | 408 | 91 |
| x=0.50 | 12.26 | 39.48 | 26.01 | 22.25 | 425 | 94 |
| x=0.75 | 10.91 | 42.53 | 24.36 | 22.20 | 437 | 97 |
| x=1 | 10.76 | 41.73 | 20.63 | 26.88 | 700 | 119 |

[a] Estimation from SEM images by ImageJ software.
[b] Calculation from XRD (*hkl*) peaks by Scherrer equation.

### *3.2. Silver-conducting properties of $CsAg_4Br_{2.5}I_{2.5}$*

In practice, ionic conductovity of solid electrolyte can be measured by electrochemical impedance spectroscopy (EIS). Hereby, the silver-conducting properties of $CsAg_4Br_{2.5}I_{2.5}$ are characterized using the setup shown in figure 1 based on the $Ag^+$ ion beams generation. The collected ion current is proportional to the temperature-dependent ionic conductivity of solid electrolyte and the potential applied on the apex of emitter [16]. In the section, ion current as functions of working temperature and accelerating voltage are recorded, as shown in figures 5(a) and 5(b). The current increases with the temperature following the Arrhenius equation:

$$I = I_0 exp^{-\frac{Q}{k_B T}} \quad (3)$$

where $Q$ indicates the surface potential barrier. A threshold feature is observed at the temperature of 150°C, which is close to the melting point of $CsAg_4Br_{2.5}I_{2.5}$. The threshold character can be attributed to the quasi-liquid film occurring on the surface of SE resulting in a formation of Taylor cone [17].

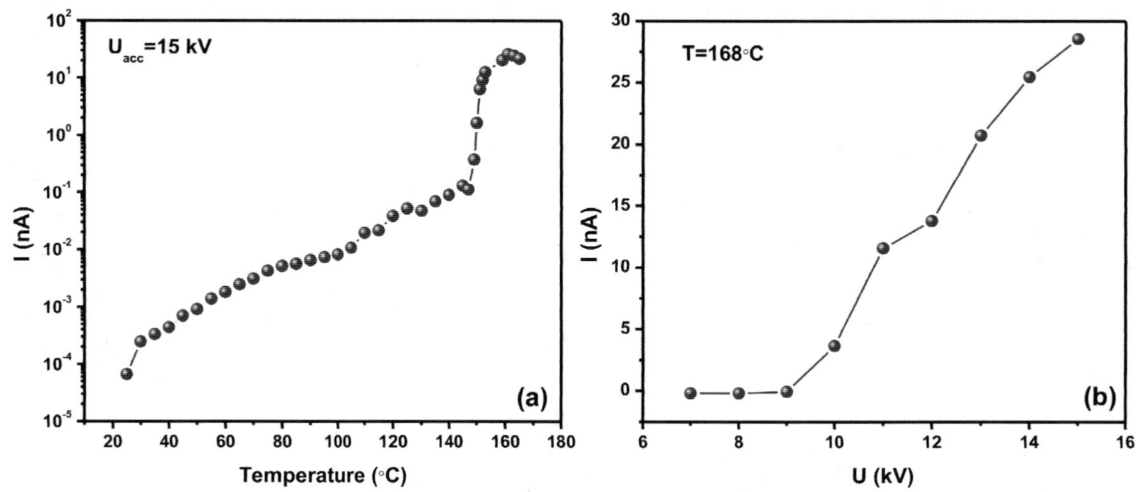


**Figure 5.** Ion current ($I$) versus (a) working temperature ($T$) obtained at 15 kV and (b) accelerating voltage ($U_{acc}$) at 168°C.

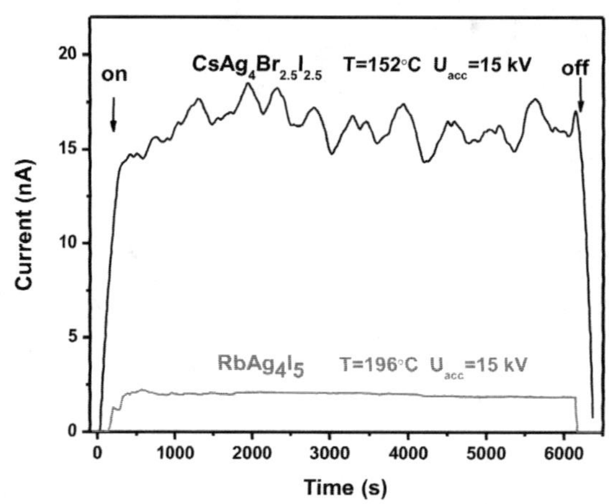


**Figure 6.** The ion emission current recorded at 152°C and 15 kV for about 2 hours.

The applied voltage on the emitter is also able to enhance the ion emission, seen in figure 5(b) due to the Schottky effect *i.e.* the applied field lowers the surface potential barrier for ion emission [18]. Figure 6 presents the stability of ion emission current for a long term (*ca.* 2 hours), where the 15 nA current is obtained at 15 kV and 152°C using $CsAg_4Br_{2.5}I_{2.5}$ film. The red line is the obtained current using $RbAg_4I_5$ film. It is found that using $CsAg_4Br_{2.5}I_{2.5}$ film results in stronger ion emission than $RbAg_4I_5$ film, which demonstrates that the silver-conducting capacity of $CsAg_4Br_{2.5}I_{2.5}$ can be higher than that of $RbAg_4I_5$.

## 4. Conclusions

In the present work, superionic conductors $CsAg_4Br_{3-x}I_{2+x}$ ($0<x<1$) were successfully synthesized by means of high-energy ball milling at rotation rate of 450 rpm for 3 hours. The expand of lattice and the grow of grain were observed with an increase of iodine content. The grain size of $CsAg_4Br_{3-x}I_{2+x}$ was distributed in the range of 0.2-1 μm. The silver-conducting properties of superionic conductor $CsAg_4Br_{2.5}I_{2.5}$ were characterized based on the generation of $Ag^+$ ion beams. The ion current of 15 nA was obtained at a temperature of 152°C and accelerating voltage of 15 kV. It was demonstrated that $CsAg_4Br_{2.5}I_{2.5}$ can possess higher ion-conducting capacity than $RbAg_4I_5$.

## Acknowledgment

This work was supported by the Fundamental Research Funds for the Central Universities under the grant number 2042019gf0073 and China Postdoctoral Science Foundation under the grant number 2019M652687. In addition, the authors are grateful to Mr. Han Chengzhen for his help in the mechano-chemical synthesis.

## References

[1] Salamon M B 1979 *Physics of Superionic Conductors* (New York: Springer-Verlag)
[2] Owens B B and Argue G R 1967 *Science* **157** 308
[3] Liang X F, Chen Y, Chen L, Yin J and Liu Z G 2007 *Appl. Phys. Lett.* **90** 022508
[4] Guo H X, Yang B, Chen L, Xia Y D, Yin K B, Lin Z G and Yin J 2007 *Appl. Phys. Lett.* **91** 243513
[5] Rossi M D, Pistoia G and Scrosati A 1969 *J. Electrochem. Soc.* **116** 1642
[6] Daiko Y, Segawa K, Honda S and Iwamoto Y 2018 *Solid State Ionics* **322** 5
[7] Daiko Y, Segawa K, Machida K, Imataka H, Honda S and Iwamoto Y 2018 *Adv. Eng. Mater.* **20** 1800198
[8] Tolstogouzov A, Aguas H, Ayouchi R, Belykh S F, Fernandes F, Gololobov G P, Moutinho A, Schwarz R, Suvorov D V and Teodoro O 2016 *Vacuum* **131** 252
[9] Zuo W B, Pelenovich V, Tolstogouzov A, Zeng X M, Wang Z G, Song X Q, Gusev S I, Tian C X and Fu D J 2019 *J. Alloy Compd.* **790** 109
[10] Despotuli A L, Lichkova N V, Minenkova N A and Nosenko S V 1990 *Elektrokhiniya* **26** 1524
[11] Yao Y F Y and Kummer J T 1967 *J. Inorg. Nucl. Chem.* **29** 2453
[12] Takahashi T, Yamamoto O, Yamada S and Hayashi S 1979 *J. Electrochem. Soc.* **126** 1654
[13] Shivam V, Basu J, Shadangsi Y, Singh M K and Mukhopadhyay N K 2018 *J. Alloy Compd.* **757** 87
[14] Hwang S H, Byun J M, Lee S, Suk M, Oh S and Kim Y D 2014 *J. Alloy Compd.* **585** 418
[15] Koch C C 1996 *Scripta Mater.* **34** 21
[16] Eremeev M A and Yureva G A 1953 *Zhurnal Tehnicheskoj Fiziki* **23** 430
[17] Taylor G 1964 *Proc. Roy. Soc.* A **280** 383
[18] Zuo W B, Pelenovich V, Tolstogouzov A, Ieshkin A, Zeng X M, Wang Z G, Gololobov G, Suvorov D, Liu C S, Fu D J and Hu D H 2019 *Vacuum* **167** 382

# Study of grain distribution during friction stir welding of Al-Zn-Mg alloys using numerical simulation

**Y F Wu[1], J Yu[1], Z Yang[1,2] and Y L Deng[1]**

[1]School of Materials Science and Engineering, Central South University, Lushan South Rd 932, Changsha,410083, PR China

E-mail: zyang@csu.edu.cn

**Abstract.** During friction stir welding (FSW) of Al-Zn-Mg alloys, the grain size of the welded material significantly affects its mechanical properties. In this study, the FSW process was simulated using the finite element method (FEM). Subsequently, the Zener-Hollomon parameters and grain size were calculated using the results of the stimulation of the temperature and strain rate fields. Also, the accuracy of numerical simulation was proved by experiments. It was found that the Zener-Hollomon relationship correctly predicted the grain size for the weld stirring zone (SZ) and thermal-mechanical affected zone (TMAZ), and we concluded that low rotating and high welding speeds resulted in smaller grain sizes for the SZ. Moreover, the high welding speed could effectively improve the grain condition at the edge of the TMAZ and the specific relationship between Zener-Hollomon parameters and the average grain diameter of SZ can be expressed as: $Z = 9.09 \times 10^{13} d^{-4.74}$.

## 1. Introduction

Recently, the worldwide demands for high-speed railways and subways have increased rapidly. There are currently thousands of high-speed trains traveling on railways in Europe, Japan, China, and other countries, at speeds that can rival propeller-driven airplanes. Aluminum alloys are increasingly used for building high-speed trains and subways to meet the lightweight, environmental safety, and other requirements for these vehicles, as indicated by Yang and Lin[1]. Currently, aluminum alloy railway car bodies are manufactured combining aluminum alloy-extruded profiles using metal inert gas (MIG) welding to achieve structural integration and body weight reduction. Generally, 6 and 7 series aluminum-extruded profiles are MIG-welded using ER5356 aluminum soldering wires to achieve good mechanical properties. However, previous reports [2-5] indicated that the aluminum alloys welded by traditional MIG fusion welding is prone to defects, such as blow holes, cracks, coarse grains, and insufficient welding joints during the welding process, which could become limiting factors in the development of aluminum alloy car bodies.

Friction stir welding (FSW) was developed by The Welding Institute in 1991[6]. As a solid phase connection process, in addition to being a non-polluting process, FSW presents advantages that cannot be achieved through conventional fusion welding of aluminum alloys, like high joint mechanical properties, cracks-free joint, low distortion, as well as low energy consumption, as indicated by many researchers[7-9]. However, the FSW process is quite complicated; therefore, it is difficult to formulate suitable process parameters for various FSW procedures. Poor process parameters increase production costs and the risk of defects, especially when manufacturing high-speed trains. Generally, the welding quality depends on parameters including the traverse speed and the rotating speed of the tool, tools shapes, and slope. However, primary studies should be carried out, and many expensive extrusion profiles would be wasted. Therefore, the numerical simulation of the FSW process has been widely studied. Recently, lots of efforts have been spent on studying the influences of the welding speed, tool shape, and slope.

Chen et al.[10] used computational fluid dynamics (CFD) to analyse the effect of the pin thread on the flow and temperature of materials and concluded that pin thread play an important role on the temperature distribution. Zhu et al.[11] predicted and analysed the effects of welding parameters and tool pin profile using the Abaqus software based on the Coupled Eulerian–Lagrangian (CEL) method and concluded that welding and rotating speeds affected hole defects. Moreover, their study indicated that the properties of the joint can be benefit from a low welding speed and high rotating speed.

A tool pin with a special feature may enhance the materials flow during welding and achieve defect-free welds, compared with smooth tool pins. Long et al.[12] focused on tool tilt angle during FSW. They employed the DEFORM-3D software which has the Computational Solid Mechanics (CSM) method to simulate the effort of tool tilt angle, and determined that using non-tilt FSW would cause worm hole defects in the advancing side of the weld line, but using 2° inclination FSW would not. Moreover, compared to FSW featuring no tool tilt angle, FSW using a tool tilt angle exhibited higher peak temperature and better material flow. Carlone et al.[13] researched the influence of the welding parameters of FSW on fatigue cracks using numerical simulation methods. Ajri and Shin[14] researched the effects of welding and rotating speeds on welding quality using experiments and numerical simulation methods.

To date, most research has focused on temperature field distribution and material flow during FSW using experiments as well as finite element simulations. For example, Tutunchilar et al.[15] predicted the shape and defects of the weld stirring zone (SZ) by simulating the flow of material during FSW. Asadi et al.[16] predicted the width of the SZ and temperature and deformation required for recrystallization at the bottom of the SZ using material flow simulations and experimental techniques. Shi et al.[17] simulated the material flow and temperature distribution. Their research indicated that the shape and size of the thermal-mechanical-affected zone (TMAZ) was greatly affected by rotating speeds of the shoulder and pin. Hasan et al.[18] studied the effects of using unworn vs. worn tools on material flow using the FLUENT CFD software, and determined that the worn tool causes a different flow behaviour around and under the pin comparing to the unworn one. However, the microstructure of the material significantly influences its mechanical properties and is one of the key FSW research tasks. Buffa et al.[19] predicted grain size by establishing two different models for continuous dynamic recrystallization. Arora et al.[20] researched the temperature histories at different locations of FSW samples to further estimate the growth rate and size of grains.

To investigate the grain size distribution of welding Al-Zn-Mg alloy under different process conditions, we investigated the temperature field and strain rate in different regions and at different times during FSW using the DEFORM-3D commercial finite element software. Then, the grain sizes of different areas were calculated using the Zener-Hollomon parameter. We analysed the microstructure of materials subjected to welding and used numerical simulations to analyse the distribution of grain size during the FSW process.

## 2. Simulation details

### *2.1. Geometric model*

In this study, we used a 120 × 100 × 12.3 mm Al-Zn-Mg alloy sample. The stirring pin was conical, featuring a conical angle of 20° and a shoulder exhibiting a concave shape with a concave angle of 8°. The diameter of the shoulder, diameter of the bottom pin, and height of the pin were 26, 7.5, and 11 mm, respectively. The length, width, and height of the back plate were 160, 160, and 3 mm, respectively. The FEM model is illustrated in figure 1.

IOP Conf. Series: Materials Science and Engineering **668** (2019) 012022 doi:10.1088/1757-899X/668/1/012022

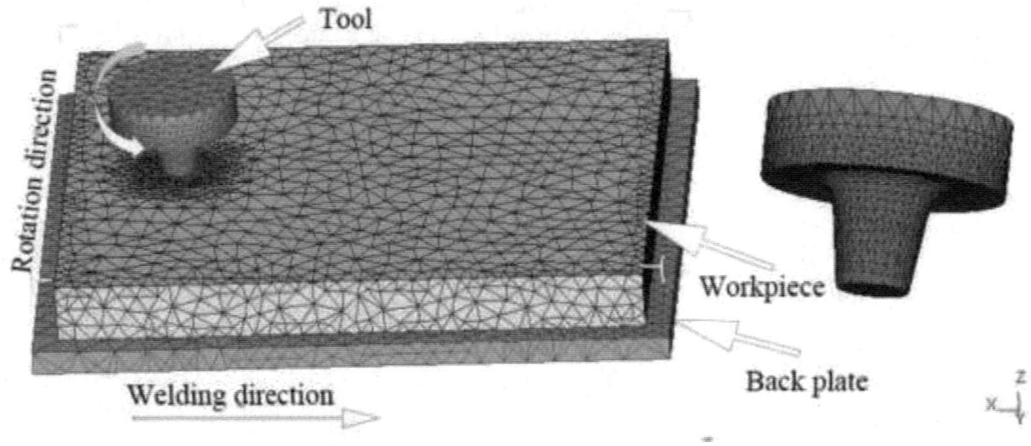


**Figure 1.** Finite element model.

In this simulation, the DEFORM-3D finite element software was employed, and the simulation used the implicit Lagrangian code proposed by Buff et al.[21]. All models adopted tetrahedral mesh cells, to improve computational efficiency and calculation accuracy. In addition, the meshes of the workpiece and tool were locally refined using a mesh window, and the mesh window followed the movements of the tool. The workpiece presented a total of 19 451 mesh elements and the mesh size transitioned from 0.7 to 6, while the tool featured 17 812 mesh elements and the mesh size transitioned from 0.33 to 4. Since the back plate only experienced heat transfer, the mesh was not locally refined and it exhibited 11 677 mesh elements while the mesh size transitioned from 1.1 to 4.

To simplify the simulation, many assumptions were made as followed: (1) the workpiece was continuous and the material model was rigid-visco-plastic; (2) the stir pin and back plate were rigid; (3) the environmental temperature was 20 °C; (4) the speeds of the workpiece along the X, Y, and Z directions were set to zero; (5) heat was exchanged between the free surfaces of the workpiece, tool, back plate, and environment, and the heat transfer coefficient was set as constant; furthermore, the contact surface between the stirring pin and workpiece, and that between the workpiece and back plate also experienced heat exchange, and the heat exchange coefficient was also set as constant; and (6) the friction coefficient between the workpiece and stirring needle was constant. The welding simulation parameters are listed in Table 1.

**Table 1.** FSW simulation parameters.

| Pin rotating speed (rpm) | 300 | 450 | 600 |
|---|---|---|---|
| Welding speed (mm/min) | 150 | 50<br>150 | 150 |

*2.2. Material model*

The workpiece was considered to be rigid-visco-plastic, and the elastic deformation of the material was neglected. The flow stress of the material followed the Arrhenius equation (equation (1)), which describes the relationship between the temperature, stress, and strain of the material at high temperature.

$$\dot{\varepsilon} = A[sinh(a\dot{\sigma})]^n \, exp\left(\frac{-\Delta H}{RT}\right) \tag{1}$$

Where $\dot{\varepsilon}$ is the strain rate, $\dot{\sigma}$ is the flow stress, $R$ is the gas constant, $T$ is the temperature of the material, and $A$, $\alpha$, and $n$ are material constants. Liu et al.[22] determined the specific parameters in equation(1) using a hot compression test. Table 2 lists the values of the material parameters.

**Table 2.** Parameters and constant of the Arrhenius equation.

| Material | A(1/s) | n | Q(kJ/mol) | α | R (J/(mol·K)) |
|---|---|---|---|---|---|
| Al-Zn-Mg alloy | $241 \times 10^8$ | 3.55 | 145 | 0.04504 | 8.314 |

*2.3. Friction model*

The friction conditions during the FSW process are quite complex, and the shear and Coulomb friction models have been commonly used to describe them. Tutunchilar et al.[15] used a shear friction model to simulate the friction stir process using DEFORM-3D software. Therefore, in this work, we also employ the shear friction model. The shear model defines the frictional force as:

$$f = mk \tag{2}$$

where $f$ is the frictional stress, $k$ is the shear yield stress, and $m$ is the friction factor. In this study, m was considered to be a constant. Frigaard et al.[23] indicated that the friction factor of the shear friction model during FSW ranged between 0.2 and 0.6, and m was finally determined to be 0.35.

During FSW, heat generation is mainly derived from two processes: the friction between the stirring tool and workpiece, which was the main source of heat, and the plastic deformation of the workpiece (90% of the energy caused by deformation is transformed into heat during the welding process). The detailed formulae of the governing equations for the thermal model and finite element simulation have been described in detail by Jain et al.[24].

## 3. Experimental

According to the process parameters given in Table 1, the welding process is carried out. The chemistry of the AL-Zn-Mg alloy was determined by ICP component analysis, and its result shows in Table 3. The welding samples were sectioned vertical to the welding line. Optical microscopy (OM) and electron backscatter diffraction (EBSD) were employed to analyse the microstructure. The preparation of OM specimens was treated by Graff Sargent Corrosion Reagent (1ml HF+16ml $HNO_3$+3g $CrO_3$+83ml $H_2O$). Meanwhile the EBSD specimens was prepared by electro-polishing in a solution (10% $HClO_4$ +90% $C_2H_5OH$) with voltage of 20V for 10s at 25 Celsius. The metallographic observation was conducting using OLYMPOUS-PMG3 metallographic microscope and the EBSD test was carried out with a field emission gun-environmental scanning electron microscopy device Zeiss EVO MA10 with NordlysMax2 Detector.

**Table 3.** Chemical composition of Al-Zn-Mg alloy (wt%).

| composition | Fe | Si | Cu | Mn | Mg | Zn | Cr | Zr | Ti | Al |
|---|---|---|---|---|---|---|---|---|---|---|
| content | 0.089 | 0.057 | 0.19 | 0.36 | 1.45 | 4.29 | 0.26 | 0.1 | 0.013 | Bal |

## 4. Results and discussion

The classic FSW model indicated that the welded material could be divided into four areas, as follows: from the centre toward the sides of the weld we identified the SZ, TMAZ, heat-affected zone (HAZ) and base material (BM). To illustrate these four areas more clearly, the point tracking method of DEFOEM-3D software was used to describe the material flow and deformation during the whole process. As shown in figure 2, the areas where the tool caused the point positions to change are the SZ and TMAZ, and the other areas are the HAZ and BM (AS: Advancing side, RS: Retreating side). Figure 3 depicts a typical temperature field and the corresponding strain rate field during FSW when the rotating speed and welding speed were 300 rpm and 150 mm/min, respectively. The HAZ and BM areas could be distinguished using their temperatures, and the temperature of the BM area was almost the same as the ambient temperature.

During FSW, the SZ and TMAZ were subjected to heat and deformation; as a result, the strain rate and deformation temperature severely affect the grain sizes of these two zones. According to the high-temperature plastic deformation theory, the relationship between the Zener-Hollomon parameters, temperature, and strain rate can be expressed as follows:

$$Z = \dot{\varepsilon} exp(Q/RT) \tag{3}$$

where $\dot{\varepsilon}$ is the strain rate, $Q$ is the activation energy, $R$ is the gas constant, and $T$ is the temperature. In addition, Heidarzadeh and Saeid[25] have indicated that the value of the Z parameter is negatively correlated with the grain size, which is shown in equation (4).

$$d = A \times Z^b \tag{4}$$

where $d$ is the diameter of the grain, and $A$ and $b$ are material constants. Usually $b$ is a negative number; therefore, smaller Z values lead to coarser grains.

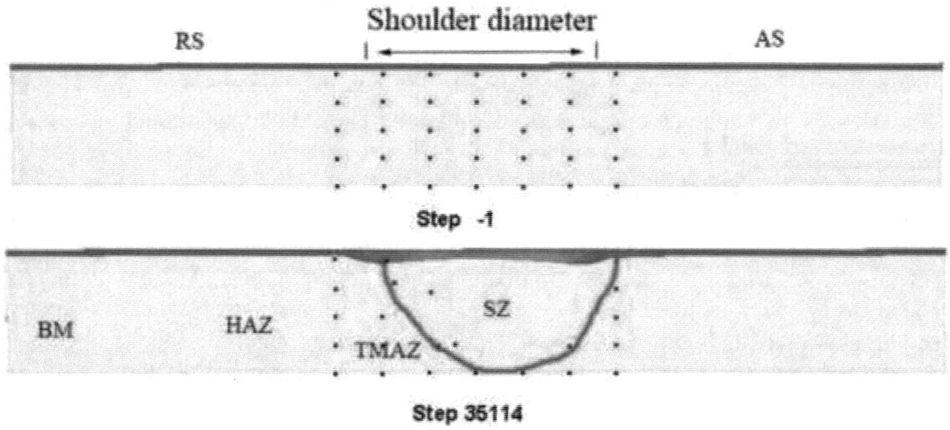


**Figure 2.** Four areas of material after FSW.

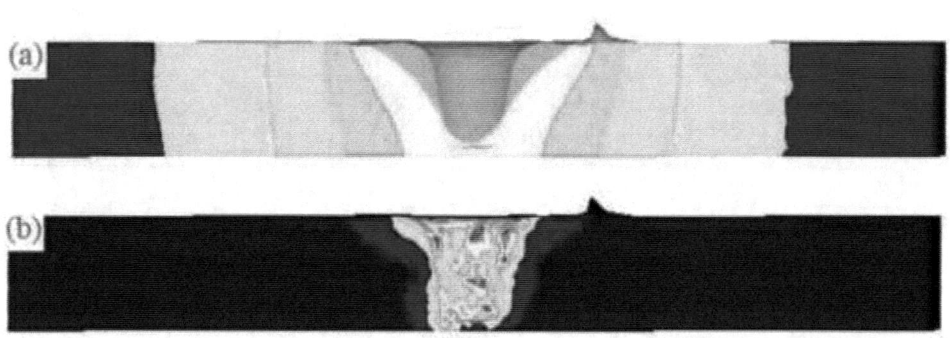


**Figure 3.** Typical (a) temperature field and (b) strain rate field of FSW for a pin rotating speed of 300 rpm and welding speed of 150 mm/min.

Figure 4(a) illustrates a typical metallurgical image of the Al-Zn-Mg alloy after FSW. From figures 3 and 4, we concluded that the SZ presented the highest temperature and strain rate of the four above-mentioned areas. Simultaneously, the grain size of the SZ was the finest of all areas. The TMAZ exhibited a lower strain rate than the SZ and much coarser grains. Ji et al.[26] reached similar conclusions through grain size analysis.

Figure 4(b)~(e) show the metallographic microstructure of the typical four areas from the welded specimen. Figure 4(b) and (c) show that the BM and HAZ has obvious processing microstructure characteristics, and the crystal grains are elongated. Additionally, affected by temperature, the grains of HAZ slightly grow. Meanwhile the TMAZ and SZ have totally different microstructure, showed in figure4(d) and (e). In these two areas, the grains undergo dynamic recrystallization under the influence of deformation and temperature, and the grain morphology changes from elongated to equiaxed. The SZ is directly affected by the stirring needle and experiences the most severe deformation. The new grains formed by recrystallization are broken by the stirring of the pin, so that the region forms a equiaxed grain which is finer than TAMZ.

### *4.1. Effect of different welding speeds on SZ grain size*

Both the temperature and strain rate fields at different welding speeds (50 and 150 mm/min) and the same rotating speed (450 rpm) were predicted using numerical simulations. Uyyuru and Kailas[27] indicated that when the tool advanced a minimum distance equal to its diameter, the process was considered to have reached a steady state. The temperature and strain rate of the YZ plane near the stirring pin in the steady state were analysed. It should be noted that the centre of the bottom surface of the plate was the origin of the coordinate axes.

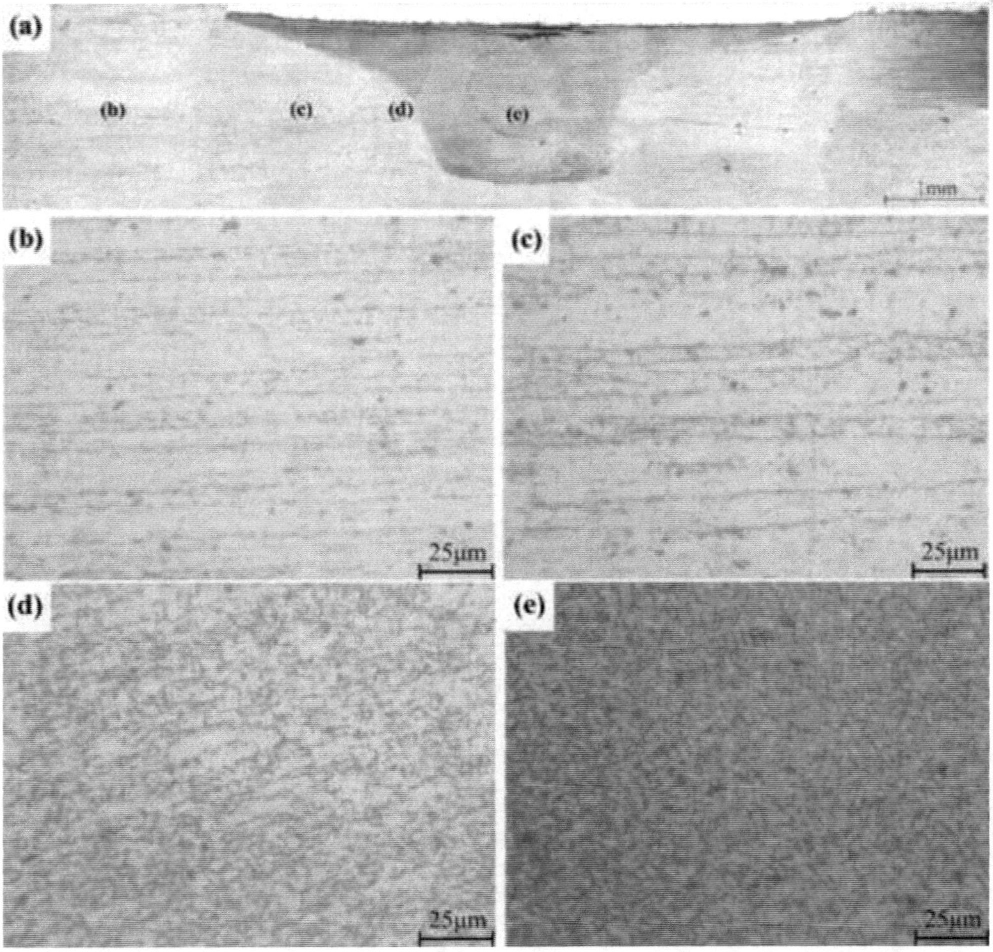

**Figure 4.** Typical metallurgical structure of Al-Zn-Mg alloy after FSW for a pin rotating speed of 300 rpm and welding speed of 150 mm/min. (a) weld joint, 50x, (b) BM, 400x, (c) HAZ, 400x, (d) TMAZ, 400x, (e) SZ, 400x

The temperature and strain rate plots of the YZ plane are showed in figure 5. As shown in figure 5(a), the welding speeds of 50 mm/min corresponded to peak temperatures of 526°C while the 150mm/min one is 513 °C. Meanwhile figure 5(b) reveals that the strain rates are almost the same at both welding speeds, thus the traverse speed of the stirring pin has a neglected influence on the strain rate of the FSW. Pashazadeh et al.[28] also reached similar conclusions through finite element simulation. That is, the strain rate was almost constant while the peak temperature during the process decreased as the welding speed increased. The Zener-Hollomon parameter was calculated according to the obtained curves of temperature and strain rate, as shown in figure 6.

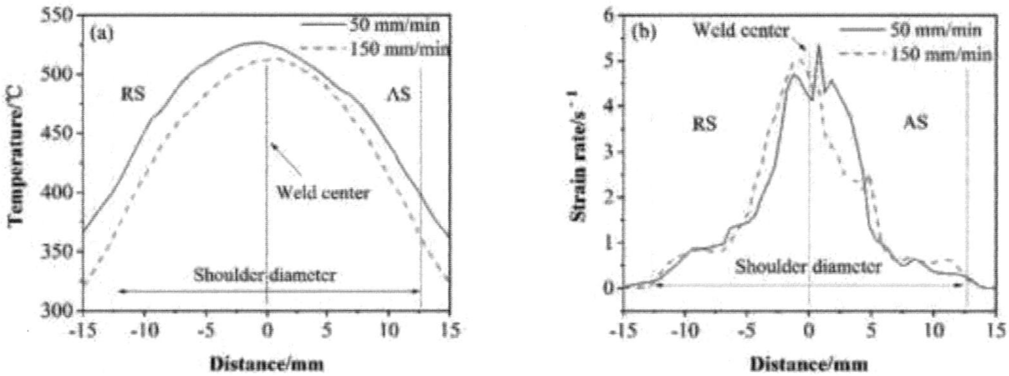


**Figure 5.** (a) Temperature and (b) strain rate plots of SZ of FSW for a rotating speed of 450 rpm and welding speeds of 50 and 150 mm/min.

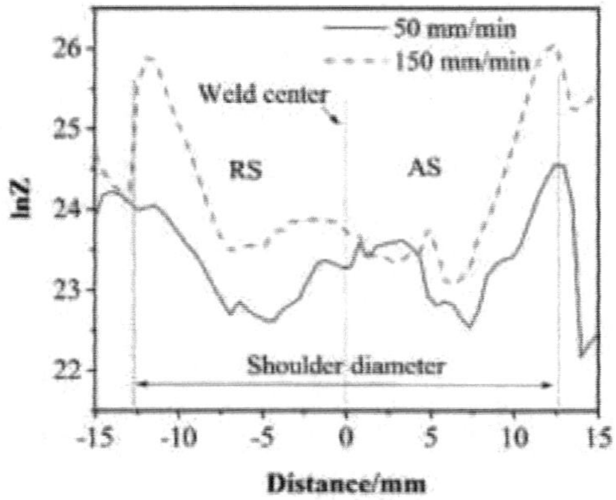


**Figure 6.** Zener-Hollomon parameter of SZ of FSW for a rotating speed of 450 rpm and welding speeds of 50 and 150 mm/min.

The Zener-Hollomon parameter increased as the welding speed increased, indicating that as the welding speed increased, the grain size of the material in the SZ became smaller. This phenomenon has been experimentally demonstrated. The EBSD technique was used to observe the grain distribution in joints after welding. The images in figure 7 represent the positions indicated by the white arrow in figure 4. Figure 7 indicated that a higher welding speed result in a smaller grain size distribution of the joint. The average grain diameters were approximately 6.49 and 5.80 μm when the welding speeds were 50 and 150 mm/min, respectively. Obviously the grain size was much smaller at the welding speed of 150mm/min.

**Figure 7.** EBSD analysis of SZ of FSW for a rotating speed of 450 rpm and welding speeds of (a) 50 and (b) 150 mm/min.

*4.2. Effect of different rotating speeds on SZ grain size*

During FSW, the grain size distribution after welding is not only related to the welding speed but also to the tool rotating speed. Keeping the same welding speed (150 mm/min), different rotating speeds (300, 450, and 600 rpm) produce different temperature and strain rate fields. Due to the presence of a pinhole at the location of the stirring pin, the points near the YZ plane of the trailing edge of the stirring pin were taken. These points were located 2 mm from the upper surface of the workpiece and equidistantly distributed on both sides along the centre of the weld. The graphs of the temperature and strain rates are illustrated in figure 8.

It can be seen in figure 8(a), the corresponding peak temperatures at 300, 450, and 600 rpm were 436, 512, and 540 °C, respectively, indicating that the temperature increased as the rotating speed increased. Meanwhile the figure 8(b) reveals that the peak strain rate increases as the rotating speed increases. That is, for same welding speed, increasing the rotating speed of the tool, the temperature and strain rate increased.

It can be seen that when the rotating speed was 300 rpm the average diameter of the grian was 3.88 μm, and the average grain size decreased as the rotating speed increased. When the rotating speed reached 600 rpm, the grain diameter reached 9.24 μm, which was much larger than that obtained at a rotating speed of 300 rpm. Wan et al.[21] used the point tracking simulation method and Heidarzadeh and Saeid[25] used experimental data to demonstrate that the increasing the tool rotating speed, the the grain grow larger.

IOP Conf. Series: Materials Science and Engineering **668** (2019) 012022 doi:10.1088/1757-899X/668/1/012022

In summary, the slow welding and high rotating speeds during FSW caused the grains of the SZ to become coarse. From figures 6, 7, 9, and 10, we concluded that the grain size of the SZ was mostly distributed within a 5 μm range when lnZ reached 24, and a relatively uniform fine grain structure was obtained. When lnZ was smaller than 24, the crystal grains tended to be coarse, and the fraction of large-sized crystal grains increased.

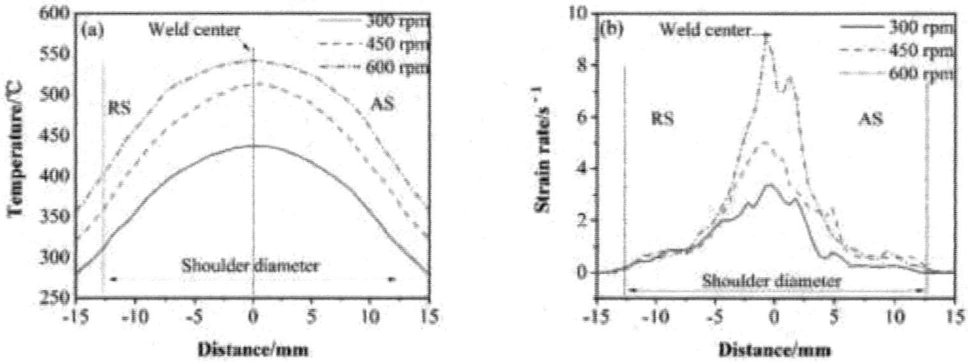


**Figure 8.** (a) Temperature and (b) strain rate plots of SZ of FSW for a welding speed of 150 mm/min and rotating speeds of 300, 450, and 600 rpm.

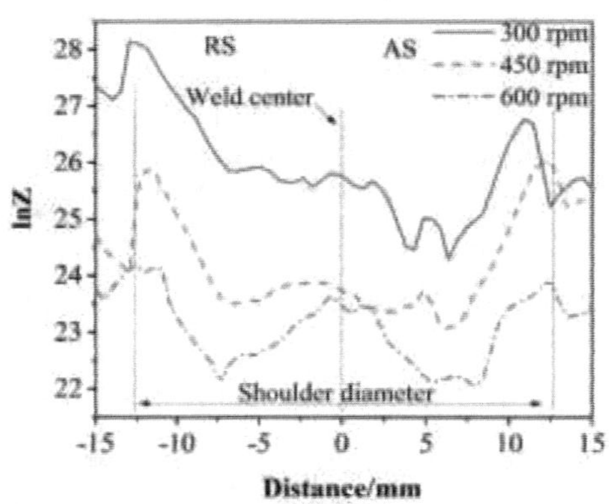


**Figure 9.** Zener-Hollomon parameter of SZ of FSW for a welding speed of 150 mm/min and rotating speeds of 300, 450, and 600 rpm.

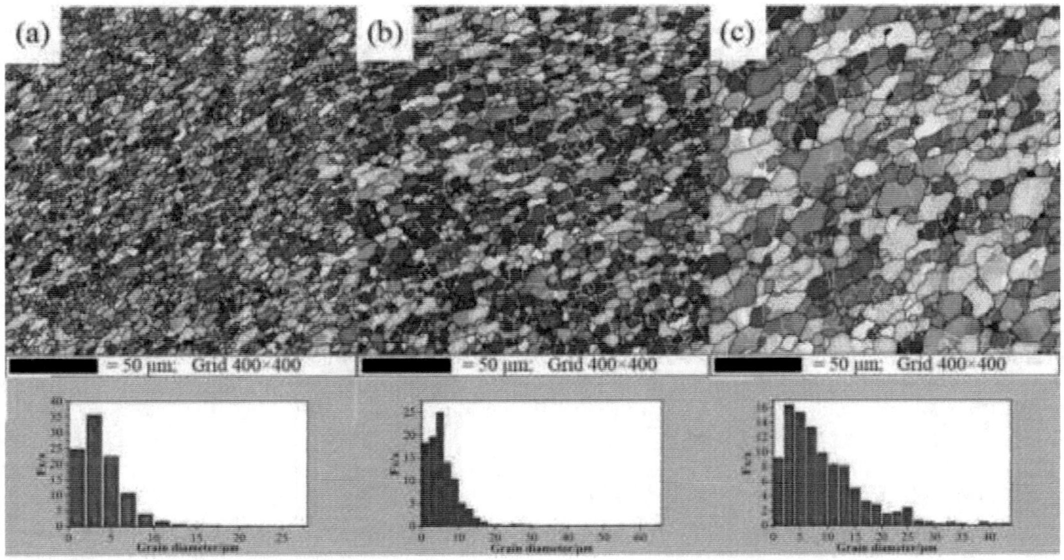


**Figure 10.** EBSD analysis of SZ of FSW for a welding speed of 150 mm/min and rotating speeds of (a) 300, (b) 450, and (c) 600 rpm.

The relationship between Z and grain size in SZ is plotted in figure 11. It indicate that the grain sizes of SZ and Z are inversely proportional. When fitting these data using a power function (equation (4)), the specific relationship between Z and the average grain diameter can be expressed as:

$$Z = 9.09 \times 10^{13} d^{-4.74} \quad (5)$$

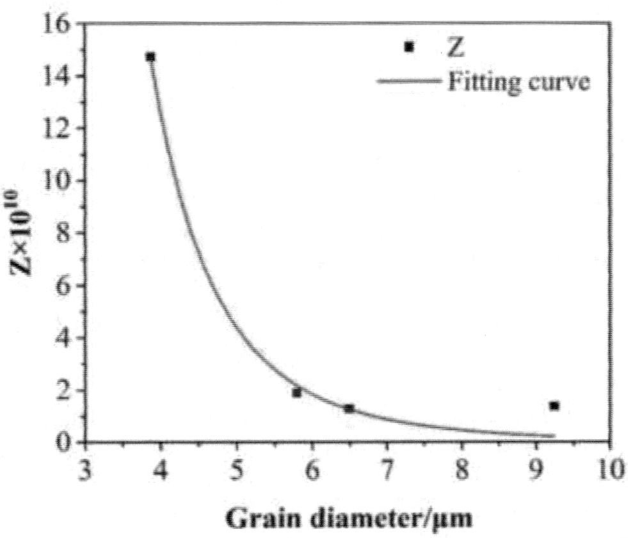


**Figure 11.** Relationship between Z and grain size of SZ.

## 5. Summary

In this study, the temperature and strain rate distributions during FSW were studied by the finite element method. The Zener-Hollomon parameter was calculated using the temperature and strain rate obtained employing numerical simulations to predict the influence of different welding parameters on the grain size. After welding experiments, the microstructure of the material was studied by OM and EBSD. The experimental data were combined with the simulation results, which further demonstrated the accuracy of the prediction. Considering the discussed results, the following conclusions were drawn:

(1) For a constant rotating speed, the strain rate during FSW was almost constant. The deformation temperature and average grain diameter of the weld nugget zone decreased as the welding speed of the tool increased.

(2) Different rotating speeds presented different effects on grain size during the FSW. The strain rate and temperature increased as the rotating speed increased. Since temperature contributed more than strain rate to grain growth, the grain size increased as the rotating speed increased.

(3) Under the designed processing parameters, the relationship between Zener-Hollomon parameter and SZ grain size can be expressed as $Z = 9.09 \times 10^{13} d^{-4.74}$ through simulation calculation and experimental analysis. As can be seen from this equation, the grain size decreases as the Zener-Hollomon parameter increases.

**Acknowledgements**

This work was supported by the National Key Research and Development Program of China (Project No. 2016YFB0300900). The authors appreciate the kind support offered by Professor Yunlai Deng, the Project leader.

**References**

[1] Yang S L and Lin Q L 2010 Microstructures and properties of the al-4.5zn-1.5mg-0.5mn aluminum alloy welding joints *Adv. Mater. Res.* **148-149** 640-3

[2] Mishra R S and Ma Z Y 2005 Friction stir welding and processing *Mater. Sci. Eng. R: Reports* **50** 1-78

[3] Oiwa N, Iwaki S, Okada T, Eguchi N, Tanaka S and Namba K 2005 Studies on characteristics of friction stir welded joints in structural thin aluminium alloys part 1: imperfections in friction stir welded zones and their precision non-destructive testing *Weld. World* **49** 76-82

[4] Okada T, Hori H, Hashimoto T, Tanikawa H, Iwaki S, Takeda J, Miyamichi T, Eguchi N, Tanaka, S, Oiwa, N and Namba K 2005 Studies on characteristics of friction stir welded joints in structural thin aluminium alloys part 2: metallurgical features and mechanical properties of friction stir welded joints *Weld. World* **49** 83-92

[5] Li D, Yang X, Cui L, He F and Zhang X 2015 Investigation of stationary shoulder friction stir welding of aluminum alloy 7075-T651 *J. Mater. Process. Tech.* **222** 391-8

[6] Wan L, Huang Y, Lv Z, Lv S and Feng J 2014 Effect of self-support friction stir welding on microstructure and microhardness of 6082-T6 aluminum alloy joint *Mater. Design* **55** 197-203

[7] Li J Q and Liu H J 2012 Design of tool system for the external nonrotational shoulder assisted friction stir welding and its experimental validations on 2219-T6 aluminum alloy *J. Int. J. Adv. Manuf. Tech.* **66** 623-4

[8] Liu H, Zhao Y, Hu Y, Chen S and Lin Z 2015 Microstructural characteristics and mechanical properties of friction stir lap welding joint of alclad 7B04-T74 aluminum alloy *Int. J. Adv. Manuf. Tech.* **78** 1415-25

[9] Zhang H J, Liu H J and Yu L 2011 effect of water cooling on the performances of friction stir welding heat-affected zone *J. Mater. Eng. Perform.* **21** 1182-7

[10] Chen G, Li H, Wang G, Guo Z, Zhang S, Dai Q, Wang X, Zhang G and Shi Q 2018 Effects of pin thread on the in-process material flow behavior during friction stir welding: a computational fluid dynamics study *Int. J. Mach. Tool. Manu.* **124** 12-21

[11] Zhu Z, Wang M, Zhang H, Zhang X, Yu T and Wu Z 2017 A finite element model to simulate defect formation during friction stir welding *Metals* **7** 256.

[12] Long L, Chen G, Zhang S, Liu T and Shi Q 2017 Finite-element analysis of the tool tilt angle

effect on the formation of friction stir welds *J. Manuf. Process.* **30** 562-9
[13] Carlone P, Citarella R, Lepore M and Palazzo G S 2014 A FEM-DBEM investigation of the influence of process parameters on crack growth in aluminum friction stir welded butt joints *Int. J. Mater. Form.* **8** 591-9
[14] Ajri A and Shin Y C 2017 Investigation on the effects of process parameters on defect formation in friction stir welded samples via predictive numerical modeling and experiments *J. Manuf. Sci. Eng.* **139**(11) V001T02A008
[15] Tutunchilar S, Haghpanahi M, Besharati Givi M K, Asadi P and Bahemmat P 2012 Simulation of material flow in friction stir processing of a cast Al–Si alloy *Mater. Design* **40** 415-26
[16] Asadi P, Mahdavinejad R A and Tutunchilar S, 2011 Simulation and experimental investigation of FSP of AZ91 magnesium alloy *Mater. Sci. Eng. A* **528** 6469-77
[17] Shi L, Wu C S and Liu H J 2015 Analysis of heat transfer and material flow in reverse dual-rotation friction stir welding *Weld. World* **59** 629-38
[18] Hasan A F, Bennett C J and Shipway P H 2015 A numerical comparison of the flow behaviour in friction stir welding (FSW) using unworn and worn tool geometries *Mater. Design* **87** 1037-46
[19] Buffa G, Fratini L and Shivpuri R 2007 CDRX modelling in friction stir welding of AA7075-T6 aluminum alloy: analytical approaches *J. Mater. Process. Tech.* **191** 356-9
[20] Arora H S, Singh H and Dhindaw B K 2012 Numerical simulation of temperature distribution using finite difference equations and estimation of the grain size during friction stir processing *Mater. Sci. Eng. A* **543** 231-42
[21] Buffa G, Hua J, Shivpuri R and Fratini L 2006 A continuum based FEM model for friction stir welding—model development. *Mater. Sci. Eng. A* **419** 389-96
[22] Liu J, Jin L, He Z, Lu L, Liu H and Zhang Y, 2011 Hot deformation behavior of 7N01 aluminum alloy *Rare Metals* **35** 812-7
[23] Frigaard O, Grong O and Midling O T 2001 A process model for friction stir welding of age hardening aluminum alloys *Metall. Mater. Trans. A* **32** 1189–200
[24] Jain R, Pal S K and Singh S B 2016 A study on the variation of forces and temperature in a friction stir welding process: a finite element approach *J. Manuf. Processes* **23** 278-86
[25] Heidarzadeh A and Saeid T 2016 Correlation between process parameters, grain size and hardness of friction-stir-welded Cu-Zn alloys *Rare Metals* **37** 388-98
[26] Ji S D, Meng X C, Li Z W, Ma L and Gao S S 2015 Investigation of vertical compensation friction stir-welded 7N01-T4 aluminum alloy *Int. J. Adv. Manuf. Tech.* **84** 2391-9
[27] Uyyuru R K and Kailas S V 2006 Numerical analysis of friction stir welding process *J. Mater. Eng. Perform.* **15** 505-18
[28] Pashazadeh H, Masoumi A and Teimournezhad J 2013 A study on material flow pattern in friction stir welding using finite element method *P. I. Mech. Eng. B- J. Eng.* **227** 1453-66

# Research on speed regulation method of winding machine based on rectangular line mode

**Y S Zhang[1,2], H W Ma[1] and H T Wang[1]**

[1]College of Mechanical Engineering, Xi'an University of Science and Technology, Xi'an, 710054, China

E-mail: 714720644@qq.com

**Abstract.** In the traditional semi-automatic winding machine winding process, the tension of the enameled copper wire changes greatly, and the cable is inaccurate, and cannot be separated from the manual machine. The emergence of automatic winding machine realizes the automation of winding production, reduces the labor intensity of winding, and greatly improves production efficiency. In order to improve the winding quality of the rectangular coil of the motor, a motor winding automatic winding machine is designed. The kinematics and dynamics analysis of the winding process of the winding machine were carried out. The kinematics of the enamelled copper wire and the working principle of the governing mechanism were studied. The kinematics mathematical model of the enamelled copper wire was simulated. The results show that the designed motor coil automatic winding machine has accurate cable routing and smooth winding process.

## 1. Introduction

In the manual winding process, the production efficiency of the winding is low due to the variety of the wound products and the proficiency of the operator. Especially in the increase of the number of turns, such as the number of individual motor turns in the millennium or 10,000-degree motor, the manual winding efficiency is even lower [1-3]. Due to the inconsistent tension of the wires, the accuracy of the number of windings is not guaranteed, and raw materials are wasted. The magnitude of the external tension applied to the magnet wire depends on the cross-sectional dimension of the magnet wire. Different specifications of the electromagnetic wire, the tension required in the coil winding process is also different. If the tension is too large, the insulation of the bread outside the electromagnetic wire is easily scratched. If the tension is too small, the coil forming is certainly not ideal [4-7].

The foreign winding industry has developed earlier and is developing rapidly [8-10]. METEOR, Switzerland, mainly produces precision winding machines, including semi-automatic and fully automatic workpiece rotary winding systems. During the winding process, both ends of the workpiece must be clamped, which will cause the magnetic properties of the workpiece to change, affecting the quality of the workpiece [11,12]. Japan's Hitachi Company uses the workpiece as the main body and uses the flying fork to wind the high-speed rotary winding. This not only ensures the quality of the rotating body is constant, but also runs smoothly, avoiding the additional installation torque to deform the workpiece and affect the workpiece quality [13]. In order to meet the needs of the industry, combined with China's national conditions, it is imperative to develop a fully automatic winding machine with excellent performance and low price [14,15]. The traditional motor coil winding machine working state and winding process are analyzed, and a motor coil automatic winding machine experimental platform with smooth winding process and precise cable routing is designed to achieve automatic winding, reduce human factors and reduce Labor intensity to increase production efficiency.

## 2. Motor coil automatic winding machine structure

The conventional winding machine is mainly composed of a clamping mechanism, a coil, an electric motor, and a wire mold. The working principle is that the motor drives the wire mold to rotate, and the copper wire is wound around the wire mold to obtain the required coil, as shown in figure 1.

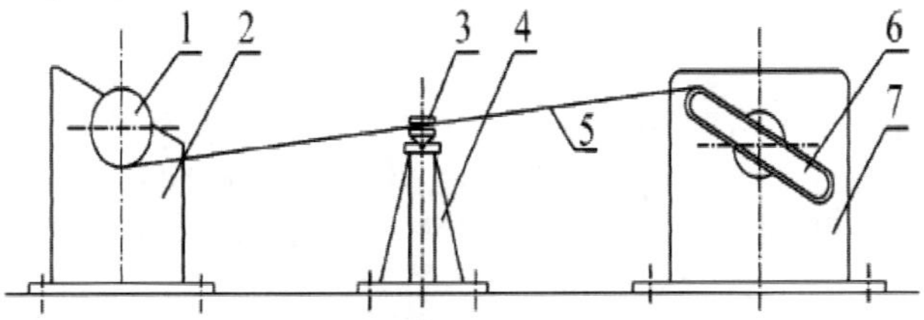


1-Reel; 2-4-Stand; 3- Tensioning device; 5- Copper wire; 6-Line mode; 7- Bracket and motor

**Figure 1.** Winding machine schematic.

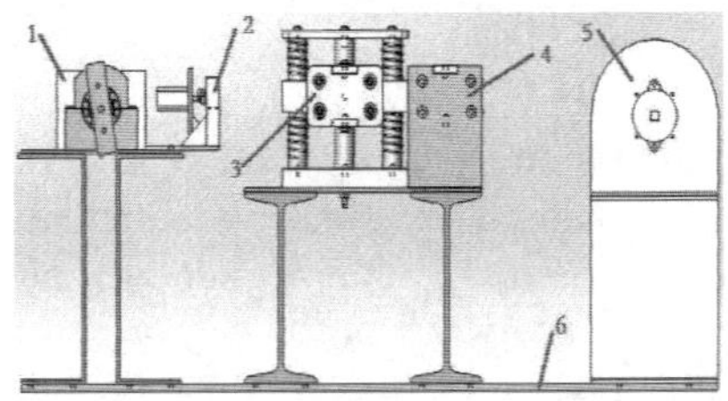

1- Winding mechanism; 2- Cable mechanism; 3- Speed governing mechanism; 4- Guiding mechanism; 5- Take-up mechanism; 6- Bottom plate

**Figure 2.** Winding machine 3D assembly main view.

In the figure, the motor rotates at a constant speed, but since the motor coil is usually rectangular and has a large volume, the wire mold may vibrate during the winding process, the winding speed is uneven, and the tension is uneven. However, the tensioning device in the figure can not adjust the clamping force at any time, so the performance of the wound coil will be affected to some extent.

The motor coil automatic winding machine system consists of six parts: control system, winding mechanism, cable arranging mechanism, speed regulating mechanism, guiding mechanism and pay-off mechanism. Its winding machine assembly is shown in figure 2.

- Winding mechanism: the winding mechanism motor is used as the power source of the whole system, and the driving line mode realizes the rotary motion of the wire mold when winding. The mechanism is composed of DC servo motor, motor bracket, coupling bracket, coupling, and wire mold. The coupling is connected with the wire mold, and the coupling bracket is mounted on the frame to remove the influence of the cantilever beam and ensure the neutrality of the motor output shaft. The wire mold uses a bolt-on clamping plate as a clamp to wind the coil. In order to prevent the coil from rotating without tension during the pay-off, a rotary damper is installed on the pay-off mechanism to adjust the tension of the wire on the lacquered copper wire during the pay-off.
- Cable arranging mechanism: The mechanism is composed of a slider, a connecting rod, an internal rail, a motor base, and a stepping motor. The stepping motor rotates at a constant speed, and the crank slider is driven to perform periodic motion. By changing the connecting rod, changing the stroke of the slider, controlling the pulse frequency received by the stepping motor to change the width of the wound coil, the enamelled copper wire passes through the long hole in the slider, and the two deep groove ball bearings are connected in the slider. The sliding friction is converted into rolling friction, thereby achieving a stable and uniform line of the mechanism.
- Speed regulating mechanism: The speed regulating mechanism adopts a combined structure of rollers and springs. When the rectangular coil is wound by a spring, the copper wire of the enamelled copper is buffered by the tension change, and the winding speed is adjusted. The utility model solves the problems that the axial length of the coil is increased due to the manual operation and the need for continuous tapping and shaping during the winding

process, and the material for the motor production is reduced, and the weight of the motor is reduced. volume.

- Guide and pay-off mechanism: The guide mechanism is composed of a roller structure, and a frame is shared with the speed-regulating mechanism. The nylon material idler is used to plan the winding path, which reduces the weight of the mechanism and reduces the rolling friction on the paint. Damage to the copper wire insulation layer. As the whole system supply mechanism, the pay-off mechanism is relatively simple in structure, and the reel is fixed by the fixed form of the tailstock. The rotary damper is installed on one side of the frame to provide damping for the rotation of the reel, which solves the problem of inertia rotation of the reel.

## 3. Principle analysis of speed regulating mechanism of winding machine

The traditional winding machine does not have a constant tension control mechanism, and the tension control is obtained by the worker continuously adjusting the manual control splint. The process cannot rely on the experience of the worker, and the constant tension cannot be achieved, thereby causing the winding coils to be inconsistent and affecting the motor winding. the quality of. In the winding process, since the wire mold is generally rectangular in cross section, the wire moving speed is uneven and the coil tension is not constant. These factors will also cause the electromagnetic wire to be damaged, the coil size is inconsistent, and the winding work is difficult. Therefore, the speed governing mechanism of the winding machine must be able to ensure the stability of the coil winding process.

### *3.1. Motion analysis without speed regulation*

First, we discuss the movement of the enamelled copper wire when the winding machine is not equipped with a speed governing mechanism. Since the wire mold is far away from the wire reel, regardless of the thickened size of the copper wire wound on the wire die during the winding process and the thinned size of the enamelled copper wire on the wire reel, it can be considered that the lacquered copper wire periodically vibrates when the wire is wound in the wire.

Kinematic analysis of the enamelled copper wire was performed during the process of intercepting the rotation of the wire mold. In this process, the enamelled copper wire passes through the four corner points of the wire mold, so the rotation motion of this week can be divided into four segments for analysis.

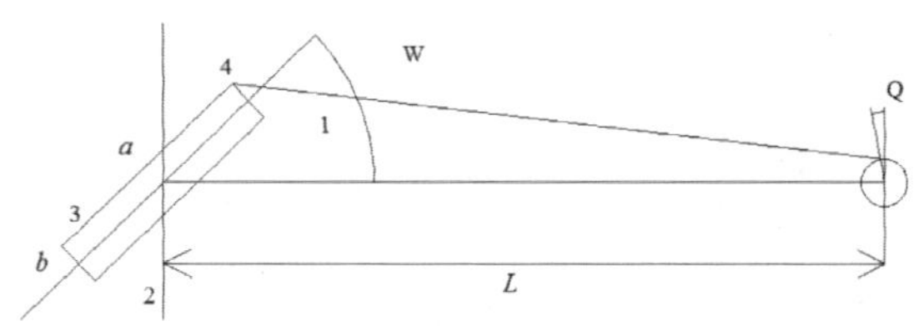


**Figure 3.** The first section of the wire mold winding.

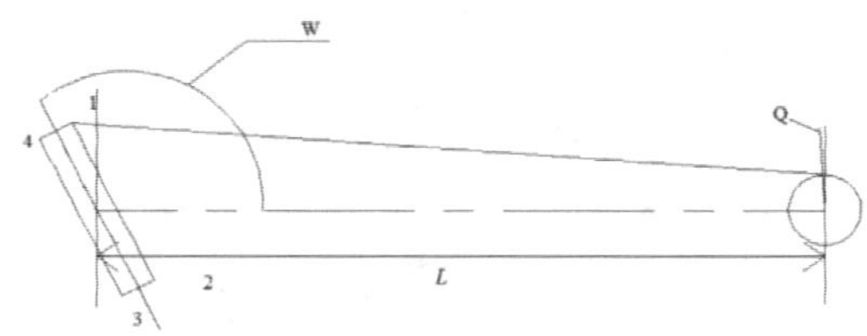


**Figure 4.** Schematic diagram of the second section of the wire mold winding.

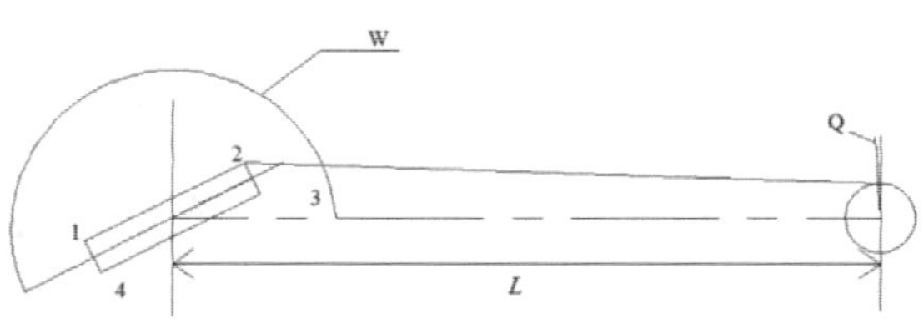


**Figure 5.** The third section of the wire mold winding.

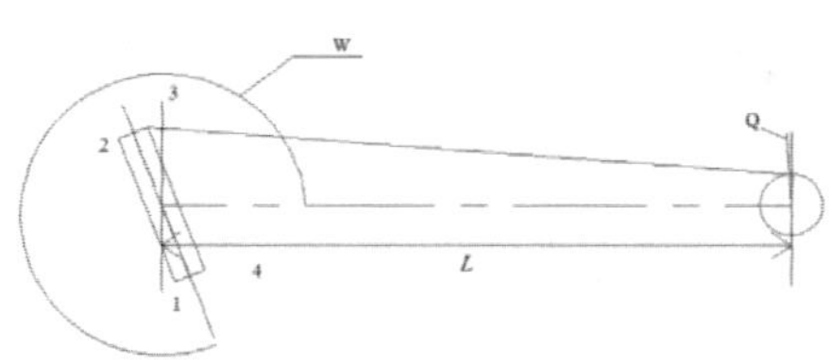


**Figure 6.** Schematic diagram of the fourth section of the wire mold winding.

The first paragraph: the wire mold starts to rotate from the starting horizontal position until the enameled wire passes the first corner point position as shown in figure 3.

The second paragraph: the wire mold continues to rotate after the enamelled copper wire passes through the first corner point until the enamelled copper wire passes through the second corner point as shown in figure 4.

The third paragraph: the wire mold continues to rotate after the enamelled copper wire passes through the second corner point until the enamelled copper wire passes through the third corner point as shown in figure 5.

The fourth paragraph: the wire mold continues to rotate after the enamelled copper wire passes through the third corner point until before passing through the fourth corner point as shown in figure 6.

When the center of the wire mold is larger than the wire spool center from the center of the wire, the angle Q can be ignored. The center of the wire mold and the center of the coil are L=2100 mm, a=200 mm, b=50 mm. Calculate the angle W to which the line mode is turned when the first and second segments are converted. The solution is W1=88 degrees and W2=91 degrees. The reason for the double solution is to ignore the angle Q. The essence of ignoring the angle Q is that the enamelled copper wire is always horizontal during the rotation of the wire die. Under the ideal state, the wire die is turned to the angle W=90 degrees, and the first segment and the second segment are converted, and the obtained result is obtained. Approximate to the ideal state.

The resulting length change $d_L$ of the four-stage enameled wire is:

The first stage: $d_L = L - \frac{\sqrt{a^2+b^2}}{2}\cos\left(W + \arctan\frac{b}{a}\right)$; The second stage: $d_L = b + L - \frac{\sqrt{a^2+b^2}}{2}\cos\left(W - \arctan\frac{b}{a}\right)$;

The third stage: $d_L = b + a + L - \frac{\sqrt{a^2+b^2}}{2}\cos\left(W - \pi + \arctan\frac{b}{a}\right)$; The fourth stage: $d_L = b + 2a + L - \frac{\sqrt{a^2+b^2}}{2}\cos\left(W - \pi - \arctan\frac{b}{a}\right)$.

The trace speed $v_L$ changes to:

The first stage: $v_L = \frac{\sqrt{a^2+b^2}}{2}\omega\sin\left(W + \arctan\frac{b}{a}\right)$; The second stage: $v_L = \frac{\sqrt{a^2+b^2}}{2}\omega\sin\left(W - \arctan\frac{b}{a}\right)$;

The third stage: $v_L = \frac{\sqrt{a^2+b^2}}{2}\omega\sin\left(W - \pi + \arctan\frac{b}{a}\right)$; The fourth stage: $v_L = \frac{\sqrt{a^2+b^2}}{2}\omega\sin\left(W - \pi - \arctan\frac{b}{a}\right)$.

The line acceleration $a_L$ changes to:

The first stage: $a_L = \frac{\sqrt{a^2+b^2}}{2}\omega^2\cos\left(W + \arctan\frac{b}{a}\right)$; The second stage: $a_L = \frac{\sqrt{a^2+b^2}}{2}\omega^2\cos\left(W - \arctan\frac{b}{a}\right)$; The third stage: $a_L = \frac{\sqrt{a^2+b^2}}{2}\omega^2\cos\left(W - \pi + \arctan\frac{b}{a}\right)$; The fourth stage: $a_L = \frac{\sqrt{a^2+b^2}}{2}\omega^2\cos\left(W - \pi - \arctan\frac{b}{a}\right)$.

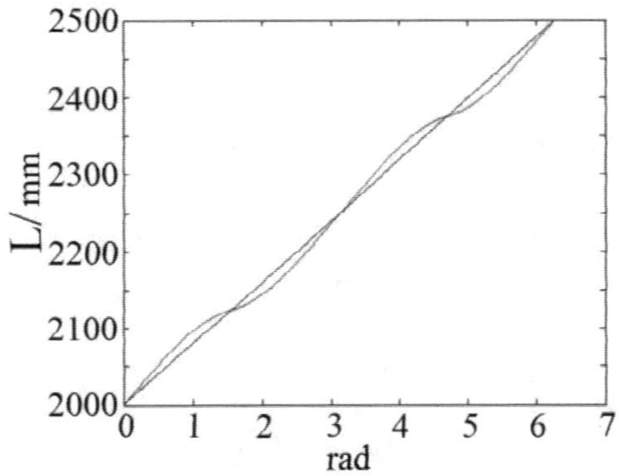


**Figure 7.** Enamelled wire length change curve.

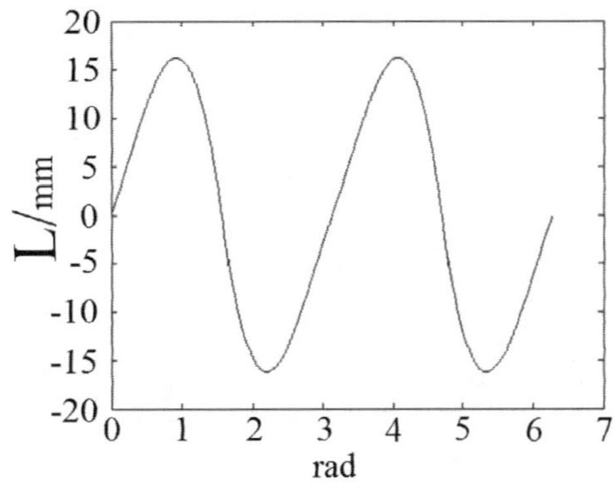


**Figure 8.** Difference curve.

Solving the kinematics equation, the calculation results are shown in figure 7. The red curve in figure 7 is the actual variation curve of the enamelled copper wire length, and the black straight line is the ideal change. From this, the difference curve between the two lines can be obtained, as shown in figure 8. The curve of the line speed is shown in figure 9. The curve of the line acceleration is shown in figure 10. From the above analysis results, the linear velocity, acceleration, and line length change periodically during the one-rotation of the rectangular line. The linear mode is cyclically shifted during the winding, and the tension on the enameled wire changes greatly.

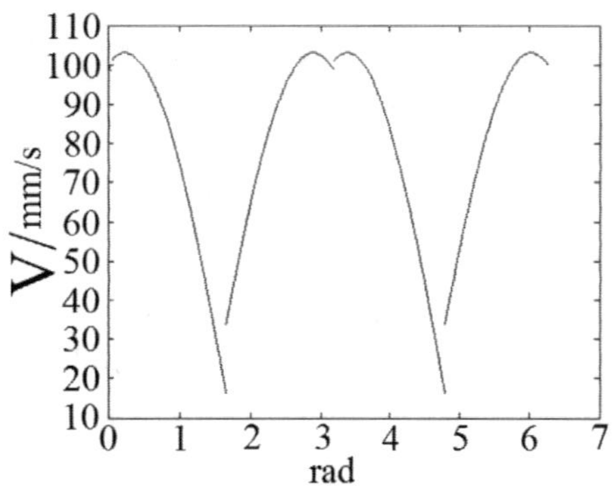


**Figure 9.** Enameled wire trace speed curve.

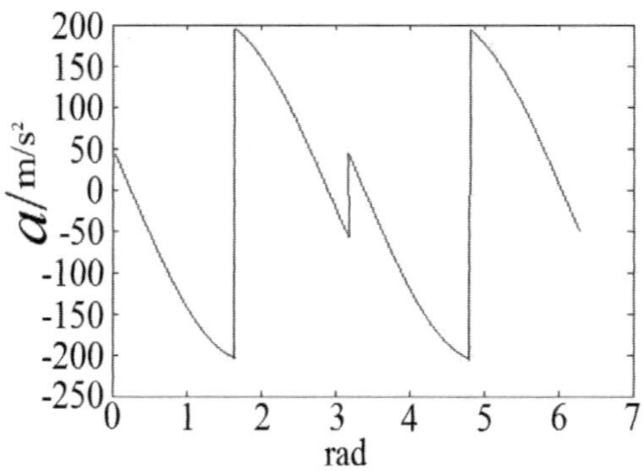


**Figure 10.** Enameled wire trace acceleration curve.

*3.2. Motion analysis during speed regulation*

The speed regulation process of the spring roller speed regulating mechanism in the space can be roughly divided into four stages: the falling force, the returning equilibrium position, the rising and the returning equilibrium position under the component force of the enamelled copper wire pulling force. Since the roller speed control mechanism uses light materials, the influence of the mechanism gravity on the speed regulation process of the winding machine is reduced. Therefore, the four stages have a symmetrical relationship when moving in the space, and the space is adjusted by the idler mechanism. The periodic reciprocating motion adjusts the wire speed of the enameled copper wire to improve the quality of the winding machine when winding the rectangular coil.

In the winding process of the winding machine, when the enamelled copper wire is subjected to the spring action in the speed regulating mechanism, the winding working process is as shown in figure 11:

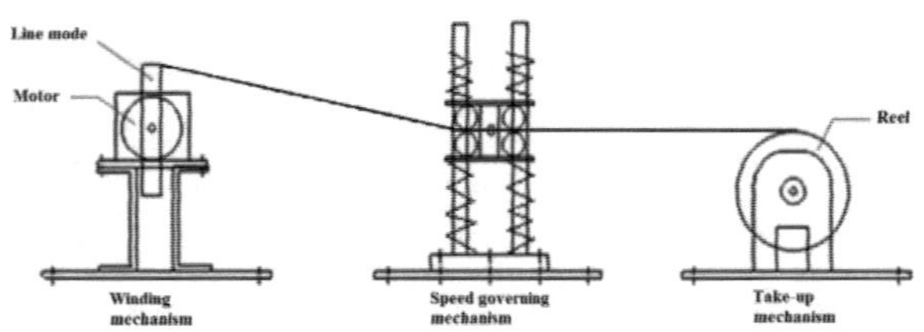


**Figure 11.** Speed regulation mechanism speed diagram.

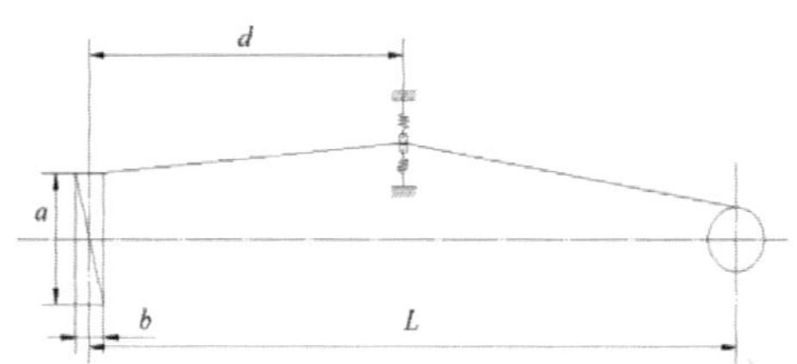


**Figure 12.** Schematic diagram of speed governing mechanism.

At this point, the working diagram of the speed governing mechanism is shown in figure 12.

For comparison, the dimensions in the calculation below are the same as above, L=2100 mm, a=200 mm, b=50 mm, taking d=1000 mm. Set the initial position of the spring, that is, the position when the spring is not deformed. In order to avoid excessive deformation of the spring, the intermediate position of the enameled wire at the upper and lower limits at the position in the natural state is taken as the initial position of the spring.

The lower limit position is shown in figure 13, where $c_1$: $c_1 = \dfrac{r\left[d + \dfrac{b}{2}\left(\dfrac{l}{r - b/2}\right)\right]}{\sqrt{\left[\dfrac{b}{2}\left(\dfrac{l}{r - b/2}\right) + l\right]^2 - r^2}}$.

The upper limit position is shown in figure 14, where $c_2$: $c_2 = \dfrac{d\left(r - \dfrac{1}{2}\sqrt{a^2 + b^2}\right)}{l} + \dfrac{1}{2}\sqrt{a^2 + b^2}$. Then the distance c of the center position of the spring from the center line of the line mold is: $c = \dfrac{c_1 + c_2}{2}$.

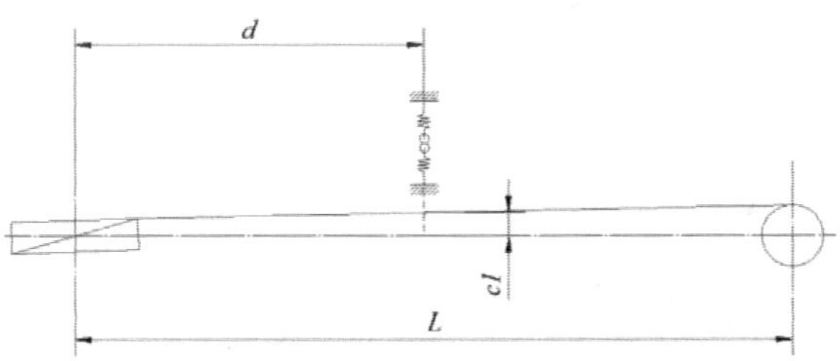


**Figure 13.** Lower limit position.

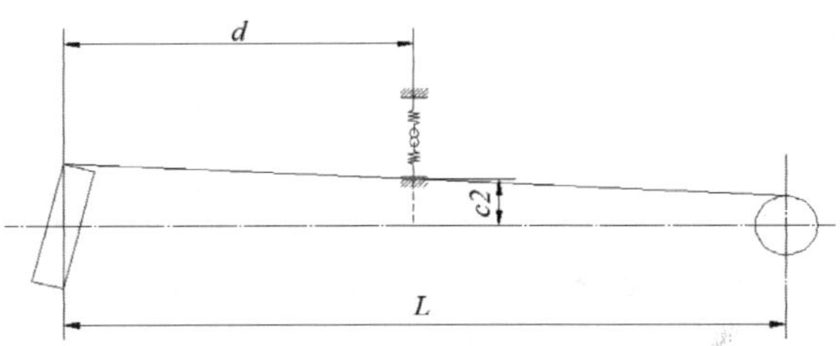


**Figure 14.** Upper limit position.

A similar method to that without the speed governing mechanism is used to model the length, trace speed and acceleration of the enameled wire during the one-turn rotation of the rectangular wire mold. The simulation results are shown in figure 15.

The red curve in the figure is the actual curve of the length of the enamelled copper wire, the black line is the ideal curve, and the blue curve is the curve of the length of the enamelled copper wire without the speed governing mechanism. It can be seen that after increasing the speed governing mechanism, the length change of the enamelled copper wire is very close to the ideal curve. From this, the difference curve of the two lines can be obtained, as shown in figure 16. The difference is very small and can be ignored.

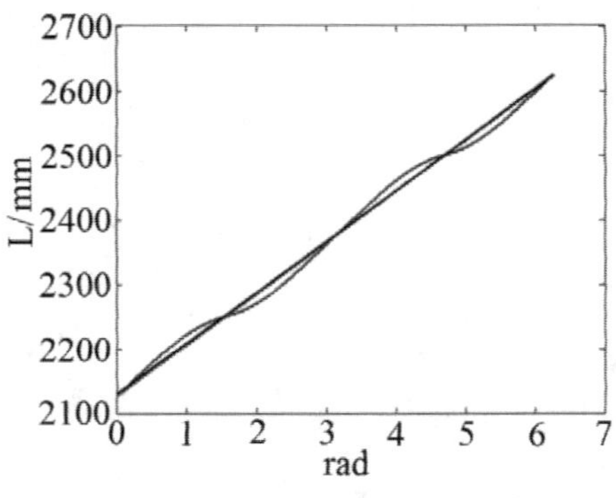


**Figure 15.** Increase the speed change diagram of the speed governing mechanism.

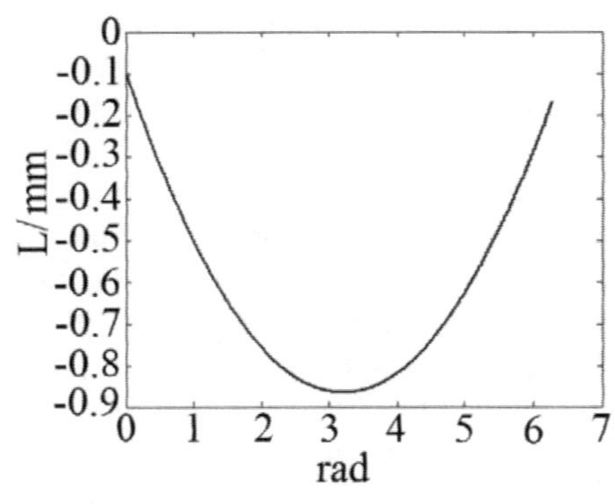


**Figure 16.** Line length difference curve.

Figure 17 shows the curve of the trace speed of the enamelled copper wire. The blue curve shows the curve of the wire speed of the enamelled copper wire without the speed governing mechanism. It can be seen that after increasing the speed governing mechanism, the wire speed of the enamelled copper wire is close to a constant speed. The curve of the line acceleration is shown in figure 18. It can be seen that the line is approximately zero, and the uniform circular motion of the line mode is realized.

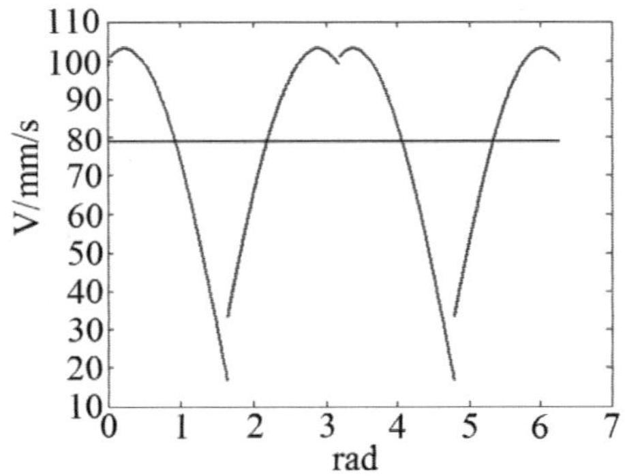


**Figure 17.** Trace speed curve.

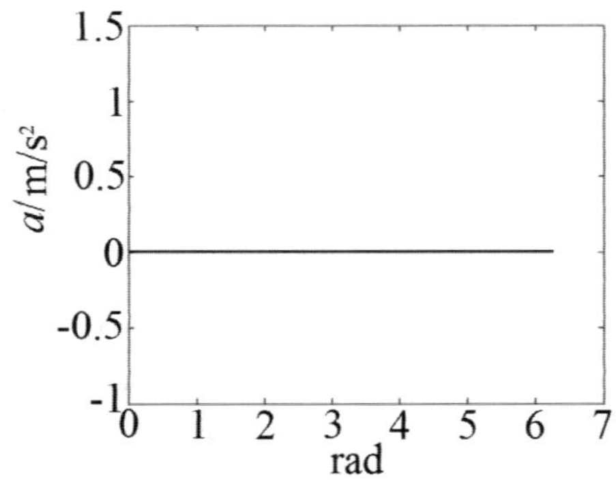


**Figure 18.** Trace acceleration change.

From the above results, it can be seen that after the speed governing mechanism designed in this paper is added to the winding machine, the tension change caused by the periodic vibration during winding is buffered by the spring roller speed regulating mechanism, thereby achieving the speed of the enamelled copper wire. The purpose of the adjustment is to change the periodic shifting circular motion to a nearly uniform circular motion, thereby improving the winding work process.

## 4. Conclusion

The motor coil is the core of the motor. The quality of the motor coil directly affects the quality and performance of the motor product. The winding of the motor coil is not tight, and the short-circuit resistance of the motor will be reduced, which directly affects the quality of the motor. Aiming at the coil speed regulation problem of the motor coil automatic winding machine, the kinematics equation of the enamelled copper wire in the winding process is established. The simulation calculation results show that the speed change mechanism changes the line length variation of the copper wire through the displacement change in the space. The value curve is compensated, and the purpose of adjusting the copper wire routing speed is achieved. The periodic shifting circular motion of the copper wire is changed to a uniform uniform circular motion, which indicates that the designed speed governing mechanism can meet the requirements of the smoothness of the winding process. The analysis method can be applied to windings of different specifications, so that the winding machine has a certain flexibility and widens the production range of the product.

## References

[1] Zhu Q P 2005 Development of automatic winding machine for transformer high voltage coil *China Science and Technology Information* **2005** 5-6

[2] Mao J X, Liu W and Zhou G F 2012 Study on control method of automatic winding machine with high precision *Computer Measurement & Control* **20** 95-7

[3] Zhang J H 2013 Development of 10 kV large-scale wound three-phase asynchronous motor *Electrical Manufacturing* **2013** 47-9

[4] Shen J X and Fan Z X 2008 Development trend and key technologies of modern electric machines *Small & Special Electrical Machines* **36** 55-7

[5] Wen J B, Wang G H and Wang C J 2011 Numerical calculation of medium-sized high-voltage induction motor rotor temperature fields based on coupled field *Explosion-Proof Electric Machine* **46** 1-3

[6] Zhang Q, Chen C J and Yan Y Z 2007 NC system based on motion controller for deflection winding machine *Machinery Design & Manufacture* **2007** 162-4

[7] Ma Z L and Gao X M 2004 Design of dedicated flat winding machine *Machine Tool & Hydraulics* **2004** 120-1

[8] Xie L F, You L R and Wang L S 2009 Design of hybrid controller base on PC104 and PLC for automatic winding-machine *Modular Machine Tool & Automatic Manufacturing Technique* **2009** 10-6

[9] Wei H B, Zhu Y H and Zheng H H 2005 Design of JR-28 automatic winding machine for metallized film capacitors *Electronics Process Technology* **2005** 1-3

[10] Barcaro M, Bianchi N and Magnussen F 2011 Six-phase supply feasibility using a pm fractional-slot dual winding machine *IEEE T. Ind. Appl.* **47** 2042-50

[11] Li H 2011 Research on vision-based error detection system for optic fiber winding *Int Conf Optical Instruments and Technology: Optoelectronic Imaging and Processing Technology* (Beijing, China) pp 40-1

[12] Djurović S, Vilchis-Rodriguez D and Smith A C 2012 Vibration monitoring for wound rotor induction machine winding fault detection *Xxth International Conference on Electrical Machines* (Marseille, France) pp 1906-1912

[13] Sauter D, Jamouli H, Keller J Y *et al* 2005 Actuator fault compensation for a winding machine *Control Eng. Pract.* **13** 1307-14

[14] Yuan X L, Tian H, Duan N *et al* Development of low-speed CNC winding machines *Sci. Technol. Innov.* **2014** 7-8

[15] Li X F 2017 Improvement of synchronization control accuracy of automatic cable winding machine *Electronic Technology & Software Engineering* **2017** 116-7

IOP Conf. Series: Materials Science and Engineering **668** (2019) 012024 doi:10.1088/1757-899X/668/1/012024

# Fabrication of Ag micro-patterns by electrohydrodynamic jet printing

**H Lv[1] and X Wang[1,2]**

[1]College of Mathematics and Physics, Qingdao University of Science and Technology, Qingdao, Shandong, 266061, China

E-mail: phwangxia@163.com

**Abstract.** Electrohydrodynamic jet printing is a rapid manufacturing method in additive manufacturing fields, which is employed to generate micro-nano patterns, functional structures, sensors and electronics in recent years. It is a rapid manufacturing, low-cost, mask-free route to manufacture one dimensional to three dimensional structures by nanoink. In this paper, silver dots, lines and designed patterns are printed by electrohydrodynamic jet printing. The results are stable and uniform, which can be adjusted by printing parameters. With high voltage, large pulse width or small stand-off distance, the size of dots increase. By increasing frequency or decreasing shifting speed, the distance between dots becomes short, then the structures are getting into lines from dots. Multi-layer silver lines are characterized by X-ray imaging and exhibit good absorption of X-ray, leading to the significant radiation attenuation effectiveness. The printed silver structures are good candidates for radiation shielding in electronics and circuit boards.

## 1. Introduction

Electrohydrodynamic jet (e-jet) printing is a rapid manufacturing method in additive manufacturing fields, which can generate many kinds of micro-nano structures from metal to polymer [1-5]. There are many advantages of e-jet printing, such as rapid manufacturing, low-cost, mask-free and pattern variation, thus, many functional structures can be fabricated on glass, flexible films and silicon [6-9]. The nanoink for printing are prepared by nanomaterials and typical solvent, resulting in the good property for printing with silver nanoparticles [9], quantum dots [10], silver fibers [11], even carbon nanotubes [12] and graphene [13]. Compared with the traditional manufacturing methods, e-jet printing provides a green, low environmental approach to generate micro-structures, which has great potentials in electronic sensors, bio-sensors and circuit boards.

The property of printed results is greatly influenced by printing parameters, such as voltage, shifting speed, frequency, stand-off distance, needle size, etc. Han *et al* [14] reported high aspect-ratio 3D structures of sub-10 μm resolution by e-jet printing, which showed high resolution of this method. AC-pulsed voltage was used to manufacture the transparent electrode [15], electrical features and connectors [16,17]. Qin *et al* [18] fabricated a functional touch sensor with high sensitivity via e-jet printing using silver nanoink. Many composite materials were employed to manufacture multi-layer structures, including $WO_3$-Ag [19,20], Ta [21], $WO_3$ [22], which were good candidates of electron devices. In addition to traditional substrate, flexible films were employed for printing, such as polydimethylsiloxane (PDMS) and Polyethylene terephthalate (PET) [23,24]. In the previous reports, the researchers focused on improving printing quality through different approaches by controlling the relevant printing parameters, but the printing mechanism and quality control were still challenges for manufacturing functional structures.

One of the potential applications for printed metal structures is radiation shielding. The electronic components and circuit board used in a high-radiation environment always cause operational errors of the system because of heavy intensity radiation. X-ray absorbed metals are good selections for radiation-hardening techniques, which requires special manufacturing processes and tailored electronic design. Furthermore, it was difficult to realize particular part radiation protection, because the micro/nano fabrication of given shape metal is complex. Fortunately, with the development of nanotechnology, many metal nanoparticles were synthesized and commercialized. These nanoparticles can be made into nanoink for printing with a particular additive solvent to generate different patterns on the substrate.

In this paper, silver micro-structures in dots, lines and other designed patterns were fabricated by e-jet printing. The influence of voltage, pulse width, frequency, shifting speed and stand-off distance on the size of printed results were discussed. With multiple printing processes, multi-layer silver lines were generated, which showed good absorption of X-ray by characterizing them with X-ray topography imaging. The printed silver structures are good candidates for radiation shielding in electronics and circuit board.

## 2. Methods and materials

### 2.1. E-jet printing system

The schematic e-jet printing system is shown in figure 1(a). A nozzle with a micro-needle in the tip is employed to contain the ink. There is a shifting platform (Aerotech A3200), which moves freely in *xyz* directions to control the designed patterns. A high voltage generated by the wave generator and amplifier is applied between the metal needle and the plate electrode. The substrate is on the electrode, which moves by program control in *xy* plan. The distance between the needle tip and substrate is adjusted by moving nozzle positions along *z* axis.

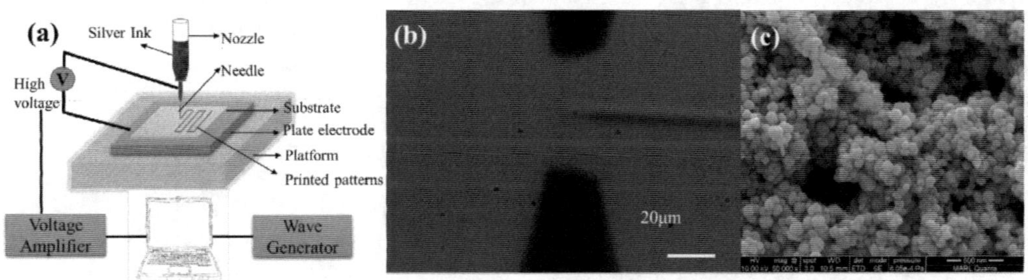


**Figure 1.** (a) Schematic of e-jet printing. (b) Optical image of the needle of the printing process. (c) Scanning electron microscope (SEM) image of printed silver dots.

The whole system is set on a flat stable platform to avoid vibration. In order to get the printing results in real-time, a charge-coupled device (CCD) camera coupled with a lens system is added in the system [25]. Figure 1(b) shows the needle shape and size, as well as the printed line. Different size patterns are generated by controlling the stand-off distance and needle size. In the printing process, the nanoink is added to the nozzle, then it is jetted by the high voltage, then deposited on the substrate, resulting in functional patterns. The nanoink used in this research is silver nanoink from Sigma Aldrich, in which the ~50 nm silver nanoparticles are solved in the solvent Triethylene glycol monomethyl ether (TGME) at 50 *wt*%.

### 2.2. Characterization

The printed patterns were characterized by optical microscope (Hirox RH-2000). The microscope obtained both the surface morphology and film thickness. The largest magnification was 5000X. A scanning electron microscope (SEM, FEI Quanta 250) was employed to estimate the details of printed patterns. With the characterization results, the printing control and patterns optimization were realized. An X-ray imaging system (figure 7(a)) was employed to characterize the radiation absorption property. It includes X-ray source, rotating platform, focus system, detector, and image process program. The X-ray topology image was recorded by the detector, which shows different gray scale to estimate the absorption ability of X-ray [26].

## 3. Results and discussions

The control of parameters in the e-jet printing process is significant to get designed patterns. By altering these parameters, the target results change in size and shape. Voltage is one of the important parameters in e-jet printing. The nano ink is jetted by electrical force in the electrical field generated by high voltage. Different from gas-driven jet printing, e-jet printing is better to control the droplet by voltage. Direct current (DC) [27] and alternating current (AC) [28] were both employed to drive the printing process in the previous reports. AC voltage is controllable by frequency, positive and negative value, pulse width, resulting in charges removing of nanoink, which benefits on printing results. An amplified AC voltage is applied with variable voltage amplitude, frequency and pulse width. Figure 1(c) shows the SEM of the printed dot, which is composed of dense silver nanoparticles, indicating that the stability of nanoparticles is not destroyed in the e-jet printing process.

Figure 2(a) shows the optical microscope image of printed silver dots with different voltages. The size of dots becomes large as the increase of voltage amplitude. By the curve of dots areas as a function of printing voltage shown in figure 2(b), the change rule meets the function of $S$=-0.0073$u^2$+17.976$u$-8598.2. In the equation, S is the mean area of the dots, $u$ is the voltage. It indicates that the dot diameter linear changes as the voltage altering. The area of the dots represents the volume of the droplet of e-jet printing. With large voltage, the electrical field force becomes large, resulting in large quantity size volume. There is another advantage of high voltage used in ink-jet printing, good stability of the printing process. The program can control the voltage and set a given value, which leads to regular and high-quality printing structures. We can see from figure 2(a) that, while the voltage reaches 900 V, there is a tip of each dot. That's because the ink split under high voltages, resulting in small satellite droplets and tip-dots on the substrate.

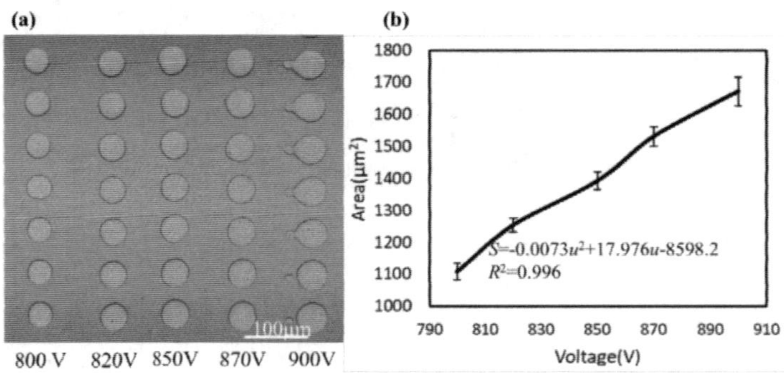


**Figure 2.** (a) Optical microscope image of printed dots with different voltages. (b) The dots areas as a function of voltages.

By changing the stand-off distance, the size of the dots is changed gradually, shown in figure 3. The stand-off distance here is the distance between the needle tip and the substrate, which is usually 10-100 μm. The mean area of dots changes from 2250 $\mu m^2$ to 420 $\mu m^2$, as the stand-off distance changes from 10 μm to 45 μm, respectively. When the stand-off distance becomes short, the electrical force applied to the droplet decreases. Thus, small volume droplet is generated, and deposited on the substrate, resulting in small dots. The needle and printing results are both in micrometer level, so that it is important to control the stand-off distance accurately. With the application of e-jet printing in flexible electronics, the substrate surface is no longer flat, such that the program should adjust the constant stand-off distance. The selection of stand-off distance is also affected by the size of the needle. In our experiment, a needle with ~28 μm is used to generate dots and lines.

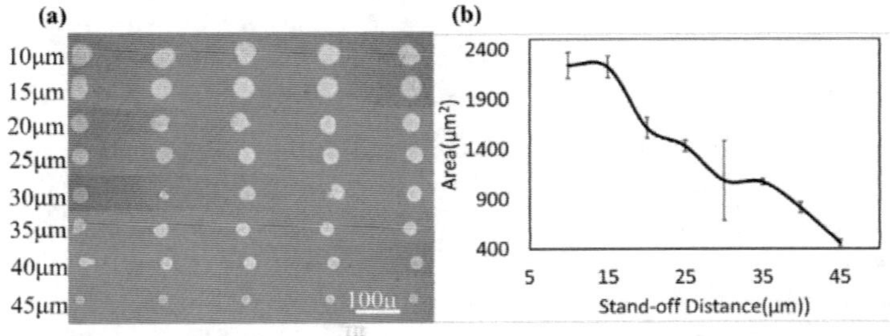


**Figure 3.** (a) Optical microscope image of printed dots with different stand-off distance. (b) The dots areas as a function of printing stand-off distance.

Pulse width is the time period in the open status of the voltage switch, which allows the supply of high voltage for printing. It affects droplet volume, resulting in size control of printed structures. The printed lines and dots at different pulse width are shown in figure 4. By controlling printing speed, we obtained lines (0.2 mm/s) and dots (4 mm/s). The results show an increase of line width and dot diameter with the increase of pulse width. With very large pulse width, there are also tip-dots patterns because of the droplet split, shown in figure 4(b).

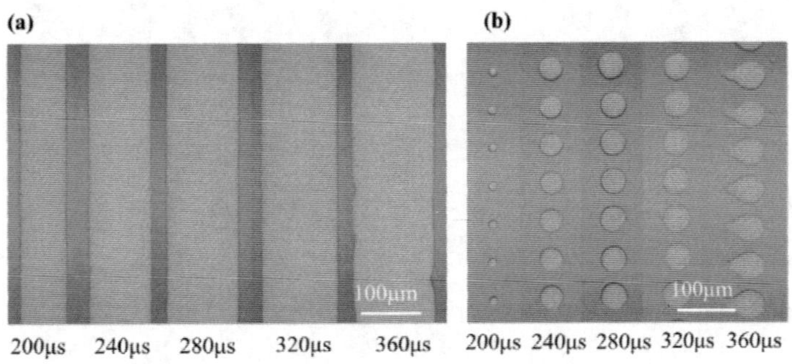


**Figure 4.** Optical microscope image of printed silver (a) lines and (b) dots with different pulse width.

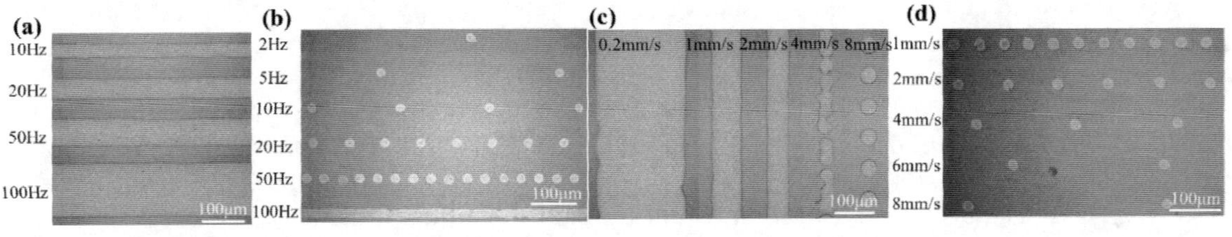


**Figure 5.** Optical microscope image of printed silver (a) lines and (b) dots with different frequency, (c) lines and (d) dots with different printing speeds.

Lines and dots are generated by changing the printing speed or frequency. It affects the distance of two nearby dots by the time period of droplet deposited onto the substrate. Once a small frequency or large speed are applied in printing, separate dots are printed, shown in figures 5(b) and 5(d). With the increase of frequency and decrease of speed, the dots become close to each other, forming a dense dots arrays. Due to the constant voltage, pulse width and stand-off distance, the size of the dots don't change. Lines are fabricated while two or more adjacent dots overlap with each other, shown in figures 5(a) and 5(c). At this condition, the width of lines is changed with frequency and speed. The droplet is large and aggregates together during the printing process, which results in different widths of lines at a low solidified rate of nanoink. The shifting speed is changed by moving the platform in our experiment, where the electrical field is also changed. Many defects will appear at high moving speed during the printing process, especially in manufacturing some complex functional patterns.

With computer design, many functional patterns are realized by e-jet printing. Figure 6(a) is a micro-glass composed of silver lines and circles. The linewidth is about 30 to 35 μm. It is noticeable the linewidth of the circle is larger than the line, which is caused by the circle route of printing at a small speed. With the spread of nanoink deposited on the substrate, the outline of the circle becomes wider. The line patterns connected together as circuit candidates in electronics are shown in figure 6(b). Both circles and lines are fabricated by e-jet printing with silver nanoink, which indicates the good printing performance in manufacturing most functional patterns on the substrate. The letters of 'QUST' grouped by micro silver dots are generated in figure 6(c). The amplified optical image is shown in figure 6(d). The diameter of the dot is 36 μm, almost the same as the diameter of the needle. The dot is very uniform and regular, which is the high-quality results of optimizing printing parameters.

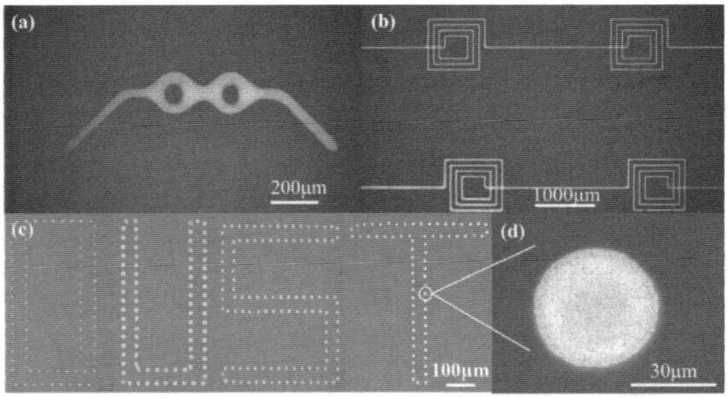


**Figure 6.** Optical microscope image of printed designed silver patterns. (a) micro-glass, (b) circuit, and (c) letters with dots. (d) is the magnified image of one dot in (c).

Multi-layer silver lines are printed for X-ray absorption analysis by multiple e-jet printing. The lines were deposited layer by layer at a small shifting speed. The previous layer became solid partly while the new layer deposited. Thus, high density and thickness silver line are generated to form the X-ray image in the system of figure 7(a). A 30 kV, 1000 μA X-ray source was employed for scanning the samples. The exposure time was 6500 ms to get a high-quality image. The detector with resolution 3072×3888 was used to get the X-ray transmitted data. Each pixel has position information with grey value to indicate the absorption ability of X-ray. Figure 7(b) shows the amplified optical image of silver lines 1, 10, and 50 layers. The mean width is 18.64 μm, 19.56 μm and 28.43 μm, respectively. The width difference is caused by the spread of printed nanoink. The more number of layers are, the wider of the lines generate. The different layers represent single layer line, medium layer line and thick layer line. Furthermore, other different layers between 1 and 200 layers can also be printed according to the requirements.

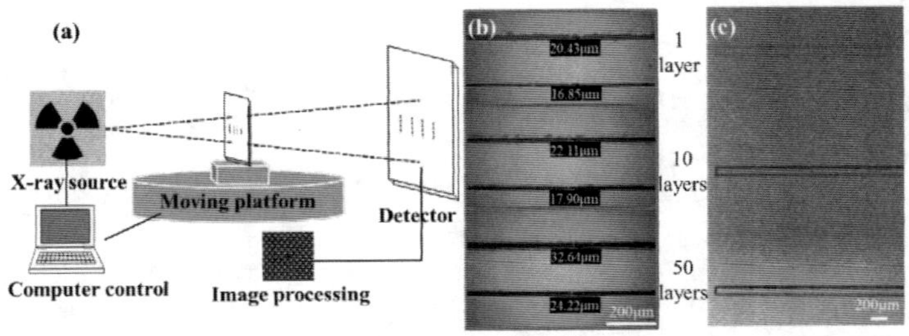


**Figure 7.** (a) Schematic of the X-ray imaging system. (b) Optical microscope and (c) X-ray image of multi-layer silver lines.

The X-ray image is shown in figure 7(c). The lines are clear with a different gray value which is obtained with image processing in Matlab. The gray value ratio $G=I/I_0$, where $I$ and $I_0$ are gray value of the substrate and printed silver lines, separately. The dark color in the image refers to the small gray value. Thus, a small gray value ratio means high X-ray absorption ability. With the printing layer increase, the color of the lines in the X-ray image becomes dark. It is due to the larger absorption rate of X-ray of 50 layers than 10 layers and a single layer. The X-ray shielding efficiency is determined by the density and thickness of the structure. In e-jet printing, larger number layers result in higher density and more thickness. They have much more atoms at the direction X-ray passed through, leading to better absorption ability. However, it is still a challenge to get sufficient thick layer for radiation shielding, because of the flow of ink during the printing process. In the future, we will focus on building several grating masks and print silver nanoink into them to enhance the X-ray shielding ability.

## 4. Conclusion

To summarize, we have printed functional silver micro-structures of different typical dots and lines using an e-jet printing. The influence of printing parameters on results was discussed. With high voltage, large pulse width or small stand-off distance, the size of dots increase. By increasing frequency or decreasing shifting speed, the distance between dots becomes short, then the structures are getting into lines from dots. The printed silver structures show good absorption of X-ray by an imaging system measuring, resulting in application potentials in electronics radiation shielding. The research provides a significant method of fabricating functional micro-structures of e-jet printing and could pave a way for the applications in electronic radiation shielding.

## Acknowledgments

Project supported by the National Natural Science Foundation of China (Grant numbers 11874232), the International Cooperation Program for Excellent Lectures of 2017 by Shandong Province Education Department, the Key research and development plan in Shandong Province (Grant number 2018GGX101008), and the Doctoral Found of QUST (Grant number 010022936).

## References

[1] Hong S, Zhu J and Mirkin C A J S 1999 *Science* **286** 523-5
[2] Sirringhaus H, Kawase T, Friend R, Shimoda T, Inbasekaran M, Wu W and Woo E 2000 *Science* **290** 2123-6
[3] Oyama S T, Clark P, Wang X, Shido T, Iwasawa Y, Hayashi S, Ramallo-Lopez J and Requejo F 2002 *J. Phys. Chem. B.* **106** 1913-20
[4] Boscherini M, Adriani O, Bongi M, Bonechi L, Castellini G, D'Alessandro R, Gabbanini A, Grandi M, Menn W, Papini P, Ricciarini S B, Simon M, Spillantini P, Straulino S, Taccetti F, Tesi M and Vannuccini E 2003 *Nucl. Instrum. Methods Phys. Res. Sect. B* **514** 112-6
[5] Maillé L, Sant C, Le Paven-Thivet C, Legrand-Buscema C and Garnier P 2003 *Thin Solid Films*

**428** 237-41
[6] Jang Y, Kim J and Byun D 2013 *J. Phys. D. Appl. Phys.* **46** 155103
[7] Yudistira H T, Tenggara A P, Nguyen V D, Kim T T, Prasetyo F D, Choi C G, Choi M and Byun D 2013 *Appl. Phys. Lett.* **103** 211106
[8] Brossard F S F, Pecunia V, Ramsay A J, Griffiths J P, Hugues M and Sirringhaus H 2017 *Adv. Mater.* **29** 1704425
[9] Lee D Y, Shin Y S, Park S E, Yu T U and Hwang J 2007 *Appl. Phys. Lett.* **90** 357-65
[10] Kim B H, Onses M S, Lim J B, Nam S, Oh N, Kim H, Yu K J, Lee J W, Kim J H, Kang S K, Lee C H, Lee J, Shin J H, Kim N H, Leal C, Shim M and Rogers J A 2015 *Nano Lett.* **15** 969-73
[11] Cui Z, Han Y, Huang Q, Dong J and Zhu Y 2018 *Nanoscale* **10** 6806-11
[12] Jeong Y J, Lee X, Bae J, Jang J, Joo S W, Lim S, Kim S H and Park C E 2016 *J. Mater. Chem. C* **4** 4912-9
[13] An B W, Kim K, Kim M, Kim S Y, Hur S H and Park J U 2015 *Small* **11** 2263-8
[14] Han Y, Wei C and Dong J J M L 2014 *Manuf. Lett.* **2** 96-9
[15] Park J and Hwang J 2014 *J. Phys. D. Appl. Phys.* **47** 405102
[16] Wei C, Qin H, Ramirez-Iglesias N A, Chiu C-P, Lee Y-S and Dong J 2014 *J. Micromech. Microeng.* **24** 045010
[17] Wei C, Qin H, Chiu C-P, Lee Y-S and Dong J 2015 *J. Manuf. Syst.* **37** 505-10
[18] Qin H, Cai Y, Dong J and Lee Y-S 2017 *J. Manuf. Sci. E-T. ASME* **139** 031011
[19] Hoseinzadeh S, Ghasemiasl R, Bahari A and Ramezani A H 2018 *J. Electron. Mater.* **47** 1-8
[20] Hoseinzadeh S, Ghasemiasl R, Bahari A and Ramezani A H 2017 *J. Mater. Sci-Mater. El.* **28** 14855-63
[21] Ramezani A H, Hoseinzadeh S and Bahari A 2018 *J. Inorg. Organomet. P.* **28** 847-53
[22] Kim Y H, Kwon S, Lee J H, Park S M, Lee Y M, Kim J W 2011 *J. Phys. Chem. C* 115 6599-6604
[23] Coppola S, Vespini V, Grilli S, Ferraro P J 2011 *Lab. Chip.* **11** 3148-52
[24] Yang S M, Lee Y S, Jang Y, Byun D and Choa S-H 2016 *Microelectron Reliab.* **65** 151-9
[25] Qin H T, Dong J Y and Lee Y S 2017 *Robot Cim-Int. Manuf.* **43** 179-87
[26] John M and Boonea J A S 1997 *Med. Phys.* **25** 1661-70
[27] Zou W H, Yu H B, Zhou P L and Liu L Q 2019 *Mater. Design* **166** 107609
[28] Nguyen V D, Byun D 2009 *Appl. Phys. Lett.* **94** 173509

IOP Conf. Series: Materials Science and Engineering **668** (2019) 012025 doi:10.1088/1757-899X/668/1/012025

# Microstructre and mechanical properties of Mg/Al joints by pulse MIG assisted induction pre-heating with composite interlayers

**H C Zhang[1,5], L Y Li[2], G D Zhang[3], Z H Hu[4] and Y L Chen[4]**

[1]School of Rongcheng, Harbin University of Science and Technology, Rongcheng 264300, Shandong Province, People's Republic of China
[2]School of Mechanical and Automotive Engineering, QiLu Univesity of Technology, Jinan 250353, Shandong Province, People's Republic of China
[3]School of Mechanical Engineering, Shandong University of Technology, Zibo 255000, Shandong Province, People's Republic of China
[4]Shandong Value Robotics Intelligent Technology Co., Ltd, Shandong Province, People's Republic of China

E-mail: hongchangzhang123@163.com

**Abstract.** Butt Joints of AZ31 magnesium alloys and Al1060 aluminium plates were produced by conventional MIG welding process with composite interlayers composed of zinc, copper and aluminium foil and induction pre-heating method was used to promote the melting degree of the high melting interlayers, which achieved the connection of dissimilar materials effectively by comparison via SEM and tensile test. The results indicated that crack-free Al/Mg butt joints were obtained. The analysis of the microstructure of the joints showed that the phase composition at the interface between base metal and fusion zone related to the sequence of the different foils. The tensile strength of the joints could reach 21 MPa and the fracture located at the interfacial zone which was mainly composed of brittle Al-Mg compounds.

## 1. Introduction

Aluminum alloys is widely used in aerospace, ship industries and automotive industries because of have its high strength and combination mass reduction [1,2]. Magnesium has a greater potential for weight reduction than other lightweight alloys due to its lower density [3,4]. Therefore, the joining of magnesium to aluminum become a key technology for industrial application of these dissimilar metals joint in the future [5,6].

The main problem of joining Mg and Al is that the very brittle Mg-Al intermetallic compounds(IMCS) formed at the interface between fusion zone and magnesium base metal, which can deteriorate the mechanical properties of the joints [7,8]. So many researchers have attempted to use brazing [9], friction stir welding [10,11], laser-TIG hybrid welding processes to joining the two different metals [12]. It is difficult to use conventional fusion welding process to join the two metals directly for the macroscopic crack usually appear after completion of the welding. However, using fusion welding process to join the two metal also is attractive for its low cost and convenient manipulation. According to Mg-Zn, Al-Cu binary phase diagrams and a series of previous researches [13-16],, Al/Cu/Zn composite interlayers was selected to reduce the brittle Al-Mg IMCS. In this paper, pulse MIG assisted induction pre-heating with Al/Cu/Zn composite interlayers was used to joint dissimilar aluminum and magnesium alloy and the microstructure and prosperities of the joints were researched.

## 2. Experiment

In order to exclude the influence of other alloying elements, the materials used for welding are AZ31 and Al1060 sheets with the thickness of 3 mm. And the chemical composition of of the experimental materials was shown in table 1. The dimensions of the plates were 50 mm×200 mm×3 mm and I-groove was selected as shown in figure 1.

**Table 1.** The chemical composition of Mg and Al alloy.

| Element(wt.%) | Al | Zn | Mn | Si | Cu | Mg |
|---|---|---|---|---|---|---|
| Al 1060 | 99.6 | 0.05 | 0.05 | 0.2 | 0.05 | 0.05 |
| AZ31 | 3 | 1 | 0.5 | 0.08 | 0.01 | 94.6 |

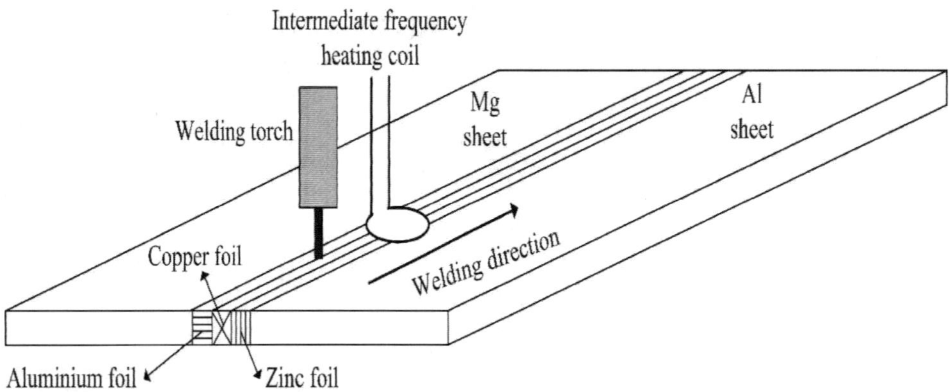


**Figure 1.** The schematic illustration of the MIG welding process

Welding process were finished with conventional pulse MIG welding power source (ADP 350) and a medium frequency induction coil was placed upon the assembled sheets up to about 150°C to preheat the base metal and interlayer, which could promote the fusion of the high melting point interlayer metal . ER4043 Al-Si wire was selected as filler metal and the welding speed was 6.6 mm/s, the welding basic current was 20 A, the voltage of induction pre-heating was 200 V. In the experiments,the sequence of zinc, copper and aluminum foil was changed to investigate its effect on the joint properties. Label "ZCA" represented joint made with the interlayer which the zinc foil contact with magnesium base metal and aluminum foil contact with aluminium base metal and copper foil was placed at the center. On the contrary, label "ACZ" represent the aluminum foil contacted with magnesium alloy during welding.

Scanning electron microscopy (SEM, TESCAN VEGA II) with energy dispersive spectrometer (EDS) were used to observe the cross-section microstructure and fracture surface of the tensile specimens. The tensile specimens were cut from weld to undergo tensile strength tests on an electronic tension machine (Instron 5967) so as to estimate the mechanical property of the different types joint.

## 3. Results and discussion

### *3.1. Weld appearances*

Figure 2 shows the surface appearances of Mg/Al joints made by this novel process. The surface appearance of the weld "ZCA" was shown in figure 2(a) and the surface appearance of the weld "ACZ" with the same processing parameters was shown in figure 2(b). The weld formation was good and obvious burning loss of magnesium element could be seen from the surface of magnesium alloy near the weld.

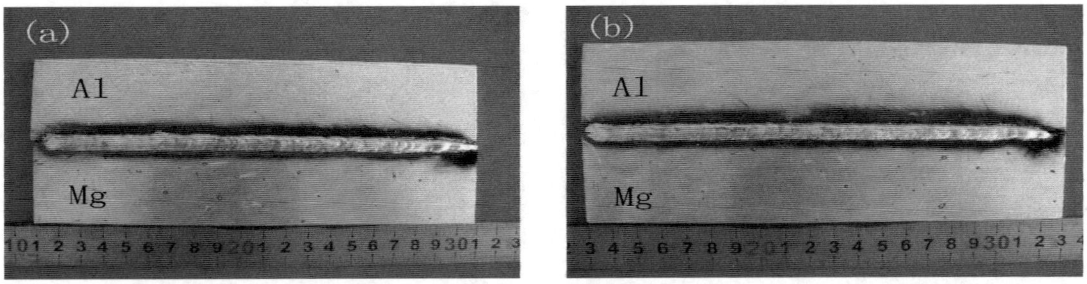


**Figure 2.** Surface appearances: (a) of the weld “ZCA” and (b) the weld “ACZ”

### *3.2. Microstructure of dissimilar metals joints*

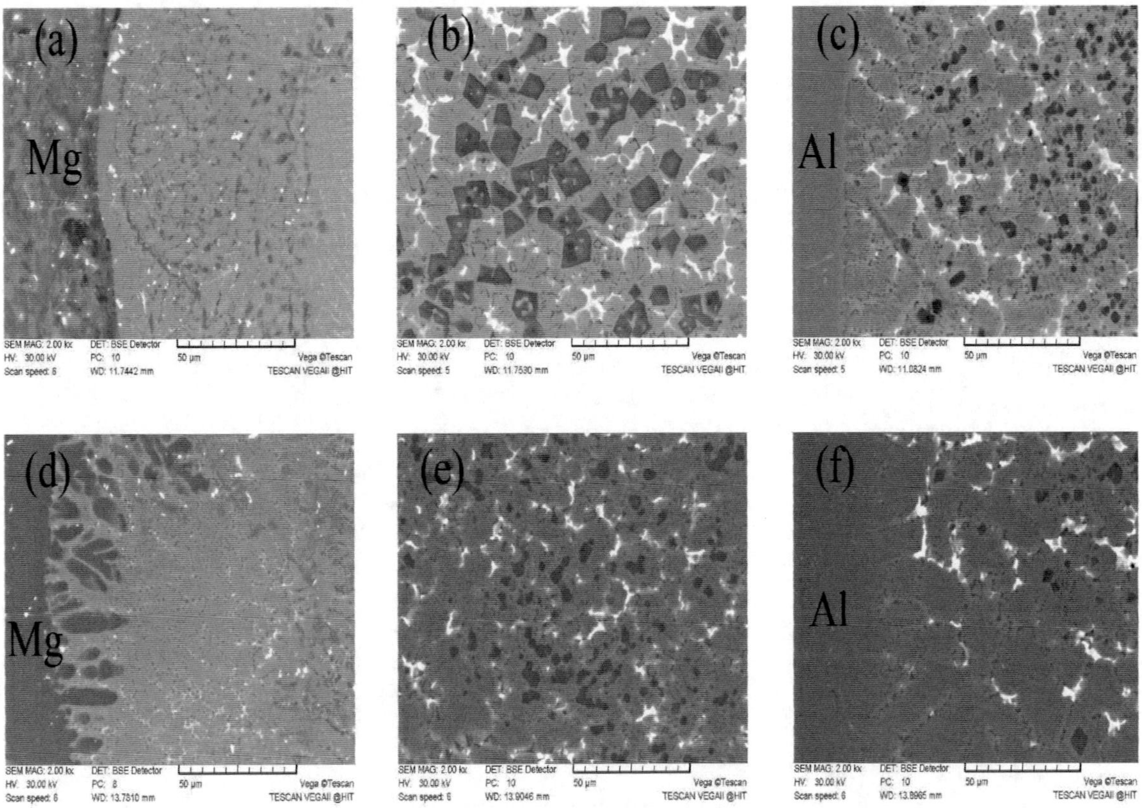


**Figure 3.** The SEM images of joints in different areas

Figure 3 shows the scanning electron microscope image (SEM) photos of the different types’ joints in different areas. As to weld “ZCA”, the SEM microstructure of joint cross sections and XRD results of it were illustrated in figures 3(a)-3(c) and figure 4, respectively. And the obvious interfacial compound layer between magnesium and the fusion zone as shown in figure 3(a); there are no obvious micro-cracks between the compound layer and the magnesium side. The chemical constitution of compound layer marked in figure 3(a) was analyzed by XRD. This intermetallic compound layer was composed of $Al_{12}Mg_{17}$ phase near magnesium base metal. Some MgO crystals with irregular shape could also be found at the center of fusion zone with white $Mg_2Si$ phase distributing in its interior as shown in figure 3(b). The main phase in the center and the zone near aluminum side was Al–Cu eutectic structure zone in figures 3(b) and 3(c).

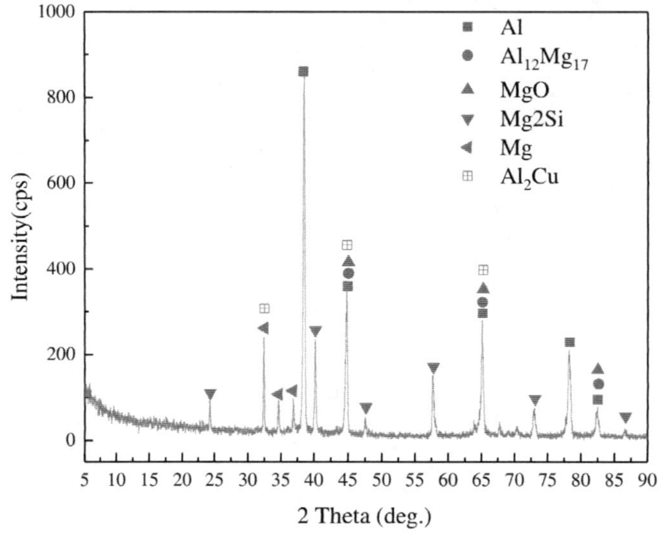


**Figure 4.** X-ray diffraction patterns of joints of ZCA

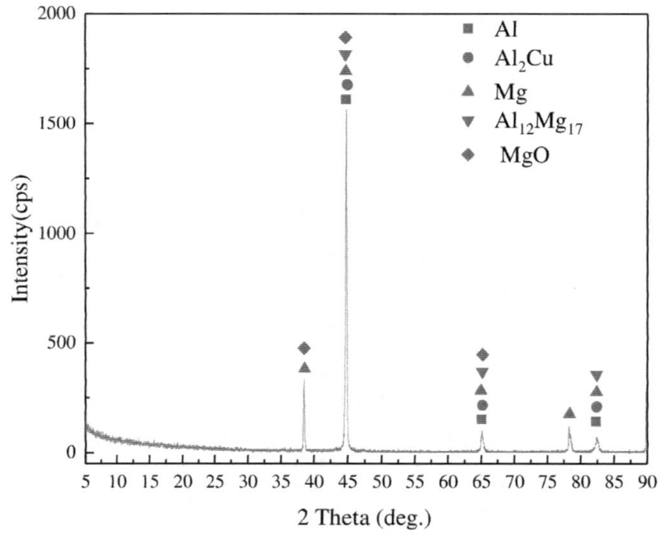


**Figure 5.** X-ray diffraction patterns of joints of ACZ

"ACZ" Joint was made when aliminum foil contacted with magnesium base metal during welding. And the SEM microstructure and XRD results of joints were sketched in figures 3(d)-3(f) and figure 5, respectively. The microstructure of fusion zone near magnesium base metal was very different with that of "ZCA" joint. Some disperse Mg base solid solution with aluminum atoms dissolved in it could be seen near magnesium side and mass of $Al_{12}Mg_{17}$ phase were distributed around the Mg base solid solution as shown in figure 3(d). In the fusion zone there was mass of Al–Cu eutectic structure and some regular shape of MgO crystals in it as shown in figure 3(e). The main phase near aluminum side was Al-Cu eutectic structure in figure 3(f).

### *3.3. Tensile property and fracture behaviour*

The fracture surface of the two different joints presents brittle fracture as shown in figure 6. In tensile tests, fracture occurred at the edge of the fusion zone which was composed of Mg–Al intermetallic and the average tensile strength of the joints could reach 21 MPa in figure 7 when the zinc foil was put near magnesium. However, the average tensile strength of the joints without interlayer reached 13 MPa.

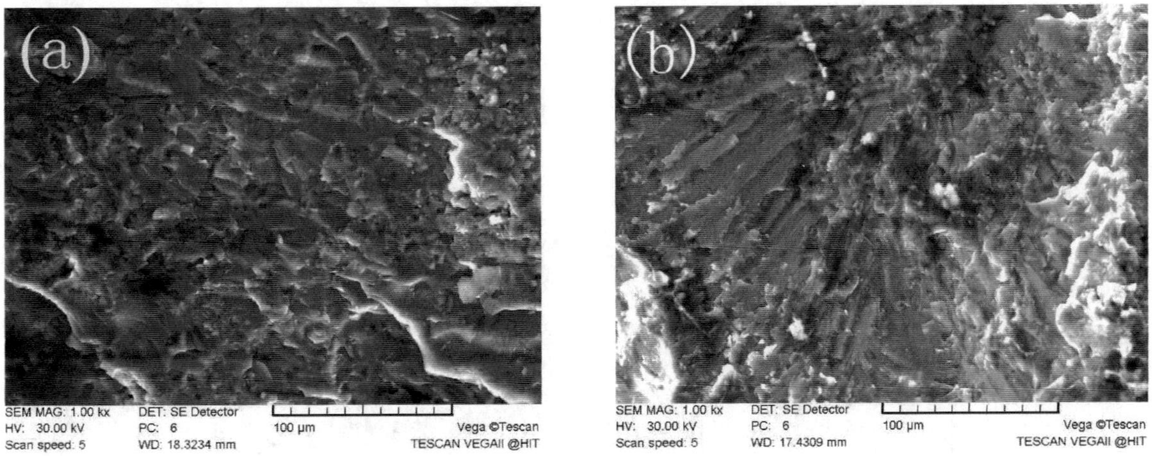


**Figure 6.** The SEM images of fracture surface

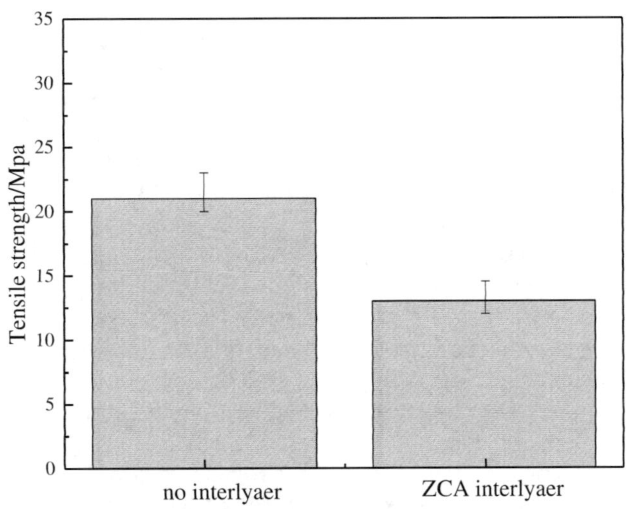


**Figure 7.** Tensile strength of welded joints with different interlayer

It was a novel technology using composite interlayers to join aluminum and magnesium with oinduction pre-heating by conventional MIG welding process. However, the brittle Mg-Al intermetallic could not be restrained effectively by the Al-Cu-Zn composite interlayer. There was no Mg-Zn intermetallic compound generated in the joint for the evaporation of zinc foil during welding by XRD and a better interlayer should be found to improve the mechanical property of the Mg-Al joints by comparison and this give a reference to join aluminum to magnesium by conventional fusion welding process.

## 4. Conclusions

Dissimilar metals joints between Al1060 and AZ31B magnesium alloy were made by conventional MIG welding process with composite interlayers assisted in inductionpre-heating method. Major conclusions of this study could be summarized as follows: butt joints between AZ31 and Al1060 can be successfully jointed by conventional MIG welding process with composite interlayers. The quantity of Al–Mg intermetallic compounds phase can be decreased obviously for the Al-Cu eutectic structure generated in the fusion zone and the tensile strength of the joint could reach 21 MPa. The existence of the brittle Mg-Al intermetallic compounds determined the tensile strength of the dissimilar metals joint and the joints all fractured at the brittle zone which was composed of Al-Mg intermetallic compound phase.

**Acknowledgments**
The research was financially supported by Yantai Key R & D Program (2018YT06030485).

**References**

[1] Jata K V, Sankaran K K and Ruschau J J 2012 Friction-stir welding effects on microstructure and fatigue of aluminum alloy 7050-T7451 *Metall. Mater. Trans. A.* **31** 2181-92
[2] Xu X, Liu D, Ni A *et al* 2017 Effects of pre-corrosion on mechanical properties of 7B50-T7751 aluminum alloy in sodium chloride solution *Mater. Corros.* **69** 870-80
[3] Wu G, Chan K, Zhu L *et al* 2017 Dual-phase nanostructuring as a route to high-strength magnesium alloys *Nature* **545** 80-3
[4] Hao F C, Cui Z Q and Wang W X 2017 Effects of spark plasma sintering temperature on microstructure and mechanical properties of ZK61 magnesium alloy *T. Mater. Heat Treat.* **38** 34-40
[5] Zhang Z, Wang S, Zhang H *et al* 2017 Corrosion behavior of joints of Mg-alloy AZ31 fabricated by friction stir welding *J. Chin. Soc. Corros. Rrot.* **37** 117-25
[6] Xu Z, Zhengwei L, Lei X *et al* 2018 Reduction of intermetallic compounds in ultrasonic-assisted semi-solid brazing of Al/Mg alloys *Sci. Technol. Weld. Join.* **24** 1-8
[7] Hwang I H, Watanabe T and Doi Y 2007 Dissimilar metal welding of steel to Al-Mg alloy by spot resistance welding *Adv. Mater. Res* **15-17** 381-6
[8] Shi H, Chen K, Liang Z *et al* 2017 Intermetallic compounds in the banded structure and their effect on mechanical properties of Al/Mg dissimilar friction stir welding joints *J. Mater. Process. Technol.* **2017** 359-66
[9] Yang T, Wang K, Zhang D *et al* 2017 Contact-reaction brazing of an AZ31 magnesium/3003 aluminum alloy using a silver-copper-zinc interlayer *J. Mater. Process. Technol.* **249** 531-7
[10] Liu Z, Ji S, Meng X *et al* 2017 A practical strategy for improving the weldability of ultra-thin Al/Mg sheets by ultrasonic-assisted friction stir welding *Can. Metall. Quart.* **2017** 1-5
[11] Mohammadi J, Behnamian Y, Mostafaei A *et al* 2015 Friction stir welding joint of dissimilar materials between AZ31B magnesium and 6061 aluminum alloys: Microstructure studies and mechanical characterizations *Mater. Charact.* **101** 189-207
[12] Liu L and Jiang J 2011 The effect of adhesive on arc behaviors of laser-TIG hybrid weld bonding process of Mg to Al alloy *IEEE T. Plasma. Sci* **39** 581-6
[13] Zhang H T and Song J Q 2011 Microstructural evolution of aluminum/magnesium lap joints welded using MIG process with zinc foil as an interlayer *Mater. Lett.* **65** 3292-4
[14] Zhang H T, Dai X Y and Feng J C 2014 Joining of aluminum and magnesium via pre-roll-assisted A-TIG welding with Zn interlayer *Mater. Lett.* **122** 49-51
[15] Dai X, Zhang H, Zhang H *et al* 2016 Arc assisted ultrasonic seam welding of Mg/Al joints with Zn interlayer *Mater. Sci. Technol.* **32** 1743284715Y
[16] Barekatain H , Kazeminezhad M , Kokabi A H . Microstructure and Mechanical 2013 Properties in Dissimilar Butt Friction Stir Welding of Severely Plastic Deformed Aluminum AA 1050 and Commercially Pure Copper Sheets *J. Mater. Sci. Technol.* **30**(8):826-834.

IOP Conf. Series: Materials Science and Engineering **668** (2019) 012026 doi:10.1088/1757-899X/668/1/012026

# The effects of $R_i/T$ ratio on residual stress of SAPH440 steel welding for auto parts

**L X Chi[1,2,3], S F Liang[1], B L Yan[1] and X Z Gao[1]**

[1]College of Materials Science and Engineering, Chongqing University of Technology, Chongqing 400050, China
[2]Chongqing Municipal Engineering Research Center of Institutions of Higher Education for Special Welding Materials and Technology, Chongqing University of Technology, Chongqing 400050, China

E-mail: chiluxin200195@163.com

**Abstract.** In order to study the effect of $R_i/T$ on welding residual stress in forging presses for Auto parts material, a welding experiment program was developed. The residual stress of thick pipe 10mm joint was measured by the blind hole drilling method. The simulation results within and near the weld joint, the transverse residual stresses are tensile on the inside surface which already close to yield strength with creasing the $R_i/T$ ratio, and compressive on the outer surface; The prediction results from FEM are in a good agreement with the experimental measurement which are important to apply data for manufacturing plant.

## 1. Introduction

Due to a combination of high strength, toughness and good weld-ability, the new material SAPH440 steel has been used for auto parts [1]. The Lower arm and the Upper arm with longitudinal welding are the main equipment in auto parts, the superiority or inferiority of welding quality will directly affect the car normal operation.

Many studies show that the intense heat concentration will make the weld line and its vicinity to undergo severe thermal cycles, which cause non-uniform heating and cooling of the material, thus generated inhomogeneous plastic deformation and residual stress in welded joint [2]. Therefore, understanding the residual stress distribution and improving the welding quality can ensure auto parts normal operation safety. Residual stress investigations at present are mostly focused on circumferential butt welding, Lee [3] and Yaghi [4] employed a three-dimension finite element to simulate the residual stresses distribution in circumferential welds of pipe and parametric studies with inside radius to wall thickness ratio have been presented to investigate the effects of diameter on residual stress, but less focused on different $R_i/T$ ratio. Owing to the residual stress state and distribution will change under own constrain conditions with the model size increasing, in this study, experimental measurement and finite element analysis are adopted to investigate the effects of $R_i/T$ ratio on the temperature filed and the residual stress distribution of SAPH440 steel welding to provide data for manufacturing plant.

## 2. Experimental procedure

### *2.1. Material properties*

The material properties especially at high temperature have been a critical issue in welding simulation, which is able to realize the nonlinear relation between physical parameters and temperatures in heating and cooling process. Thus, thermo-physical properties close to melting point were adopted according to the report [5, 6], the properties of the weld filler material $H08Mn_2Si$ (φ1.2 mm) were assumed to be the same as the base metal. Table 1 shows the chemical composition of SAPH440 steel, the Metal Active Gas Arc Welding conditions in experiment as in table 2, frequency was 143 HZ, and Gas flow was 15 L/min.

**Table 1.** Chemical composition of SAPH440 steel.

| C | Si | Mn | P | S | $Al_x$ |
|---|---|---|---|---|---|
| 0.21 | 0.30 | 1.5 | 0.03 | 0.025 | 0.01 |

**Table 2.** Welding condition in the experiment.

| voltage V | Current (A) | Filter Speed (mm/s) | Protection gas |
|---|---|---|---|
| 40 | 80 | 8 | $80Ar+20CO_2$ |

*2.2. Experiment data measurement*

After the completion of the welding, the thermal cycles were measured by K type thermocouples with spotting welded to the pipe. The strain gauges were used to measure the welding residual stress, the drilled holes diameter was 1.5 mm, and the holes were as deep as 2 mm. The locations and the strain gauges are shown in figure 1.

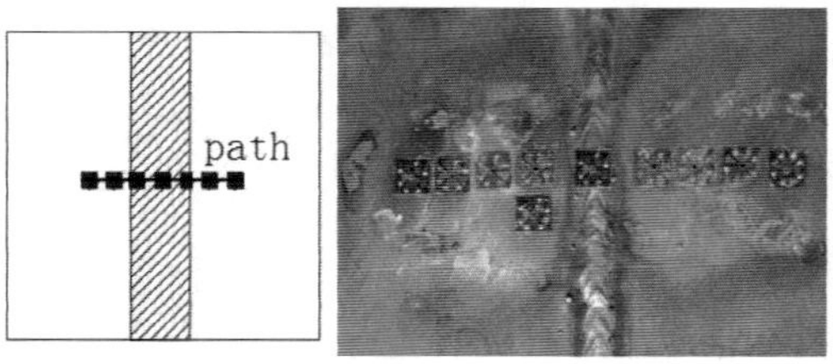


**Figure 1.** Locations of the strain gauges. (a) Location of strain gauges and (b) Rosettes bounded on the outer surface of the pipe.

After measuring released strains along the vertical welded joint, the longitudinal residual stress and the transverse residual stress were calculated according to the following equations:

$$\sigma_{1,2} = \frac{E(\varepsilon_1+\varepsilon_3)}{4A} \mp \frac{E}{4B}\sqrt{(\varepsilon_3-\varepsilon_1)^2+(2\varepsilon_2-\varepsilon_1-\varepsilon_3)^2}$$

$$tg2\theta = \frac{2\varepsilon_2-\varepsilon_1-\varepsilon_3}{\varepsilon_3-\varepsilon_1}$$

$$\sigma_{x,y} = \frac{\sigma_1+\sigma_2}{2} + \frac{\sigma_1-\sigma_2}{2}\cos(2\theta)$$

Where $\varepsilon_1$, $\varepsilon_2$, $\varepsilon_3$ are the released strains of strain gage rosette in three directions, respectively. $\sigma_1$, $\sigma_2$ are the maximum principal stress and minimum principal stress. E is Young's modulus of elasticity, A and B are the strain release coefficient, which can be get from the equipment manufacturer, $E$=2.04*$10^{11}$ Pa, $A$=-0.07255, $B$=-0.1514. $\theta$ is the angle contained by $\sigma_1$ and $\varepsilon_1$, along clockwise, if $\varepsilon_3 \geq \varepsilon_1$, $\theta=\theta$, and $\varepsilon_3 < \varepsilon_1$, $\theta=\theta+90°$.

## 3. Finite element analysis

Welding residual stress was calculated by three-dimension thermal and mechanical finite element method. First, the temperature contours were computed according to the given welding parameters and the boundary conditions. Second, the temperature contours were taken as load in the structural analysis. Meanwhile, the improved substructure algorithm method was adopted to condense internal node for reducing calculation freedom, which was aimed to simplify loading and improve simulation efficiency for pipe.

### *3.1. Finite element model of pipe*

Due to the intense heat concentration, a relatively fine mesh was designed at the weld zone, and element size increased progressively with distance from the weld center line, which reduced total number of nodes. In this study, based on ANSYS software, the simulation of the longitudinal welding has been performed on SAPH440 steel pipe with length of 1000 mm, thickness of 10 mm with different $R_i/T$ ratio is listed in table 3, which have the same mesh density. The pipe was welded by four pass as shown in figure 2.

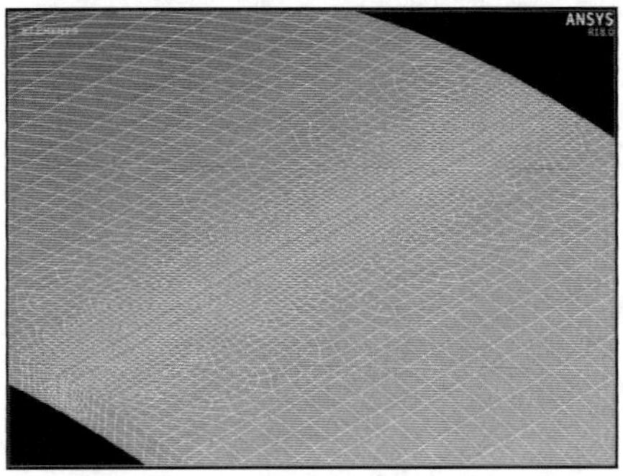


**Figure 2.** Finite element model near welding center line.

**Table 3.** Radial size and ratios in the FE analysis.

| Cylinder wall Thickness, $T$ (mm) | Cylinder inside Radius, $R_i$ (mm) | Inside radius to wall Thickness ratio, $R_i/T$ |
|---|---|---|
| 10 | 100 | 10 |
| | 200 | 20 |
| | 300 | 30 |
| | 400 | 40 |
| | 500 | 50 |

### *3.2. Thermal analysis*

Element birth and death technique was used in FE model to simulate deposit filling, which activated element before analysis by multiplying their stiffness using a severe reduction factor, similarly, mass, damping, specific heat were set to zero for deactivated elements [5]. The heat flux from the moving welding arc was given by:

$$Q = UI\eta / V$$

Where $U$ is the welding voltage, $I$ is the welding current, $\eta$ is the arc efficiency, $V$ is the activates element volume of heat source. The efficiency factor is assumed as 0.76 for the MAG process.

## 4. Results and discussion

### *4.1. Temperature contours*

The distribution and the evolution of the temperature contours were almost the same, therefore, the fourth weld pass of 30 ratio are selected to be introduced as figure 3 with the time at 1s, 50 s, 100 s and 500 s.

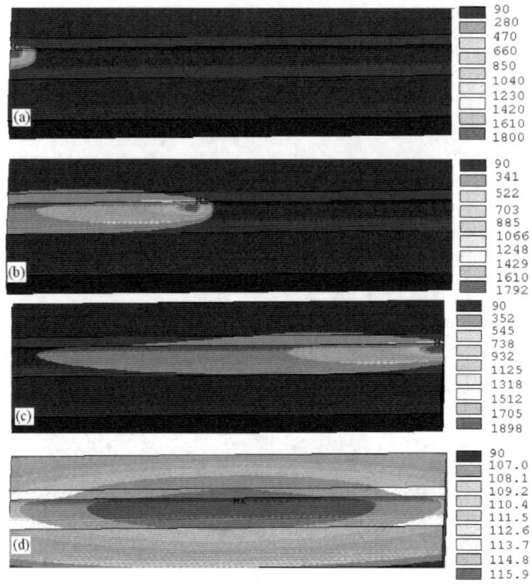


**Figure 3.** Temperature contours of the fourth weld pass. (a) t=1 s; (b) t=50 s; (c) t=100 s; (d) t=500 s.

It can be seen that the transient temperature distribution with the moving arc at different time are almost identical, the maximum temperature in weld pool is assumed to experience 1800℃, the HAZ between 825～1125℃, away from the heat source the temperature is rapidly decreased, it can be concluded that the temperature field is steady along welding direction, as figures 3(a)-3(c). The maximum temperature at the start/stop is higher than other place, which influences the residual stress. The inter-pass temperature is controlled to be lower than 120℃ at the 400s as figure 3(d).

*4.2. Thermal cycles*

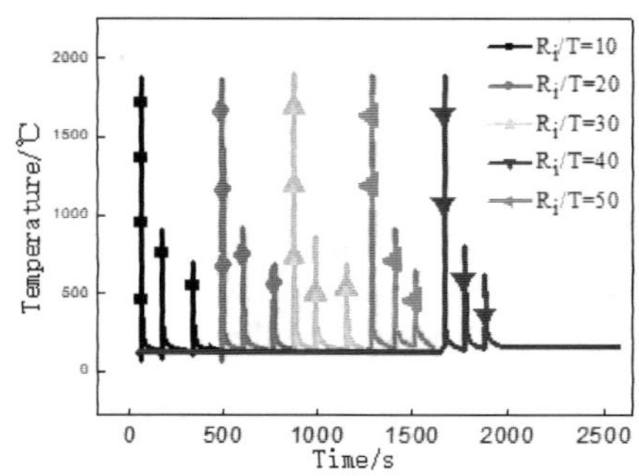


**Figure 4.** Thermal cycle simulation with $R_i/T$ ratio.

The thermal cycle simulation at welded joint is presented for the model with $R_i/T$ ratio to 50 as figure 4.

It is known that the thermal cycles of the five models have a similar trend since the same welding parameters are given to the models along the welding direction.

IOP Conf. Series: Materials Science and Engineering **668** (2019) 012026 doi:10.1088/1757-899X/668/1/012026

### *4.3. Residual stress distribution*

It is critical to establish four paths on the FE model to thorough analysis residual stress distribution as shown in figure 5. Path 1 is vertical weld center line on the outer surface of the pipe. Path 2 is vertical weld center line on the inside surface of the pipe. Path 3 is the weld centre line along the welding direction on the outer surface of the pipe. Path 4 is the weld centre line from the inside to the outer surface of the pipe along thickness. Residual stress distributions of the four paths with different $R_i/T$ ratios are shown in figure 6.

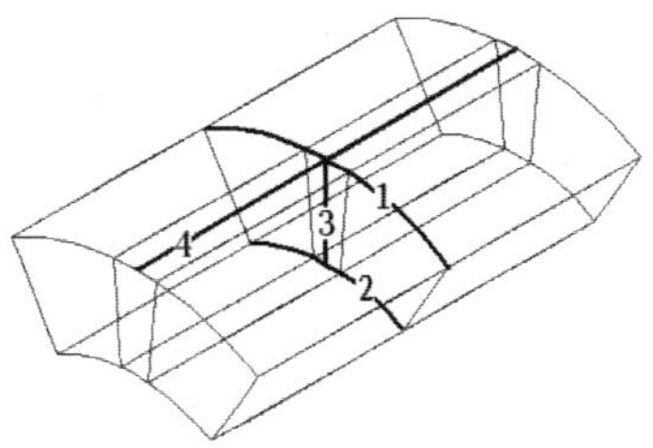


**Figure 5.** Path laying diagram.

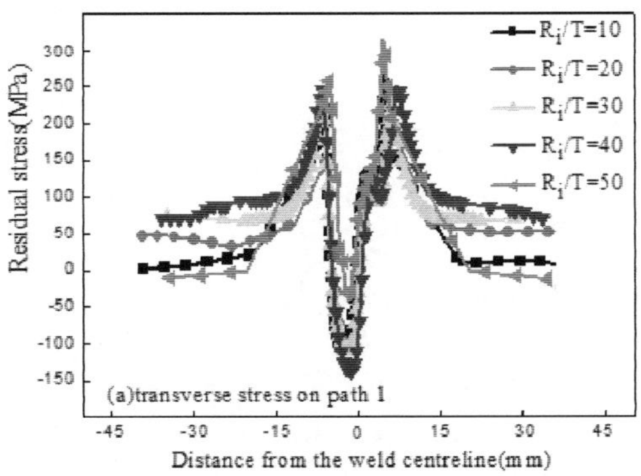


(a)

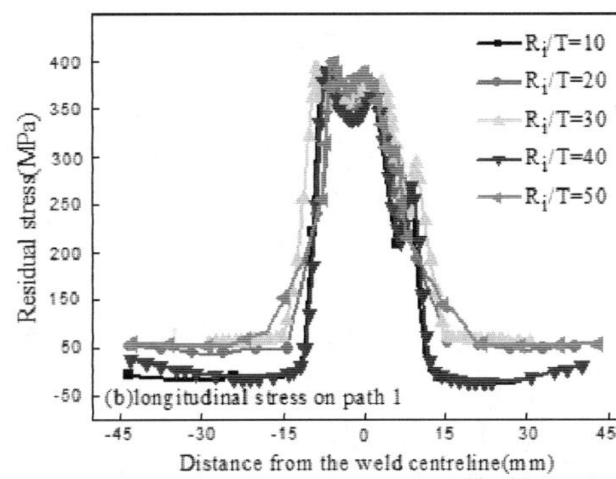


(b)

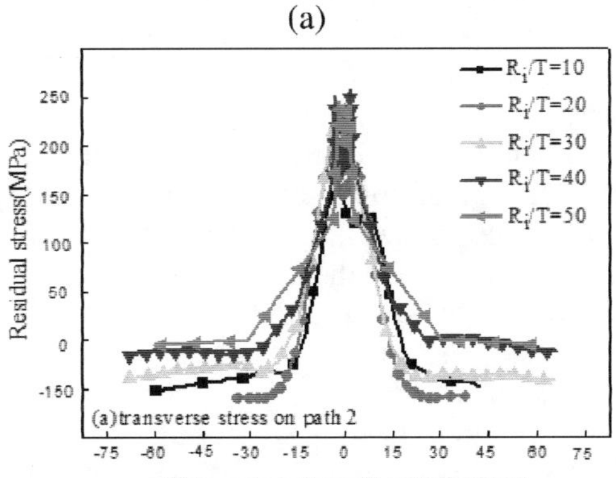


(c)

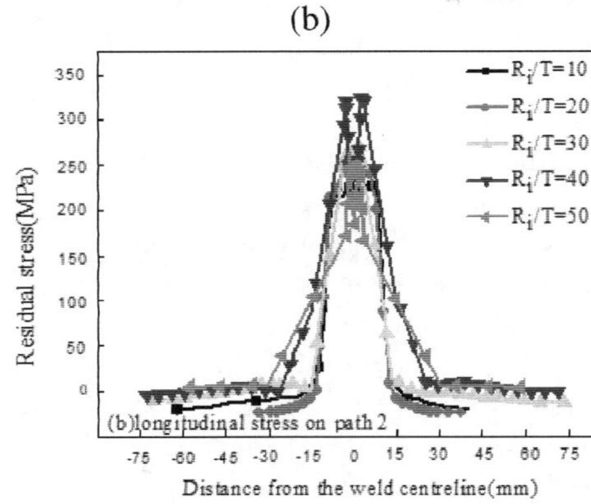


(d)

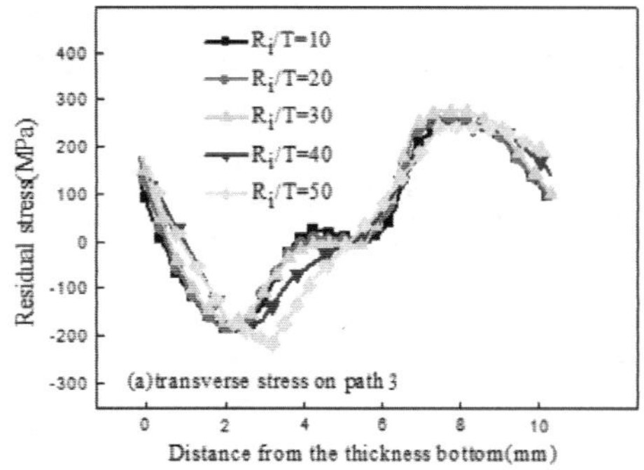


(e)

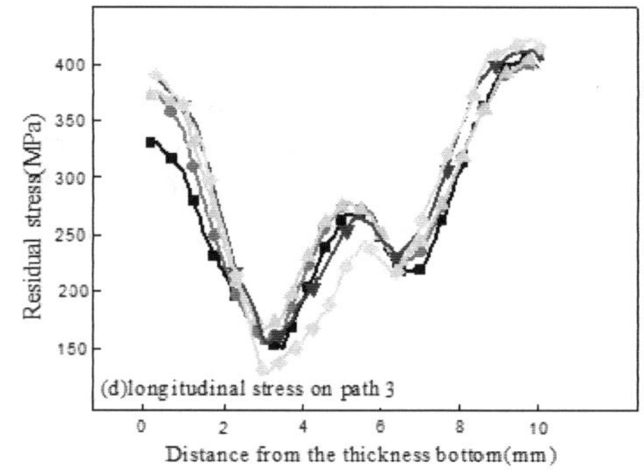


(f)

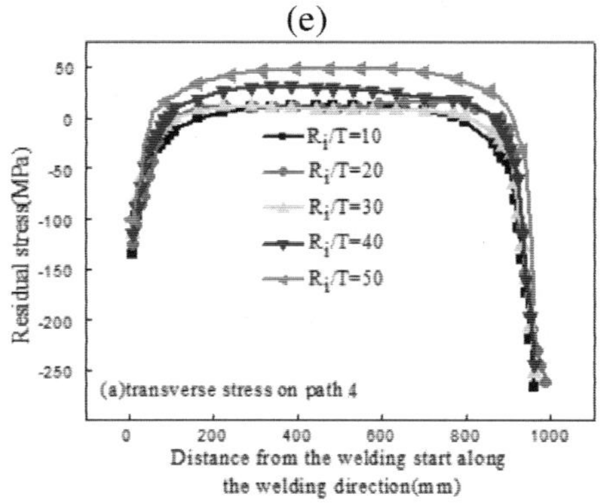


(g)

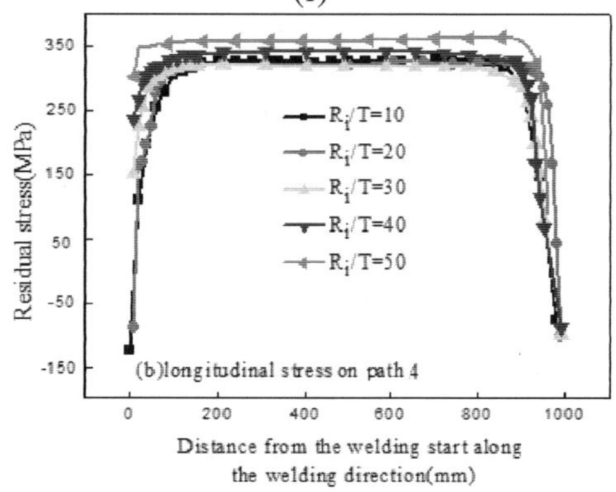


(h)

**Figure 6.** Simulated residual stress along the four paths.

It can be observed that, the transverse residual stresses in the weld and its vicinity have slightly increased with increasing $R_i/T$ to 50, and are all compressive stress on the outer surface, the longitudinal residual stress are mainly determined by the radial expansion and subsequent contraction, tensile stress at outer surface have the same trend when the $R_i/T$ increase to 50 as shown in figures 6(a) and 6(b). From the weld joint to the metal on the inside surface, the transverse residual stress and the longitudinal residual stress have been maintained the maximum tensile stress, which already close to the material yield strength as figures 6(c) and 6(d). Spatial variations are present at start/stop positions due to the moving arc effect along the welding direction. Increasing the Ri/T to 50 lead to a significant decrease of transverse stress from thickness 3 to 6mm, but there is a more spectacular increase at the bottom with the maximum compressive stress, and the maximum tensile stress is about 423 MPa at the position of 2 mm from the outer surface as figures 6(e) and 6(f). From the weld joint to the metal on the outer surface, the transverse residual stresses are changed from compressive stress, tensile stress to compressive stress, and then reache the maximum value 300MPa at the fusion zone, longitudinal residual stresses are changed from tensile stress to compressive stress and achieve the maximum value 532 MPa at weld joint as figures 6(g) and 6(h).

## 5. Verification of FE modeling

To verify the simulation result, the same geometrical model, welding conditions were used in the experiment, the thermal curves and the residual stresses were measured on the outer surface, and the results as figure 7.

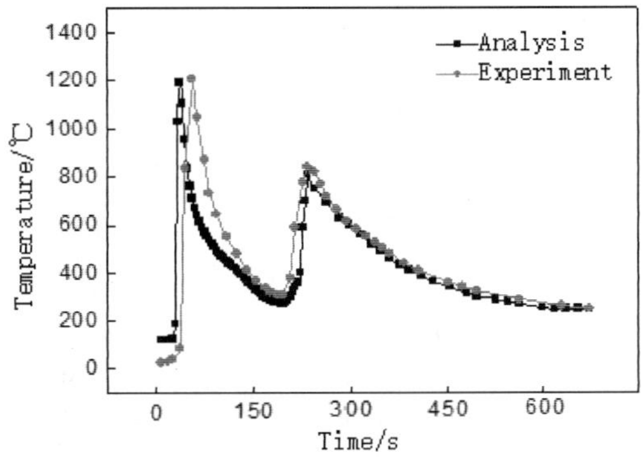


(a)

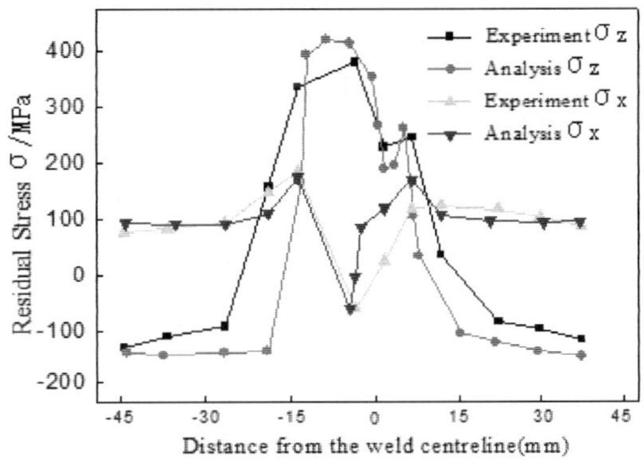


(b)

**Figure 7.** Comparison of experiment and analysis. (a) Comparison of thermal curves and (b) Comparison of residual stresses.

It is clear that the prediction results from FEM are in a good agreement with the experimental measurement. Although there is an error in the lower temperature which has no effect on the organization of the metal.

## 6. Conclusions

In this study, the simulation of residual stress on SAPH440 steel pipe were gained by three-dimensional uncoupled thermo-mechanical FE analysis, meanwhile, the influence of the $R_i/T$ ratio on residual stress was clarified. The following conclusions can be drawn:

- It can be observed that within and near the welding joint, the transverse residual stresses are tensile on the inside surface, and compressive on the outer surface, and the longitudinal residual stresses from the outer surface to inner surface are all tensile with gradual increasing which already close to yield strength.
- The residual stresses in the weld and its vicinity are increased with increasing the $R_i/T$ ratio.
- These observations are important to apply the residual stress results for manufacturing plant.

**Acknowledgments**
This work is financially supported by the Science and Technology Research Program of Chongqing Municipal Education Commission (Grant No. KJ1709197).

**References**

[1] Lee B S, Kim M C, Yoon J H *et al* 2009 Characterization of high strength and high toughness Ni-Mo-Cr low alloy steels for nuclear application *Int. J. Pres. Ves. Pip.* **2009** 1-7
[2] Liu C, Zhang J X and Xue C B 2011 Numerical investigation on residual stress distribution and

evolution during multi-pass narrow gap welding of thick-walled stainless steel pipes *Fusion Eng. Des.* **86** 288-95

[3] Lee C H and Chang K H 2008 Three-dimensional finite element simulation of residual stresses in circumferential welds of steel pipe including pipe diameter effects *Mater.Sci.Eng., A* **487** 864-74

[4] Yaghi A, Hyde T H, Becker A A and Sun W 2006 Residual stress simulation in thin and thick-wall stainless steel pipe welds including pipe diameter effects *Int. J. Pres. Ves. Pip.* **83** 210-8

[5] Sattari-Far L and Farahani M R 2009 Effect of the weld groove shape and pass number on residual stresses in but-welded pipes *Int. J. Pres. Ves. Pip.* **86** 723-31

[6] Deng D, Murakawa H and Liang W 2007 Numerical simulation of welding distortion in large structures *Comput. Method Appl. Mech. Eng.* **196** 4613-27

# Constitutive modeling of dynamic recrystallization and precipitation behavior of Nb and Ti of Q1030 high strength steel

**J J Wang[1,2,3], Y L Kang [1], Y L Liu[2] and H Yu[1]**

[1]University of Science and Technology Beijing, China
[2]Laiwu Branch of Shandong Iron & Steel Co., Ltd., China

E-mail: bandaisuo@126.com

**Abstract:** The thermal deformation and precipitation behavior at 900-1100°C and strain rate of 0.1-5s$^{-1}$ were studied by Gleeble-3800 thermal simulator of Q1030 high strength steel. The activation energy of hot deformation in austenite region was determined by regression method, and the hot deformation equation of the Q1030 high strength steel was established. The critical strain and peak strain of dynamic recrystallization were predicted accurately by fitting the inflection point with cubic polynomial of  curve of Q1030 high strength steel, and relationship between critical strain and Z parameter was established. Finally, the precipitation behavior of Nb and Ti particles during low strain rate deformation was studied, the results show that the precipitated phases in steel are rectangular TiN, quadratic (Nb, Ti) (C, N) carbonitride, elliptical (Nb, Ti) C carbide and NbC. Thermodynamic calculation shows that the order of precipitation of the second phase in steel is TiN, TiC, NbC and NbN.

## 1. introduction

Since the 1970s, the development of low-alloy high-strength steel in the world has entered a new period. Based on the control of rolling technology and microalloying metallurgy, a new concept of modern High Strenth Low Alloy Steels has been formed[1-4].In recent years, a new generation of controlled rolling and controlled cooling(NG-TMCP) technology has been widely used in industrial production. In the control rolling stage, steel is used to undergo large-scale downward deformation in the austenite recrystallization area, and then tiny austenite recrystallized grains are obtained., The deformation is carried out in the austenite non-recrystallization area, resulting in a large number of deformation bands inside the grains. These deformation bands and austenite grain boundaries can provide nucleating points for controlling the phase transition of the cooling process. In the phase transition process, smaller ferrite or bainite tissues are produced, thereby increasing the strength and toughness of the steel[ 5, 6]..

Dynamic recrystallization can change the microstructure of materials and affect their mechanical properties. It is an effective method to control the structure and performance in the process of metal thermal processing. When deformation occurs at high temperatures and high strain rates, the changes in the internal tissues of the metal are very complex, including hardening, dynamic recovery, and dynamic crystallization of austenite during deformation. These behaviors are important factors that determine the microstructure, deformation resistance, and organizational properties of the final material of the deformed austenite[7-8].

For the high-strength structural steel with low cost and high added value, the deformation behavior of austenite becomes more complicated due to the addition of Nb, V and Ti microalloy elements [9-10] .In this paper, the dynamic crystallization of high strength Q1030 steel is studied by means of the thermal simulation tester and the austenite crystallization thermal deformation equation is established. The critical condition of dynamic recrystallization is studied by the cubic polynomial fitting method of the curve. The precipitation behavior of Nb and Ti nitrogen carbides at low strain rate is studied by TEM and EDX, and the theoretical basis is provided for the establishment of a reasonable hot working process.

## 2. Experimental Materials and Methods

The test material was taken from the Q1030 high-strength steel produced by Laiwu Iron and Steel Company's wide and heavy plate production line. The chemical composition of the continuous casting slab after heating by the heating furnace is shown in Table 1. In order to preserve the state of the slabs after heating and prevent the new precipitation during the slow cooling process, the slabs are quenched into water immediately after being cut.

**Table 1.** Chemical composition of test steel (wt%).

| C | Si | Mn | P | S | Al | Cr | Ti | Nb | Mo |
|---|---|---|---|---|---|---|---|---|---|
| 0.055 | 0.18 | 1.18 | 0.016 | 0.003 | 0.042 | 0.51 | 0.120 | 0.042 | 0.21 |

Slab was made into many Φ 8 × 12 mm sample. They were tested on the Gleebe3800 thermal simulator and protected throughout by high purity argon. In order to make the alloy elements contained in the experimental steel as far as possible solid solution, the specimen is heated at a rate of 10°C/S to 1200°C and then keep 3 min warm to get a complete austenitic,then cooled to a different temperature at 10 °C/s. A single pass compression test of true strain 0.7 was performed, and the deformation temperature was 900°C, 950°C, 1000°C, 1050°C, 1100°C, the strain rate was 0.1 s-1, 1 s-1, 5 s-1. In order to prevent the specimen from oxidizing during the deformation process, high purity nitrogen is used as a protective gas, and graphite is coated at both ends of the specimen to reduce friction during the deformation process and prevent the occurrence of belly. The automatic sampling system records the stress and strain data of the compression deformation process. Immediately after the deformation of the sample, it is cooled with water to freeze the high-temperature austenite and precipitates. The deformed specimen was cut along the cross-sectional line to make the metallographic sample, corroded with saturated picric acid solution, and then the tissue was observed under a metallographic microscope. The samples were prepared into thin films and extracted compound specimens, and the microstructure was observed at high multiples by the H800 transmission electron microscope, and the precipitated phase components were analyzed by EDS.

## 3. Experimental results and analysis

### *3.1. Stress-strain curve*

Figure 1 shows the true stress(σ)-strain(ε) curve of Q1030 high strength steel at different deformation temperatures and different strain rates. Figure 1(a) is the rheological stress curve of high strength Q1030 steel at different deformation temperatures when $\bar{\varepsilon}$ =0.1s-1. It can be seen that at this strain rate, the rheological stress curve at a temperature of 950 ~ 1150°C has the characteristics of dynamic crystallized rheological stress curve. In addition, it can be seen that at the same strain rate, the peak strain and peak stress increase with the decrease of temperature.

Figure 1(b) is the rheological stress curve of high strength Q1030 steel at different strain rates at 1000°C. When the strain rate is 0.1 s-1 and 1s-1, the rheological stress curve is still of dynamic recrystallization type; When the strain rate is 5s-1, the rheological stress immediately enters a stable rheological state after reaching its peak, indicating that a dynamic recovery process occurs l at this time. It can be found that for Q1030 high strength steel, when the strain rate is small at the same deformation temperature, dynamic crystallization is easy to occur. With the increase of strain rate, peak stress and peak strain increase, the characteristics of rheological stress curve transition from dynamic recrystallization to dynamic recovery.

In the process of thermal deformation, the two mechanisms of material hardening and dynamic softening work simultaneously. At the beginning of deformation, machining hardening plays a leading role; As the deformation progresses, when the material begins to undergo dynamic crystallization, σ-ε curve is going down, and the dynamic softening mechanism plays a major role at this time; As the crystallization continues, when the hardening and dynamic softening reach equilibrium, the rheological stress σ is a certain stable value, and the material enters a complete recrystallization state.

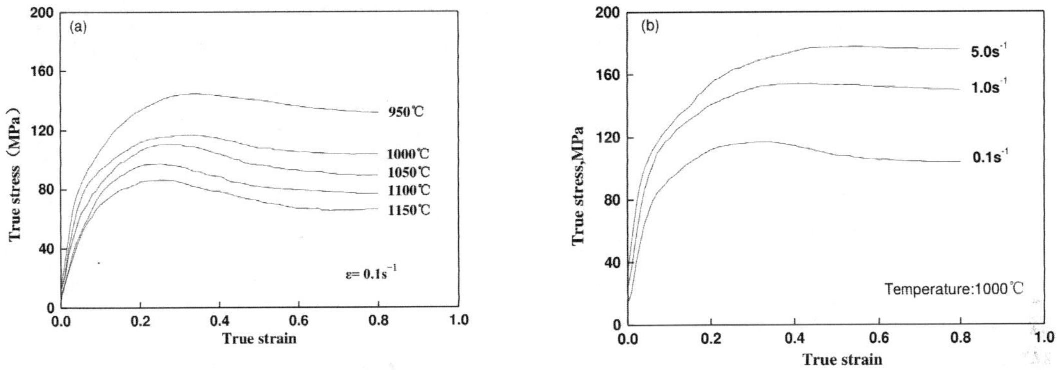


**Figure 1.** Rheological stress curve of Q1030 strength steel at (a) 1s-1 and at (b) 1000°C.

### *3.2. Thermal deformation equation*

Thermal activation energy of dynamic recrystallization of metal materials is related to peak stress ($\sigma_p$), deformation temperature (T) and strain rate ( $\bar{\varepsilon}$). According to the Zener-Hollomon parameter model, during the thermal deformation of nitrogen-microalloyed high-strength steel, the Z(Zener-Hollomon) parameter and the deformation activation energy Qdef, the temperature T and the strain rate $\bar{\varepsilon}$ have the following relationship[11] :

$$Z=\dot{\varepsilon}\exp\left(\frac{Q_{def}}{RT}\right) \tag{1}$$

In the formula, $Q_{def}$ is the thermal deformation activation energy(kJ/mol), reflecting the ease of thermal deformation of the material; R is the gas constant with a value of 8.314J/(mol·K), $\dot{\varepsilon}$ is the strain rate(s-1), and T is the deformation temperature(K). The following relationship exists between peak stress and Z parameters[12] :

$$Z=A[sinh(\alpha\sigma_p)]^n \tag{2}$$

In the formula: A is the material constant and N is the stress index; $\sigma_p$ is peak stress, Single bit is MPa, α is the stress factor, and in carbon steel material, the α value is 0.012[13], and Combine the equations(1) and(2) to get the following equation:

$$\dot{\varepsilon}=A[\sinh(\alpha\sigma_p)]^n\exp\left(\frac{Q_{def}}{RT}\right) \tag{3}$$

Equation (3) describes the effects of deformation temperature and strain rate on the rheological stress of a metal during thermal deformation. Take the logarithm of both sides of equation (3) and take the partial derivative, The expression of heat deformation activation energy $Q_{def}$ can be obtained:

$$Q_{def} = Rn\frac{\partial\{ln[sinh(\alpha\sigma_p)]\}}{\partial(1/T)}\bigg|_{\dot{\varepsilon}}\frac{\partial\, ln\,\dot{\varepsilon}}{\partial\, ln[sinh(\alpha\sigma_p)]}\bigg|_{T} = Rnb \quad (4)$$

When the deformation temperature(T) is certain, $\ln\dot{\varepsilon}$ and $ln[sinh(\alpha\sigma_p)]$ should satisfy the linear relationship, The relation curve between $ln\,\dot{\varepsilon}$ and $ln[sinh(\alpha\sigma_p)]$ is shown in Figure 2(a); Similarly, when the strain rate ( $\bar{\varepsilon}$) is certain, $10^4T^{-1}$ and $ln[sinh(\alpha\sigma_p)]$ should satisfy the linear relationship, and the slope is b/10000, and the relationship curve between it is shown in Figure 2(b). Using Origin software for linear fitting, B, N and values can be obtained.

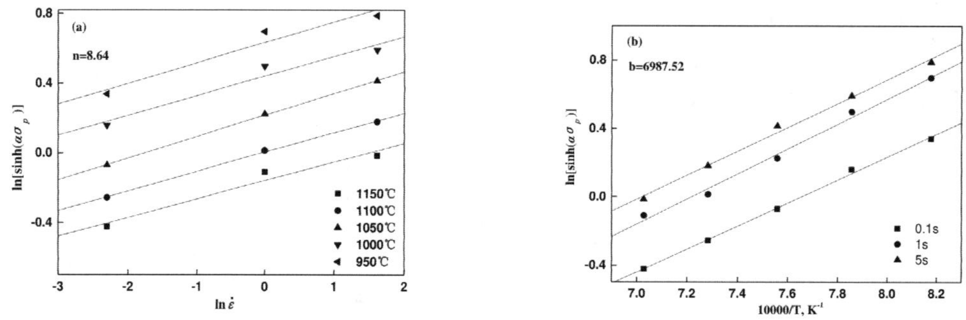


**Figure 2.** Relationships between strain rate (a), deformation temperature (b) and stress of Q1030 steel.

After calculation, the deformation activation energy of Q1030 high strength steel is 380.96 kJ/mol, B is 6987.52, N is 8.64. In the process of thermal deformation, by linear fitting lnZ and $ln[sinh(\alpha\sigma_p)]$, as shown in figure 3, lnA = 42.678 can be obtained.

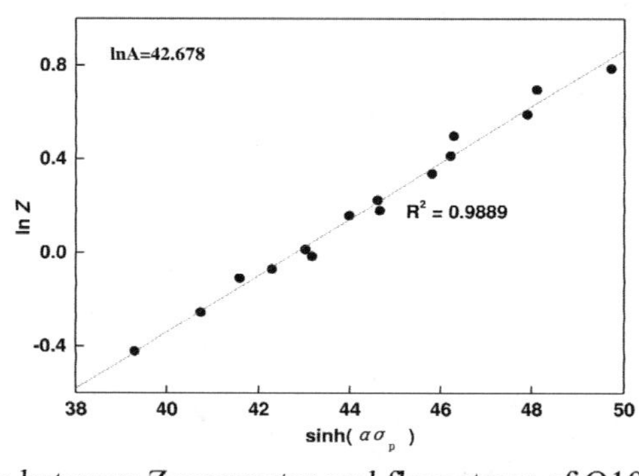


**Figure 3.** Relationship between Z parameter and flow stress of Q1030 high strength steel.

Then get the Z parameter:

$$Z=3.42\times 10^{18}[sinh(0.012\sigma_p)]^{8.64} \quad (5)$$

The thermal deformation equation of Q1030 high-strength steel is:

$$\dot{\varepsilon} =3.42\times 10^{18}[\sinh(0.012\sigma_p)]^{8.64}\exp(-\frac{380960}{8.314T}) \quad (6)$$

*3.3. Critical conditions for dynamic recrystallization*

The critical strain $\varepsilon_c$ is the key to determine whether dynamic recrystallization occurs. Only when the tangible variable is greater than $\varepsilon_c$, the austenite will undergo dynamic crystallization[ 12] . And the size of $\varepsilon_c$ indicates the difficulty of dynamic recrystallization of austenite, so the accurate determination of $\varepsilon_c$ is very important for the study of thermal deformation process parameters. In a general way, usually =(0.60 ~ 0.85). In previous scientific research, researchers proposed many mathematical models to predict the critical conditions for the beginning of dynamic recrystallization[13], Polyak et al[14] believed that the occurrence of dynamic recrystallization needs to satisfy two conditions: the maximum storage energy and the minimum dissipation rate, which can be expressed by the following formula:

$$\frac{d}{d\theta}\left(-\frac{d\theta}{\sigma}\right) \tag{7}$$

In the formula, $\theta$ is the processing hardening rate, $\theta = {d\sigma}/{d\varepsilon}$.The critical point at which dynamic recrystallization begins can be determined directly by the relationship curve with $\sigma$. And Najafizadeh-Jonas[15] simplified the above method, A cubic polynomial is used to fit the data of work hardening, and the critical stress and critical strain are determined by curve analysis method. In this research, a method similar to Najafizadeh-Jonas was used to determine the critical stress $\sigma_p$ and critical strain $\varepsilon_c$ of dynamic recrystallization, and the inflection point was determined by the cubic polynomial fitting of the pair and curve[14-16]. When fitting, the strain value is limited to the peak strain, the polynomial is as follows:

$$\theta = A_1\sigma^3 + A_2\sigma^2 + A_3\sigma + A_4 \tag{8}$$

$$ln\,\theta = B_1\varepsilon^3 + B_2\varepsilon^2 + B_3\varepsilon + B_4 \tag{9}$$

In the formula, A1-A4 and B1-B4 are constants under the corresponding deformation conditions.

The second derivative of both sides of equation (8) to $\sigma$ can be obtained:

$$\frac{d^2\, ln\theta}{d\sigma^2} = 6A_1\sigma + 2A_2 \tag{10}$$

At the critical stress at the beginning of dynamic recrystallization, the following formula holds:

$$\frac{d^2\, ln\theta}{d\sigma^2} = 6A_1\sigma_c + 2A_2 = 0 \tag{11}$$

Available:

$$\sigma_c\text{=-}A_2/(3A_1) \tag{12}$$

Similarly, it can be obtained:

$$\varepsilon_c\text{=-}B_2/(3B_1) \tag{13}$$

Take the stress-strain with a deformation temperature of 1000 °C and a strain rate of 0.1 s -1 as an example. Since the actual rheological stress curve is not a digital smooth curve, it will cause large fluctuations during analysis. Therefore, the stress-strain curve is first fitted with nine polynomial functions and then differential operations are performed. The fitting curve and its cubic polynomial fitting calculation values are shown in Figure 4 It can be seen that the calculated value and the experimental value have a high degree of quasi-coincidence, indicating that the results of the cubic polynomial fitting can be used to describe the relationship between the hardening rate and the stress and strain.

By fitting the cubic polynomial of the curve, we can calculate: $\varepsilon_c = 0.22$ ; According to peak strain, it can be obtained $\frac{\varepsilon_c}{\varepsilon_p} = 0.64$ . With the same method, the $\varepsilon_c$ values can be calculated at different temperatures and at different strain rates values of different temperatures and different strain rates can be calculated. According to the critical strain and peak strain data and the calculation results of Z parameters, the relationship graph of Q1030 high strength steel dynamic crystallization is drawn, as shown in figure 5.

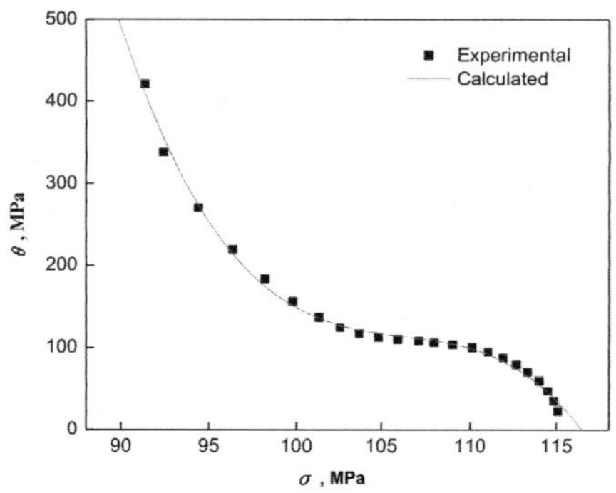


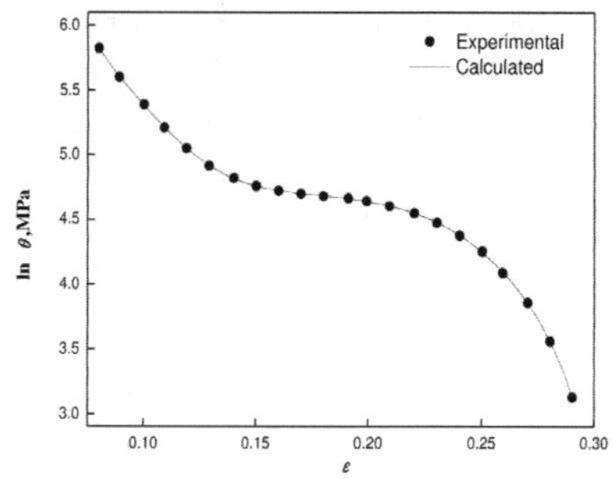


**Figure 4.** Curve at 1000°C and 0.1s-1 with their corresponding third order polynomial.

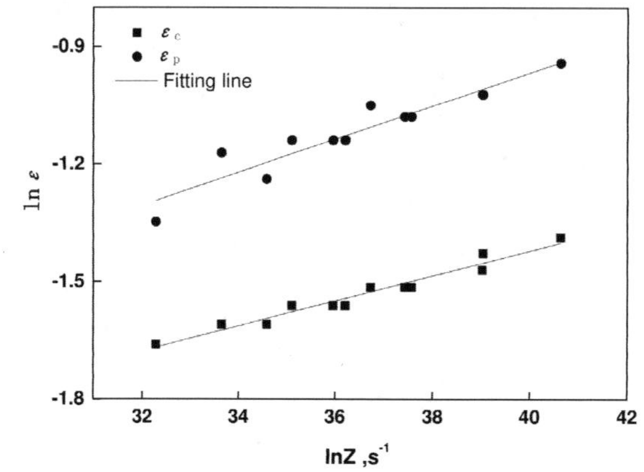


**Figure 5.** Dynamic recrystallization critical strain ($\varepsilon_c$ ) and its relationship with Z parameters of Q1030 steel.

It can be observed from figure 6 that the Z parameters can reflect the change laws of critical strain($\ln\varepsilon_p$ ) and deformation conditions. The critical strain is closely related to the heat processing parameter Z value, and $\ln\varepsilon_p$ has a good linear agreement with the lnZ: The fitting function relation is obtained by linear fitting through the least square method, that is, the critical strain model:

$$\varepsilon_p=6.88\times10^{-2}Z^{0.041} \quad (14)$$

$$\varepsilon_c=6.67\times10^{-2}Z^{0.029} \quad (15)$$

*3.4. Morphology and composition of Nb and Ti precipitated particles*

Nb and Ti particles in Q1030 steel precipitated at the strain rate of 0.1 deformation temperature of 1000°C were studied. According to the morphology of the extracted particles, there were four main types. The first type of precipitated phase is shown in Figure 6(a). Its shape is rectangular, about 150nm long and about 100nm wide. The composition of the precipitate is mainly TiN by EDS spectroscopy. The larger precipitated particles are observed by transmission electron microscopy. (100 nm or more), the ratio of the rectangular precipitated particles is between 20 and 30%. The second type of precipitated phase is as shown in Figure 6(b), and the entire precipitated particle is composed of two parts: an approximately square base (A) and an attachment (B) attached to the base in an elliptical shape.

The matrix A is Ti-rich (Ti, Nb)(C, N) is analyzed by the EDS component, and the N content is slightly higher than the C content; The Elliptic attachment B is rich in Nb(Nb, Ti)(C, N), and the C content is much higher than the N content. After observation by transmission electron microscope, in larger precipitated particles (more than 100 nm), composite precipitated particles account for about 70-80%.The third precipitation phase type is shown in Figure 6(c). There is a large number of elliptical precipitated phase particles between 50nm and 80nm. After EDS analysis, this precipitated particle is(Nb, Ti) C, but Ti content is less. The fourth type of precipitation phase type is shown in Figures 6(d) and(e). There is a kind of NbC precipitates with small size (10~20nm) and some small precipitates precipitate along dislocation.

According to the above experimental results, precipitates can be roughly divided into three categories based on the size, composition and morphology of microalloy elements: the first and second species precipitate particles are one type, most of them are over 100nm, mainly TiN, (Ti, Nb) (C, N) as shown in figure 6(a) and (b). The second type of precipitated particles is 50nm to 80nm in size, the oval in shape, and chemical composition is(Nb, Ti) C, but Ti content is small, as shown in Figure 6(c); The size of the third precipitates is 10nm~20nm , and the shape is also elliptical. The composition is NbC, and some particles precipitate along dislocation, as shown in Figure 6(d) and (e).

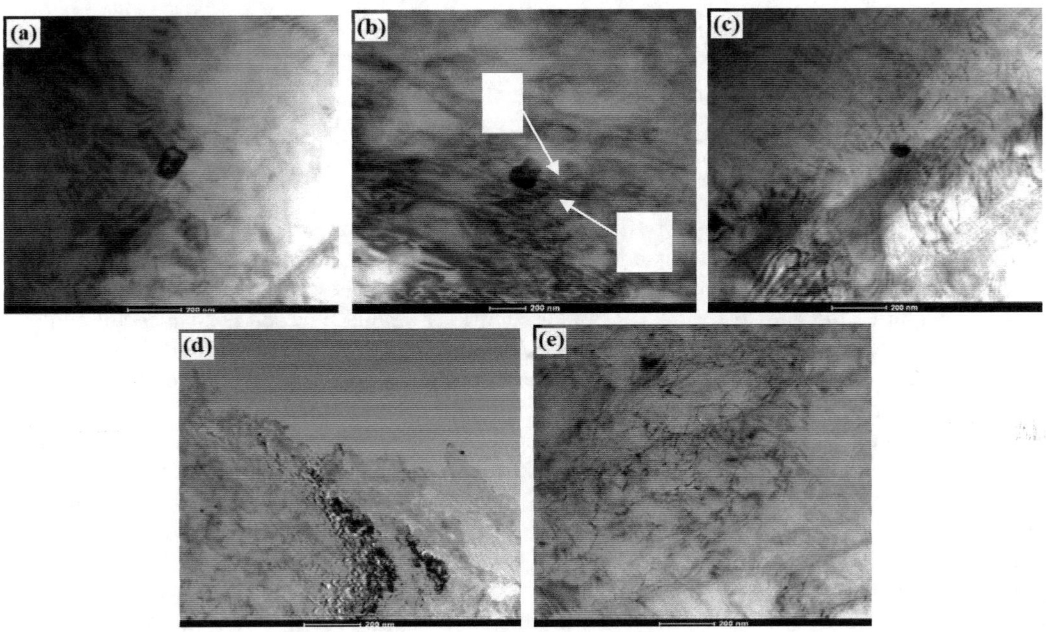


**Figure 6.** Morphology of precipitated particles.

*3.5. Thermodynamic analysis of precipitation behavior*

The precipitation of microalloy elements will lead to the decline of the free energy of the system, so the precipitation is a spontaneous process. The free energy of the system can be expressed as follows:

$$F = U - TS \tag{16}$$

In the formula, U is the internal energy of the system; T is the absolute temperature; S is the entropy of the system.

In the formula, since entropy S is multiplied by absolute temperature T, the contribution of entropy to free energy is relatively large at high temperature, and microalloy elements tend to dissolve. At low temperature, internal energy has a great influence on free energy while entropy is relatively small, and microalloy elements tend to precipitate to reduce internal energy [17]. Therefore, with the gradual decrease of temperature, the precipitation tendency of microalloying elements gradually increases. When the temperature drops to a certain extent, microalloying elements will be precipitated in the form of carbonnitride.

Generally speaking, the precipitation of carbon nitride of microalloy elements in steel can be divided into the following three stages: the first stage is the liquid phase precipitation in the solidification process of steel billet; The second stage is the austenite precipitation after solidification and during rolling. The third stage is ferrite precipitation after γ-α phase transition after rolling [18].

Microalloy elements niobium and titanium have different affinities with carbon and nitrogen atoms, and their carbonitrides have different solubility in austenite. Their sizes can be expressed by solubility products, as shown in the following formula [19] :

$$Lg([Ti][N] = 0.32 - 8000/T \tag{17}$$

$$Lg([Nb][N] = 3.79 - 10150/T \tag{18}$$

$$Lg([Ti][C] = 2.75 - 7500/T \tag{19}$$

$$Lg([Nb][C] = 2.96 - 7510/T \tag{20}$$

$$Lg([Nb][C + 12/14N] = 226 - 6700/T \tag{21}$$

According to the actual composition of the steel, the equilibrium precipitation temperature of each precipitated particle was calculated as follows: TiN-1538°C, NbN -1013°C, TiC- 1164°C, NbC -1110°C, Nb(C, N) -1146°C. It can be seen that the precipitation temperature of TiN was much higher than that of other carbon nitrides, and TiN was the first precipitated in the cooling process. Due to the high precipitation temperature of TiN, TiN has enough time to grow in the subsequent cooling process, and finally forms square TiN particles with larger size (more than 100nm), as shown in figure 6 (a).

As the temperature continues to decrease, Nb carbonitrides gradually reach the supersaturated state and begin to precipitate. Due to the similar structure of microalloy elements Nb and Ti carbonitride, therefore, the increased strain energy and interface energy of the precipitated particles on the TiN matrix are less, the nucleation resistance is less, and the precipitation is easier to achieve[20]. The particles of the new nucleation on the matrix grow up and have a certain diffusion miscible with the TiN Matrix, and finally form a compound precipitation particle, as shown in figure 6(b). Therefore, in the controlled rolling and controlled cooling Q1030 steel, the first type of precipitated particles are composite precipitates composed of TiN which starts to nucleate in the solidification process and the particles precipitated from the upper nucleation.

In the low-temperature austenite region of steel, the content of Ti in precipitated particles is relatively low due to the large amount of Ti precipitated in the early stage. Ti is a strong solid N element, which has a strong binding force with N. Almost all the N elements are fixed by TiN formed at high temperature, and there is no N element in the particles precipitated from nucleation of austenite region at low temperature. Therefore, in the controlled rolling and controlled cooling Q1030 steel, the second type of precipitated particle is (Nb,Ti) C with separate nucleation in austenite region.

The solubility of microalloying elements in austenite is much higher than that in ferrite, therefore, after the γ-α phase transition, microalloy elements will continue to precipitate from the matrix. Due to the low temperature at this time, the resulting particles are smaller in size. as shown in figure 6(d) and (e). The third type of precipitates is NbC in the ferritic region.

## 4. Conclusion

(1) dynamic recrystallization activation energy of Q1030 high-strength steel is 380.96kJ/mol, and the thermal deformation equation can be expressed as:

$$\dot{\varepsilon} = 3.42 \times 10^{18}[\sinh(0.012\sigma_p)]^{8.64} \exp\left(-\frac{380960}{8.314T}\right) \quad (22)$$

(2) the critical strain was obtained by using the cubic polynomial fitting of the sum curve of Q1030 high-strength steel to determine the inflection point, and the critical strain model of dynamic recrystallization was established, which can be expressed as:

$$\varepsilon_p = 6.88 \times 10^{-2} Z^{0.041} \quad (23)$$

$$\varepsilon_c = 6.67 \times 10^{-2} Z^{0.029} \quad (24)$$

(3) There are three types of precipitates of microalloy elements in Q1030 high-strength steel: the first type has a size of more than 100nm and a shape of regular rectangular or complex composed of rectangular matrix and elliptic attachment; The size of the second type is 50nm~80nm, the shape is oval, and the chemical composition is mainly (Nb, Ti) C. The third type is 10nm~20nm in size, oval in shape and NbC in chemical composition. Some particles precipitate along dislocation.

(4) Three types of microalloy element precipitates in Q1030 steel were formed at different stages. The first type of precipitates were TiN nucleating in the solidification process of steel, and the square TiN gradually grew to more than 100nm in the cooling process or formed composite precipitates with the particles precipitated from the upper nucleation. The second type of precipitates were (Nb, Ti) C nucleating alone in the austenite region, and the third type of precipitates were NbC nucleating in the ferritic region.

**References**

[1] Tan Z L and Xiang S 2013 Behavior of hot deformation and critical strain for dynamic recrystallization of a Q690 low carbon microalloyed steel *Trans. Mater. Heat Treat.* **34**(5) 42-6

[2] Feng R, Li L L and Li Z S 2013 Nb-V-Ti microalloy steel complex precipitation characteristics *J. Mater. Heat Treat.* **2** 37-42

[3] Liang H Q, Nan Y, Ning Y Q and Guo H Z 2015 Correlation between strain-rate sensitivity and dynamic softening behavior during hot processing *J. Alloy. Compd.* **632** 478-85

[4] Kang Y X, Cai D Y and Zhang C H *et al* 2012 Thermal Deformation Equation and Thermal Processing Diagram of Microcarbon Steel *J. Mater. Heat Treat.* **33**(6) 74-9

[5] Jang J H, Lee C H and Heo Y U *et al* 2012 Stability of (Ti, M) C(M=Nb, V, Mo and W) carbide in steels using first-principles calculations *Acta Mater.* **60**(1) 208-17

[6] Li X L and Wang Z D 2015 Interphase precipitation behavior of nanoparticles in low carbon microalloy steels containing Nb-Ti *JOM-J. Met.* **51**(4) 417-24

[7] Wei Y, Wang Z D and Han B *et al* 1997 High-temperature deformation behavior of austenite in Nb /Ti-microalloyed steel *J. North Eastern Univ.- Nat. Sci.* **18**(2) 165-8

[8] Lan L Y, Qiu C L and Zhao D W *et al* 2011 Dynamic and static recrystallization behavior of low carbon high niobium microalloyed steel *J. Iron Steel Res. Int.* **18**(1) 55-60

[9] Medina S F and Mancilia J E 1993 Determination of static recrystallization critical temperature of austenite in microalloyed steels *ISIJ Int.* **33**(12) 125-7

[10] Luo Y S and Zhang J M *et al* 2012 Evolution of precipitates in Nb-Ti binary low-carbon microalloyed steels *J. Univ. Sci. Technol. B.* **34**(7) 775-82

[11] Ferdowsi M R G, Nakhaie D, Benhangi P H and Ebrahimi G R 2014 Modeling the high temperature flow behavior and dynamic recrystallization kinetics of a medium carbon microalloyed steel *J. Mater. Eng. Perform.* **23**(3) 1077-87

[12] Mirzadeh H, Cabrera J M and Najafizadeh A 2012 Modeling and prediction of hot deformation

flow curves *Metall. Mater. Trans. A* **43**(1) 108-23
[13] Fulvio S and John J J 2000 Mathematical Modeling of the Hot Strip Rolling of MicroalloyedNb Multiple-Alloyed Cr-Mo and Plain C-Mn Steels *Metall. Mater. Trans. A* **31A**(2) 511-30
[14] Poliak E I and Jonas J J 2003 Initiation of dynamic recrystallization in constant strain rate hot deformation *ISIJ Int.* **43**(5) 684 −91
[15] Najafizadeh A and Jonas J J 2008 comparison between the methods of determing the critical stress for initiation of dynamic recrystallization in 316 stainless *J. Master. Sci. Technol***24**(6) 840-4
[16] Dong J X and Siciliano F Jr 2000 Effect of Silicon on the Kinetics of Nb(C, N) Precipitation During the Hot Working of Nb- Bearing Steels *ISIJ Int.* **40**(6) 613
[17] Xu Y B, Yu Y M and Wu D *et al* 2006 Thermodynamic Calculation of Exhalation Behavior of Nb Microalloy Steel *J. Mater. Res.* **20**(1) 104-8
[18] Meng Z B, Wu G L and Liu X B *et al* 2014 Ti precipitates in Nb-Ti microalloyed high strength steel *J. Mater. Heat Treat.* **35**(4) 106-10
[19] Kang Y L, Fu J and Liu D L *et al* 2006 *Microstructure and properties control of thin slab continuous casting and rolling steel* (Beijing: Metallurgical Industry Press)
[20] Guo H J 2013 *Metallurgical Physical Chemistry Course* (Beijing: Metallurgical Industry Press) pp 171-6

IOP Conf. Series: Materials Science and Engineering **668** (2019) 012028 doi:10.1088/1757-899X/668/1/012028

# Anomalous concentration dependence of the thermal conductivity of refractory sintered binary nano-composites TiN-ALN and $ZrC-ZrB_2$

**Y Zarichniak [1,3] and S Ordanyan [2]**

[1]National Research University of Information Technologies, Mechanics and Optics, St. Petersburg, Russia
[2]Technological University, St. Petersburg, Russia

E-mail: zarich4@gmail.com

**Abstract.** An abnormal concentration dependence of the thermal conductivity of nanocomposites Titanium Nitride-Aluminum Nitride and zirconium carbide-zirconium diboride has been experimentally established from the volume fraction of components. An explanation of the nature of the anomalous dependence and a technique for an approximate estimation of the dependence of the thermal conductivity of a nanocomposite on the volume fraction of components are proposed. Oxygen-free refractory compounds - covalent and metal-like active investigated in the development of new ceramic materials used for achieving higher physical and mechanical characteristics as well as for use in extreme conditions in mechanical engineering, nuclear engineering, aviation, special equipment. The development of technology is constantly stimulates the creation of new materials, which must manifest a complex of properties rarely show by one substance, such decisions cannot find the formation of composite (hetero) materials. In the case of eutectic systems with diffusive non-interacting refractory components, used for the synthesis of composite ceramic materials in a highly thermodynamic state, the possibility of co-building a new group of ceramics with interphase boundary (of eutectics "coarse conglomerate") with anomalous concentration dependences of the physical and mechanical characteristics of composite materials (flexural strength and modulus of elasticity ,super plasticity – elongation more than 500%, tensile, strength and fracture toughness) and thermal properties (thermal conductivity and thermal diffusivity) [1-3].

## 1. Introduction

In the literature over the past decades, was information about the patterns of interaction between the actions of refractory compounds belonging to different classes, in other words you, about the structure of polythermal sections $Me^{d'}X$ '-$Me^{d'}X$ "in the ternary systems $Me^{d'}X$ 'X "(for example, the phase diagram of type-$Me^dC$- $Me^dB2$, $Me^{IV}N$-$Me^{IV}B2$, $Me^dC$-SiC, etc.) [1-3].

It is important to note that most of these systems are described by the description diagrams of the eutectic type. Since the system Me'C(N)-Me" [2], $Me'B_2$-Me" [3], $Me^dB_2$ [4], $Me^dN$-$Me^dB_2$ [5], $LnB_6$-$Me^dB_2$, $B_4C$-$Me^dB_2$, SiC-$Me^dB_2$, SiC-$Me^dC$ [6] - are eutectic.

The authors of these publications established regularities of the eutectic composition and temperature according to the most important parameters of the electronic structure of components. These systems formed the basis for a large variety of composition ceramics and directionally crystallized materials, so-called "ceramic eutectics".

IOP Conf. Series: Materials Science and Engineering **668** (2019) 012028 doi:10.1088/1757-899X/668/1/012028

## 2. Technology of synthesis of binary composites from ultradispersed powders

When the eutectic crystallization process interphase particle boundaries must be characterized by the lowest energy, which is determined by their specific structure. In particular, for systems MeC (MeN) - MeB2 presented by cubic and hexagonal crystals can form a nearly ideal (coherent) interphase boundary in the conjugation of planes {111} MeC and {001} MeB2, the planes with a maximum surface energy $\sigma_{hkl}$.

The existing method of plasma-chemical synthesis (PHS), refractory compounds can produce powders with dispersion, sufficient to create a eutectic structure of "coarse conglomerate"

It should highlight the system TiN - AlN, in which both individual components and mixtures can be obtained by means of plasma chemical synthesis. This system is of interest to a large group of ceramics for various purposes - structural, wear-resistant (including rolling elements for bearings) and resistive (as represented by highly conductive TiN and AlN insulator AlN to the different components of the conductivity at 19 orders of magnitude).

Despite a significant difference in the nature of chemical bonding and the probable structure of the phase diagram of a system of no interacting components (non-eutectic), and sintering the mixture of powders with d < 0,1 mm is also possible to the formation of special boundaries in the resulting pair of planes {111} TiN and {0001}-typical type of dependence of conductivity on concentration of components AlN, in which the interatomic distances Ti-Ti (0.303 nm) and Al - Al (0,309 nm), respectively, differ only slightly.

The simplest and most common method of making composites of refractory components is the method of powder technology - ceramic. To date, created a variety of options for this technology, allowing to receive advanced ceramic products from micro - and nano-powders of different substances in a wide range of particle sizes from 10 nm to 10**6 nm [1,2].

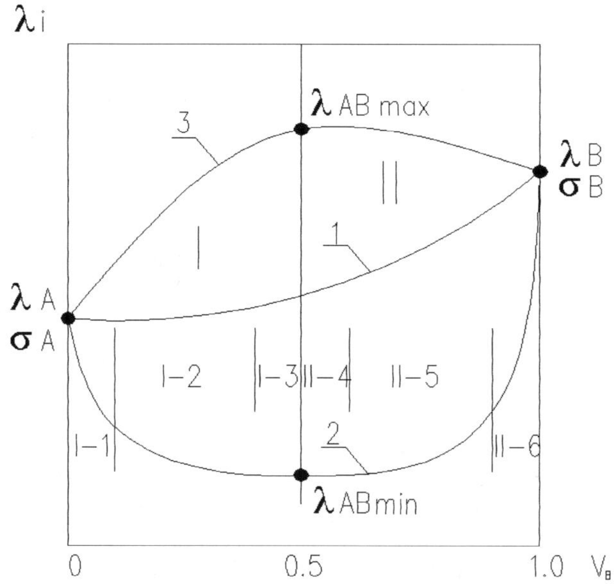


**Figure 1.** Concentration dependences of thermal conductivity $\lambda(V_B)$ and electric conductivity $\sigma(V_B)$ of binary heterogeneous systems of different nature, the nature of the structure and interaction of components: 1 - mechanical mixtures of macro- and micro-level, 2- continuous solid solutions - components are mixed in the sub-nano level (< 1 nm ), 3 – eutectics of "coarse conglomerate" with an anomalous concentration dependence.

IOP Conf. Series: Materials Science and Engineering **668** (2019) 012028 doi:10.1088/1757-899X/668/1/012028

Figure 1 shows the typical type of dependence of conductivity on concentration of components of macro- (d > 1 mm) and micro-(1 < d <100 microns) heterogeneous systems (mechanical mixes) - line 1, sub-nanodispersed system (< 1 nm) solid solution allows - line 2, and an unexpected (anomalous) type of concentration dependence for some ultradisperse (mesoscopic level) composite materials system (10 < d < 300 nm) - line 3, [7-14].

In the works [1-4,6] shows that the type of concentration dependence of aggregate size is determinate by physic-chemical factors and the technology of synthesis of composite material or alloy (static or dynamic compaction of powder components, sintering in the solid phase or with the participation of the molten phase, the melting)

Heterogeneous systems with refractory non-interacting components of the diffusion in the solid state and eutectic alloys are a group of "mechanical mixtures" with the dimensions of the in homogeneities in the macro- (d > 1 mm) and micro- (10 <d <500 microns) range. The dependence of thermal conductivity and electrical conductivity of such "mechanical mixtures" is monotonic concave to the axis of the volume concentration of the components $V_B$ (curve 1 in figure 1).

The components of heterogeneous systems belonging to the group of continuous or limited (in the region of solubility) solid solutions are mixed at the level of atoms and chemical compounds (sub-nanometer range d <10 nm). The dependence of thermal conductivity and electrical conductivity of such solid solutions has extreme character with a pronounced minimum near equiatomic or equimolar concentrations of the components Xi (curve 2 in figure 1).

One would expect that the concentration dependence of thermal conductivity of heterogeneous structures whose components are mixed at the intermediate level will be located between the extreme cases discussed above, i.e, between curves 1 and 2 in figure 1.

However, contrary to expectations, in the papers [2,6] concentration dependence of thermal conductivity of the investigated ceramics with developed interphase boundary (eutectic "coarse conglomerate ") differed from the discussed above (line 3 in figure 1) with a local or a general excess of the maximum effective thermal conductivity with respect to composites of a group of mechanical mixtures.

A qualitative explanation of the physical nature of the anomalous type of con-traditional relationships proposed in the papers [1,2].

Features compression heterophase systems - the composite ceramics in eutectic systems with the initial particle size of 0.1 - 0.3 micron at $T > 0.6\ T_{eutectic}$ manifested in the fact that close to the eutectic becoming demonstrate significantly greater rate of compaction.

Such particles obtained by vibrogrinding are characterized not only high level of specific surface area, but also by the presence of a significant number of dislocations, depending on the nature of the compounds, i.e, increased surface "activity" [2]. This is explained by the fact that different contact name, small particles are developing a much more powerful diffusive flows - providing sliding particles (seal) and accommodation (adjustment) of contacting lattices of two different particle components.

Given the difference in interatomic distances of Me-Me in these compounds can be expected , that interphase boundaries will characterized by the presence of so-called misfit dislocations. At such small sizes of grains in the composite (<1 micron) internal volumes of these compounds with predominant share of the covalent bond should be characterized by a low dislocation density (or their absence) and high hardness.

It should be noted that the eutectic metallic systems at low temperatures show a higher hardness in comparison with pure metals, which is associated with the manifestation in metals primarily intragranular dislocation slip and braking at the last of the grain boundaries [6].

This effect allows you to specifically use it when developing materials for use in conditions of rapid temperature change (especially regulated heating rate), suggesting that the processes of microplastic deformation of the boundary will contribute to the relaxation of thermal stresses occur.

Consequently, the structure of the planned heterophase sintered ceramic material must be close to that of the eutectic, which is difficult to implement. However, comparable in sophistication interfaces sintered composition (heterophase ceramics) can be created, focusing on the use of ultrafine (nano-) powders with the size of particles d = 20 ÷ 60 nm

Naturally, the simulation of eutectic structures following a convention to consider ¬ blows, but the approach to the "coarse eutectic conglomerate" is possible.

For example, earlier in [1-4,6], using submicron powders of carbides and borides MeC–(MeB2,LaB6) received sintered at T=2000°C composites in the system TiC-TiB2 ($T_{eutectic}$ = 2650°C), ZrC-ZrB2 ($T_{eutectic}$=2850°C), ZrB2-LaB6 ($T_{eutectic}$=2470°C) and in other systems throughout the concentration range studied, and their most important structure-sensitive physical and mechanical characteristics.

A characteristic feature of the structure to equal the porosity of the sintered heterophase composites (porosity ~ 2-3%) is an approximate equality of the components of particle sizes in the range of 25-75% (2-4 microns).

During sintering mono-powders each source component the particle size of both phases during sintering increases markedly - to 12-16 microns. Mutual "obstacle screening" of the phases plays the role of preventing the growth of grains of each of them is obvious. Sintering heterophase (two-component) mixtures in this concentration range (when the initial particle size of the phase components) against a background of advanced network interfaces, which are vacancy sinks.

In the end, with the free sintering of the structure is formed summarizes the properties of two closely "matched" frames, this stage and perhaps formation of a special "special" (coherent or semi-coherent) boundaries.

An important role in their formation processes are the relative rotation of the particles upon application of external forces, it is difficult implemented with increasing particle size ($d_{initial}$ >> 1 microns)

The growth of the size of the contacting particles requires a substantial increase in the time required for diffusive mass transport in the contact zone and the reversal of the particles. External forces prevent such a turn, providing the appearance of large-conventional boundaries [4].

Consequently, the "force" (pulse, dynamic), compaction (without the necessary adjustment of the diffusion boundary) and "free" sintering, there may be different in structure, almost " perfect contact at the interface (“interpenetrating”, “mutually penetrating”) lattices of the two dissimilar components of the structure and properties of each sub lattice is close to the structure and properties of monocrystalline components.

## 3. The model of the composite structure and the method of calculating the effective thermal conductivity of binary composites

To develop a model of the structure and the approximate method of calculating/predicting the thermal conductivity we use the experiences and recommendations of operations [11-14] and represent the considered binary heterogeneous system A-B in figure 1 as a quasi-ternary composition with a variable components A, B and AB component of the corresponding structure of two interpenetrating sublattices with coherent boundaries and coherent or semi- coherent phase microstructure which is close to single crystal of each of the initial components at $V_A$=$V_B$ =0.5.

The third "quasi-component" in the studied systems is the frame structure is formed in an equal volume concentrations of components $V_A$=$V_B$=0.5 as a result of diffusion of the active manifestation of the interaction at the interfaces of contacting nanoparticles crossing the starting components in a particular state - with a coherent, or semi-coherent phases border, microstructure which is close to monocrystalline of each of the initial components.

Approximation to the single crystal of each of the components reduces the scattering of phonons in the bulk of the particles, and the boundaries of different components and leads to an increase in thermal conductivity of each of the contacting the components and, consequently, a local excess of the thermal conductivity of ordinary mechanical mixture or a general maximum effective thermal conductivity of the composite material.

Then, in the range of volumetric concentrations of I (0 <$V_B$ <0.5) heterogeneous system will be considered as a binary mechanical mixture with a thermal conductivity $\lambda_A$ of component A and quasi component AB with a thermal conductivity $\lambda_{AB}$ corresponding thermal conductivity of the system with $V_A=V_B=0.5$.

In the range II ($0.5 < V_B < 1$) heterogeneous system will be considered as a binary mechanical mixture of component AB with $\lambda_{AB}$ with a thermal conductivity and thermal conductivity $\lambda_B$ of component B.

Thermal conductivity of components A and B are taken from the reference data. Thermal conductivity of the component AB is taken either from experiment (a more reliable value) [2], or calculated in the presence of background data, the thermal conductivity of $\lambda_{AB}$ max max (maximum maximorum) as the mechanical mixture of two monocrystalline components A and B with an equivolume concentration.

We now consider the method of calculation in the range of I-1 ($0 < V_B <$ (0.075 - 0.1)). In the process of sintering of components A and B at the interface between them is formed of a coherent (semi coherent) boundaries are at a low volume concentration distribution in the matrix component A in the form of droplets non contacting quasicomponent AB. Volume fraction of inclusions quasicomponent AB (reduced volume concentration of $V_{AB\,rel}$) is determinate by ratio.

$$V_{AB\,rel}= 2\,V_B \tag{1}$$

Effective thermal conductivity $\lambda_{eff}$ structures interspersed with concentrations in the range above $0 < V_{AB}$ rel < (0.15-0.2) is calculated according to any one set of relations for a given structure, for example, the formula proposed by V. Odelevsky [9].

$$\lambda_{eff} =\lambda_{matr}(1 - V_{ABrel}/((1 - \lambda_{rel})^{-1} - (1 - V_{ABrel})/3)) \tag{2}$$

where: $\lambda_{matr}=\lambda_A$, $\lambda_{incl} =\lambda_{AB}$, $\lambda_{rel} =\lambda_{incl} /\lambda_{matr}$

The matrix component is structurally dominant (in effect on the effective thermal/electric conductivity) with respect to the inclusions.

In the range I-3 (0.8-0.85) $< V_{AB\,rel} < 1$ and $0.4 < V_B < 0.425$ in figure 1 composite material has heterogeneous structure of the matrix quasicomponent AB with droplets non contacting original component A.

Effective thermal conductivity of the structure $\lambda_{eff}$ with a splash of A is also calculated by the formula (2) with an appropriate substitute index components - the dominant component of the matrix and inclusions, namely,

$$\lambda_{matr}=\lambda_{AB},\ \lambda_{incl}=\lambda_A,\ \lambda_{rel} =\lambda_{incl}/\lambda_{matr}$$

In the range of I-2 (0.15-0.2) $< V_{AB\,rel} <$ (0.8-0.85) in Fig.1 heterogeneous structure is a chaotic binary mixture of geometrically equal components A and AB which is the effective thermal conductivity can be calculated by the relation [14].

$$\lambda_{eff}=\lambda_A (1- V_{AB\,rel})^2 + \lambda_{AB} (V_{AB\,rel})^2 + 4\, V_{AB\,rel} (1- V_{AB\,rel})\, \lambda_A \lambda_{AB} / (\lambda_A + \lambda_{AB}) \tag{3}$$

In the range of II-1 in figure 1 ($0.5 < V_B < 1$) the heterogeneous system is considered as a binary mechanical mixture of component AB with a thermal conductivity $\lambda_{AB}$ and component B with thermal conductivity $\lambda_B$.

In the range of I-4 $0.5 < V_B <$ (0.575 - 0.6) in figure 1 heterogeneous structure is a quasicomponent AB matrix with droplets non contacting original component B.

The effective thermal conductivity of the $\lambda_{eff}$ structures interspersed with B is also calculated by the formula (2) with an appropriate substitute index components - the pre-dominating component of the matrix and inclusions, namely,

$$\lambda_{matr}=\lambda_{AB},\ \lambda_{incl}= \lambda_B,\ \lambda_{rel} =\lambda_{incl} /\lambda_{matr}\ V_{B\,rel}=2(V_B-0.5)$$
$$\lambda_{eff} = \lambda_{matr}(1 - V_{Brel} /( (1 - \lambda_{rel})^{-1} - ( 1 - V_{Brel}) /3) \tag{4}$$

In the range of II-3 in figure 1 (0.9 - 0.925) $< V_B < 1$ a heterogeneous system is a matrix of component B with a thermal conductivity $\lambda_B$ with non-contacting droplets of quasicomponent AB with a thermal conductivity $\lambda_{AB}$. Effective thermal conductivity structures interspersed with AB as calculated by the formula (2), with an appropriate substitute index components - the dominant component of the matrix and inclusions, namely,

$$\lambda_{matr} = \lambda_B,\ \lambda_{incl} = \lambda_{AB},\ \lambda_{rel} = \lambda_{incl}/\lambda_{matr},\ V_{AB\,rel} = 2(1 - V_B)$$
$$\lambda_{eff} = \lambda_{matr}\,(1 - V_{ABrel}/((1 - \lambda_{rel})^{-1} - (1 - V_{ABrel})/3)) \quad (5)$$

Finally, in the range of II-2 (0.575-0.6 < $V_B$ < (0.8-0.85) in figure 1 hetero-structure is a binary mixture of chaotic geometrically equal components AB and B, effective thermal conductivity which can be calculated by the relations [14].

$$V_{Brel} = 2\,(V_B - 0.5)$$
$$\lambda_{eff} = \lambda_{AB}\,(1 - V_{Brel})^2 + \lambda_B\,(V_{Brel})^2 + 4\,V_{Brel}\,(1 - V_{Brel})\,\lambda_{AB}\,\lambda_B/(\lambda_{AB} + \lambda_B) \quad (6)$$

In accordance with the principles of the theory of generalized conductivity checked the relations obtained in the limit transitions.

If $V_B = 0$, then for the range of I-1 by the formula (3) the apparent $\lambda_{eff} = \lambda_A$; when $V_B = 0.5$ for the range I-3 and I-4, we $\lambda_{eff} = \lambda_{AB}$, when $V_B = 1.0$ for range II-6, $\lambda_{eff} = \lambda_B$

Below in figure 2 shows a comparison of calculation results with experimental data obtained in the papers [1,2].

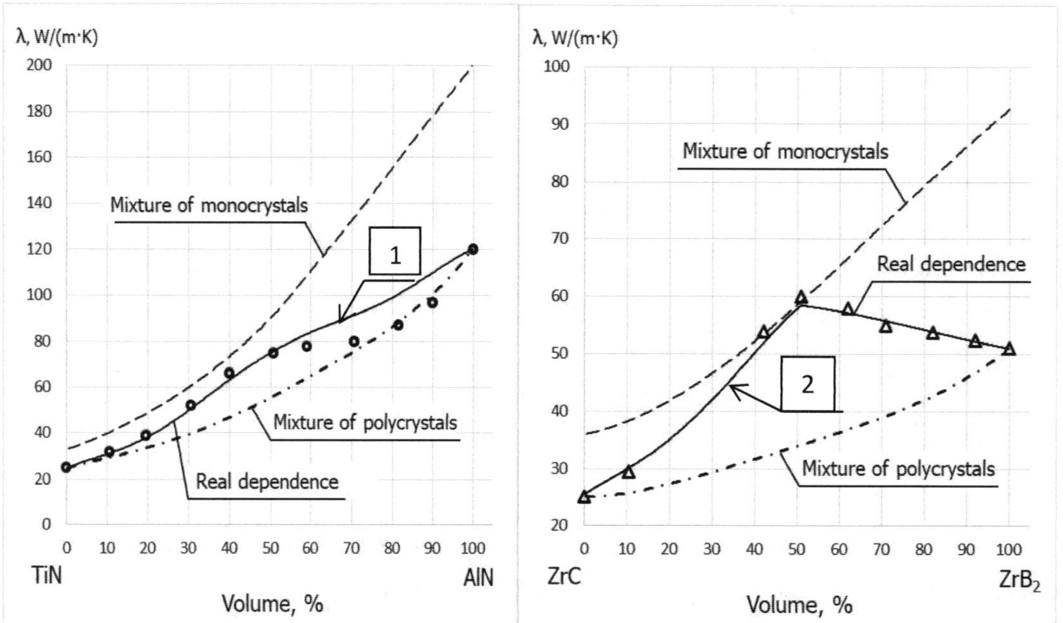


**Figure 2.** Abnormal concentration dependences of the thermal conductivity of eutectic structures of the "coarse conglomerate"-sintered nanocomposites: **O** TiN – AlN - experimental data, 1 - calculation; **Δ** ZrC-ZrB2 - experimental data, 2 - calculation.

## 4. Conclusion

The explanation for the experimental established type of abnormal concentration dependence of thermal conductivity of the binary composites sintered from ultra-dispersed (mesoskopic level 10 < d < 300 nm) powders and method of the calculation/forecasting of the thermal conductivity of such composites are offered.

The method allows to predict thermal conductivity of such composites with the minimum volume of basic data at a development stage of composite material with the required thermal conductivity that reduces expenses of time and funds for synthesis of samples and carrying out measurements.

## 5. Symbols

IOP Conf. Series: Materials Science and Engineering **668** (2019) 012028 doi:10.1088/1757-899X/668/1/012028

$V_A, V_B, V_{AB}$ - the volume fractions of components A and B and the quasi-component AB; $V_{AB\,rel}$ - reduced volume concentration of the quasi-component of the AB in the range I in figure 2 ($0 < V_B < 0.5$ ), $\lambda_A$, $\lambda_B$, $\lambda_{AB}$ - thermal conductivity of the initial components A and B and the quasi-component AB; $\lambda_{eff}$ - the effective thermal conductivity of binary nanocomposite; $\lambda_{matr}$, $\lambda_{incl}$ - thermal conductivity of the matrix component and inclusions; $\lambda_{rel} = (\lambda_{incl} / \lambda_{matr})$ - the relative thermal conductivity of the inclusions.

## References

[1] Ordanyan S 1992 *Refractory Materials* **N9-10** 10-4
[2] Ordanyan S and Unrod V 2005 *New Refractory Materials* **N7** 42-6
[3] Katz S,Ordanjyan S and Unrod V 1981 *Powder Metall.* **N12** 70-6
[4] Ordanyan S, Stepanenko E and Sokolov I 1986 *Powder Metall.* **N6** 24-8
[5] Herman G, Gleytper G and Bero G 1978 *Atomic Structure of Intergrain Boundaries* (Moscow: Mir)
[6] Fridlyander B, Neshpor V and Ordanjyan S 1979 *Thermophysics of High Temperatures* **N17** 210-6
[7] Wal A 1962 *Building and Properties of Binary Metal Systems* (Moscow: Fizmatgiz)
[8] Vertogradsky V 1972 Thermal conductivity of binary alloys,(Moscow Power Institute)
[9] Odelevsky V 1951 *Magazine of Experimental and Theoretical Physics* **N6** 667-85
[10] Duljnev G and Zarichnyak Y 1974 *Thermal Conductivity of Mixtures and Composite Materials* (Leningrad: Energy)
[11] Zarichnyak Y 1976 *News of Higher Education Institutions, Instrument Making* **N19** 11-6
[12] Duljnev G and Zarichnyak Y 1976 *Proc. V European Conf. on Thermophysical Properties of Solids* (Moscow)
[13] Zarichnyak Y 1977 *All-Union Conf. on Metallurgy* (Moscow: Atomic Energy and Engineering, Academy of Sciences of the USSR)
[14] Zarichnyak Y 1989 *The structure and thermophysical properties of composite materials and alloys* Thesis for the degree of Doctor of Physical and Mathematical Sciences (Novosibirsk: Siberian Branch Academy of Sciences of the USSR, Thermal Physics Institute)

IOP Conf. Series: Materials Science and Engineering **668** (2019) 012029 doi:10.1088/1757-899X/668/1/012029

# Functionalization of $Fe_3O_4/TiO_2/BiOCl$ nanocomposites using *Sargassum crassifolium* extract as magnetic nanophotocatalyst for cadmium sequestration

**R M T Cumba[1,2,5], R Y Capangpangan[3,4] and A C Alguno[1,2]**

[1] Materials Science Laboratory, Department of Physics, Philippines
[2] Premier Research Institute of Science and Mathematics (PRISM), MSU-Iligan Institute of Technology, A. Bonifacio Ave., Tibanga, 9200 Iligan City, Philippines
[3] Chemistry Division, Caraga State University, Philippines
[4] Materials Science and Polymer Chemistry Laboratory, Caraga State University Ampayon, Butuan City, 8600 Agusan Del Norte, Philippines

E-mail: reymarc.cumba@g.msuiit.edu.ph

**Abstract:** We report on the synthesis of $Fe_3O_4/TiO_2/BiOCl$ nanocomposites using *Sargassum crassifolium* extract as magnetic nanophotocatalyst for cadmium sequestration. Utilization of nanotechnology via surface engineering and functionalization has been considered for heavy metals sequestration. In this work, we are going to synthesize $Fe_3O_4/TiO_2/BiOCl$ nanocomposites via green synthesis for cadmium heavy metal sequestration on wastewater. Synthesis of nanocomposites includes the fabrication of functionalized $Fe_3O_4$ magnetic nanoparticles using *Sargassum crassifolium* extract followed by the fabrication of $TiO_2$ and finally capped with BiOCl to complete the $Fe_3O_4/TiO_2/BiOCl$ nanocomposites. Experimental results revealed that successful green synthesis of magnetic $Fe_3O_4$ via was observed in the Fast Fourier transform infrared spectra with the presence of Fe-O vibrations at 580 $cm^{-1}$. In addition, the band around 1000 $cm^{-1}$ was observed suggesting the $TiO_2$ was successfully formed the peak around 1155 $cm^{-1}$ can be attributed to Bi-Cl bond for the formation of BiOCl. Moreover, the presence of 370 nm peak in the ultraviolet-visible spectra suggests that $Fe_3O_4/TiO_2/BiOCl$ nanocomposites were successfully synthesized. Varied amount of cadmium on wastewater was efficiently sequestered via photocatalysis with an efficiency of about 83%. This straight forward green synthesis of $Fe_3O_4/TiO_2/BiOCl$ nanocomposites is a potential photocatalysts for cadmium sequestration.

## 1. Introduction

Gold mining has played a tremendous role in the economic development in many countries with drawback of large volume of toxic waste products which have profound impacts in the ecosystem. Proper waste management and disposal of heavy metals in mining industries can be mitigated that would result to preservation of healthy ecosystem. Most of the gold mining industries will undergo intensive mineral processing and metallurgical extraction to efficiently recover high valued minerals such as gold, nickel and copper [1]. The processing methods for minerals involve grinding of rock and ores, recovery of the desired fraction and disposal of the wastes, often as slurry, to mine tailings [2]. One such method, called leaching, is a low-cost method of removing ore from waste.

During the process, cyanide or mercury was used to dissolve gold from ores [3] and large quantities of toxic waste are produced such as cadmium which is usually present in mine tailings [4]. Thus, it is very important to have a periodic and systematic waste treatment of mine tailings to reduce the hazardous impact to the environment. There were reports that mine tailings of mining industries was able to bring environmental contaminations [5]. In fact, cadmium from mine tailing were able to contaminate drinking water [6]. Cadmium heavy metals are serious threat for living organisms, due to mutagenesis carcinogenesis and accumulation properties, even at very low concentrations [7]. The conventional mine tailings wastewater treatments include chemical precipitation, adsorption, membrane filtration, and electrodialysis [8]. The drawback in the application of these methods includes high costs, continuous input of chemicals; incomplete removal of certain metals, producing toxic sludge while the processes need high degree of operation skill. The development of innovative, cost-effective and easily engineered nanomaterials to meet the global efforts for the heavy metal sequestration has been intensively considered [9].

It has been reported that utilization of titanium dioxide ($TiO_2$) nanoparticles was utilized to excellently sequester lead and cadmium with higher efficiency [10]. However, the drawback of utilizing $TiO_2$ nanoparticles is the separation of the by-products after the wastewater treatment [11]. In addition, titanium waste is extremely acidic and its disposal methods create numerous environmental problems [12]. Moreover, zinc oxide (ZnO) nanoparticles was also utilized for cadmium wastewater treatment. High exciton binding energy of ZnO supports its activity in water purification applications. Their ability to absorb visible light and segregation of photo-excited electron/hole pair leads to surface plasmon resonance [13]. Unfortunately, toxicological studies carried out on zinc oxide nanoparticles show that potential health hazard as well as environmental risks that can have serious toxicity to bacteria, freshwater microalga, and even human cells [14].

In order to address the problem of gathering by-products after wastewater treatment, it is highly desirable to integrate magnetic nanoparticles for easy disposal of by-products using external magnet only [15]. Previous reports utilized magnetic nanoparticles such as $Fe_3O_4$ in wastewater treatment with remarkable efficiency [16]. In fact, $Fe_3O_4$ magnetic nanoparticles was synthesized using carob leaves extracts [17], *Glycosmis mauritiana* leaves extracts [18], *Sargassum muticum* extract [19], *Hordeum vulgare* and *Rumex acetosa* extracts [20], *Juglans regia* green husk extract [21] via green synthesis. But there are no reports on the utilization of *Sargassum crassifolium* which is very abundant in the Philippines and contains a potent array of antioxidants such as polyphenols, reducing sugars, nitrogenous bases, and amino acids, which is very important for reducing and stabilizing metal ions [22].

Combining $TiO_2$ nanoparticles to $Fe_3O_4$ magnetic nanoparticles can significantly increase the degradation efficiency of heavy metal sequestration. In fact, there were reports that utilizing $Fe_3O_4/TiO_2$ nanostructures can significantly increase heavy metal sequestration efficiency [23]. To the best of our knowledge, there are no reports that utilization of *Sargassum crassifolium* extract during synthesis of $Fe_3O_4$ magnetic nanoparticles was carried out and subsequently synthesis of $TiO_2$ nanoparticles was done to form $Fe_3O_4/TiO_2$ nanocomposites. Enhancing the photoelectrochemical performance of $TiO_2$ nanostructure, Bismuth Oxychloride (BiOCl) was added in the nanocomposites [24] to increase the sequestration activity of heavy metal. In this work, we are going to synthesize $Fe_3O_4$ magnetic nanoparticles using *Sargassum crassifolium* extract and subsequent synthesis of $TiO_2$ and BiOCl nanoparticles to form $Fe_3O_4/TiO_2$/BiOCl nanocomposites for cadmium wastewater sequestration via photocatalytic reaction using ultraviolet light.

## 2. Experimental methods

The seaweed *Sargassum crassifolium* were collected in Initao, Misamis Oriental, Philippines and properly washed with deionized water and then air dried at room temperature for 5 days. The dried seaweeds were cut into small pieces and then pulverized using an analytical mill. An aqueous extract was obtained by boiling 1.0 g pulverized seaweed powder with 100 mL deionized water for 6 hours at 360 rpm. The extract solution then filtered and stored at room temperature. The core magnetic nanoparticles were prepared using 2.16 g of $FeCl_3{\bullet}6H_2O$ dissolved in 200 mL deionized water and 0.4 g of $FeCl_2{\bullet}4H_2O$ dissolved in 100 mL deionized water. A 10 mL of seaweed extract was then added to the 100 mL iron salt solution while continuously stirred at 360 rpm for 15 minutes. Dropwise addition of NaOH solution was done to maintain at pH 11. The resulting magnetic nanoparticles (MNP) precipitate was washed with deionized water.

For the synthesis of MNP@$TiO_2$ nanocomposite, 2 mL of titanium butoxide was added to the 20 mL isopropanol in a 250 mL beaker and followed by the addition of distilled water and isopropanol while stirred at 360 rpm. The resulting nanocomposites were washed with isopropanol and subsequently dried in vacuum. The $Fe_3O_4/TiO_2$/BiOCl nanocomposites were prepared through a precipitation process at a room temperature. $Bi(NO_3)_3{\cdot}5H_2O$ was dissolved in 20 mL of 4M $HNO_3$ aqueous solution and mixed with NaCl and $Fe_3O_4/TiO_2$ nanocomposites while stirring until precipitate was observed. The precipitate was separated and washed using isopropanol and distilled water and subsequently dried. The $Fe_3O_4/TiO_2$/BiOCl nanocomposites were characterization using Fast Fourier transform infrared spectroscopy (FTIR) to determine the vibrational mode. Ultraviolet-Visible (UV-Vis) spectroscopy was done to determine the optical property of the synthesized nanocomposites. The photocatalytic activity of the prepared nanocomposites for cadmium sequestration under UV light irradiation was investigated by varying amount of cadmium using the synthesized $Fe_3O_4/TiO_2$/BiOCl nanocomposites.

A diluted cadmium solution was prepared to desired concentrations such as 2.5, 5.0, 15.0, 30.0 and 50.0 ppm while maintaining at pH 6.0. The actual initial concentration of cadmium was measured using an ion selective electrode (ISE) meter and recorded. A 50.0 mg of $Fe_3O_4/TiO_2$/BiOCl nanocomposites used as the nanophotocatalyst. These nanocomposites were added to a 50 mL of different diluted cadmium solutions for photocatalytic degradation activity. The mixture solution was irradiated with ultraviolet light for 40 min and measured the amount of remaining cadmium using ISE meter every after irradiation.

## 3. Results and discussion

The FTIR spectra of the synthesized $Fe_3O_4/TiO_2$/BiOCl nanocomposites using varied amount of *Sargassum crassifolium* extract is shown in figure 1. For comparison, the FTIR spectra for seaweed extract, $Fe_3O_4$ and Fe3O4/$TiO_2$ nanocomposite are also included. It is observed the presence of characteristic peak at around 580 $cm^{-1}$ correspond to Fe-O bending suggesting the formation of core $Fe_3O_4$ magnetic nanoparticles. This is due to the stretching vibration bands related to the metal in the octahedral and tetrahedral sites of the oxide structure. Obviously, this peak is absent for the *Sargassum crassifolium* extract only. On the other hand, vibrational peak at around 3300 $cm^{-1}$ for seaweed extract indicates the presence of O-H bonds while vibrational mode at 1610 $cm^{-1}$ is attributed to the conjugated carbonyl (–C=O) group [25]. In addition, $TiO_2$ nanoparticles was successfully formed due to the presence of the intense broad band below 1000 $cm^{-1}$ corresponds to Ti-O-Ti vibrations which imply to the formation of $TiO_2$ layer [26]. Likewise, the peak at 1458 $cm^{-1}$ indicates the presence of Bi-Cl bonds.

Figure 2 shows the UV-Vis spectra of $Fe_3O_4/TiO_2$/BiOCl nanocomposites. It is observed that the peaks from 320 to 350 nm can be attributed to the local vacancies present in the $Fe_3O_4$ nanoparticles due to surface plasmon resonance of electrons of metal which consists of a collective oscillation of conduction electrons excited by the incident light. The peak around 354 to 357 nm can be attributed to $TiO_2$ while peaks around 360-380 nm is attributed to the formation of pure BiOCl.

The transmission electron micrograph of synthesized $Fe_3O_4/TiO_2$/BiOCl nanocomposites is shown in figure 3. It is observed the presence of nanoparticles with an average size of 15 nm. Moreover, aggregated nanoparticles are seen in the TEM image. This is a manifestation that we successfully synthesized $Fe_3O_4/TiO_2$/BiOCl nanocomposites. The TEM measurements of $Fe_3O_4$ and $Fe_3O_4/TiO_2$ nanoparticles are underway to compare the morphology and particles size and will be reported elsewhere.

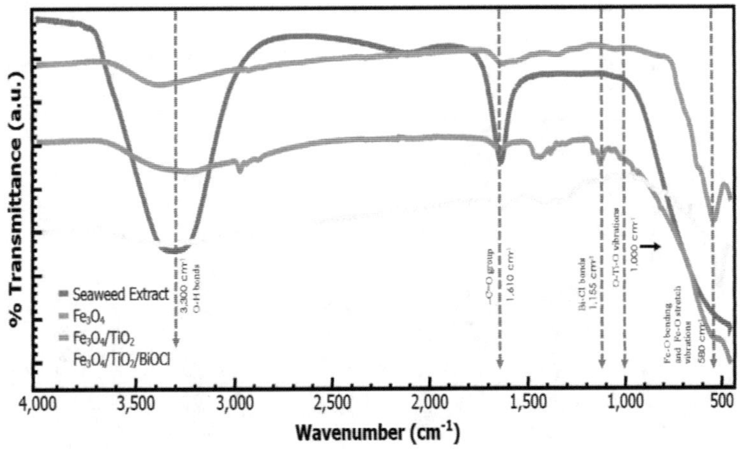


**Figure 1.** FTIR spectra of seaweed extract, $Fe_3O_4$, $Fe_3O_4/TiO_2$, and $Fe_3O_4/TiO_2/BiOCl$ nanocomposites.

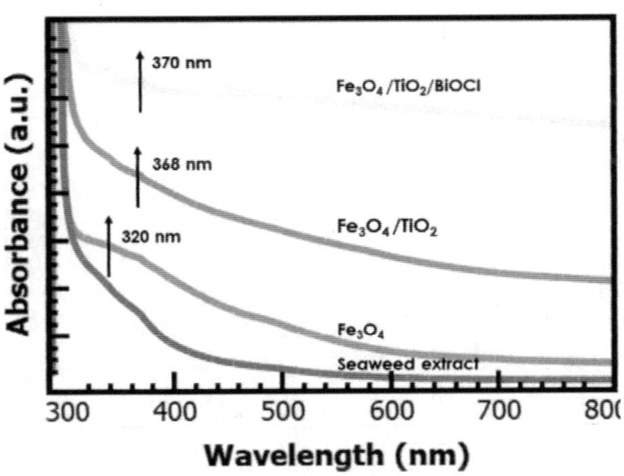


**Figure 2.** UV-Vis spectra of seaweed extract, $Fe_3O_4$, $Fe_3O_4/TiO_2$, and $Fe_3O_4/TiO_2/BiOCl$ nanocomposites.

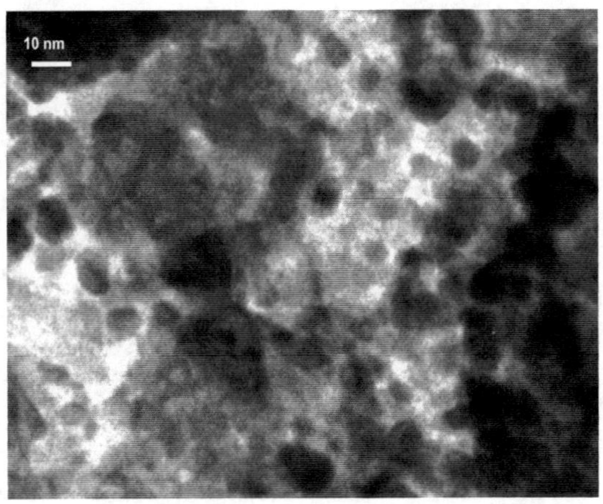


**Figure 3.** Transmission electron micrograph of $Fe_3O_4/TiO_2/BiOCl$ nanocomposites.

Cadmium sequestration was tested using $Fe_3O_4/TiO_2/BiOCl$ nanocomposites with varying amount of cadmium concentrations as shown in figure 4 in the presence of ultraviolet light. It is very apparent that about 83.48% of cadmium has been sequestered using the $Fe_3O_4/TiO_2/BiOCl$ nanocomposites. Moreover, the sequestration efficiency of cadmium in wastewater decreases as the amount of cadmium increases. This might be due to the saturation of sequestration capability of $Fe_3O_4/TiO_2/BiOCl$ nanocomposites. It is also observed that as the amount of cadmium in wastewater reaches to 50.0 ppm, the sequestration efficiency drastically drops below 50%. This might be due to the lower amount of $Fe_3O_4/TiO_2/BiOCl$ nanocomposites added in the treated cadmium wastewater. It is suggested that there must be a balance between the amount of $Fe_3O_4/TiO_2/BiOCl$ nanocomposites and the amount of cadmium concentration in wastewater treatment.

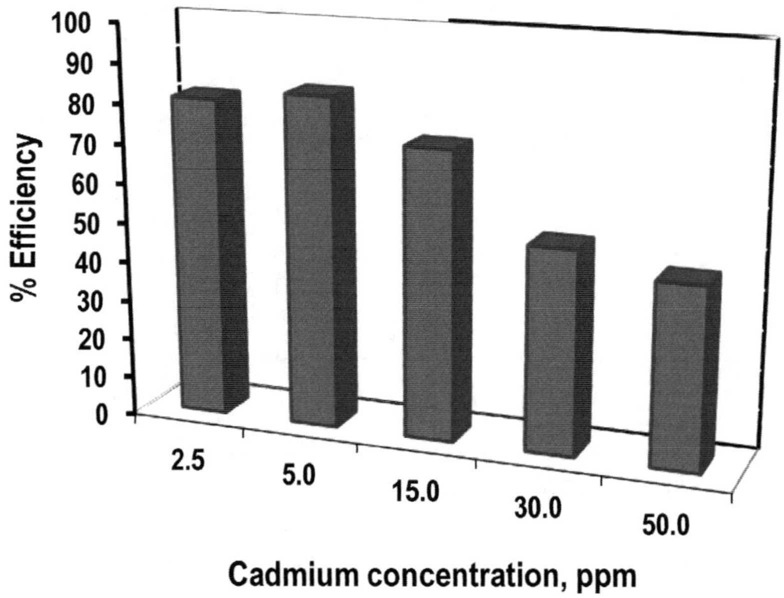


**Figure 4.** Cadmium sequestration efficiency of $Fe_3O_4/TiO_2/BiOCl$ nanocomposites with varying cadmium concentrations.

## 4. Conclusions

The magnetic $Fe_3O_4/TiO_2/BiOCl$ nanocomposite was successfully synthesized via green synthesis route using *Sargassum crassifolium* extract through a precipitation process. FTIR and UV-Vis spectra revealed that successfully formation of $Fe_3O_4/TiO_2/BiOCl$ nanocomposites and confirmed in the TEM micrograph. Higher efficiency of cadmium sequestration was done and found out that there must be a balance between the amount of $Fe_3O_4/TiO_2/BiOCl$ nanocomposites and cadmium concentration. It is highly recommended that optimization must be done in order to obtain higher cadmium sequestration efficiency. Likewise, XRD measurement is highly recommended to for $Fe_3O_4/TiO_2/BiOCl$ nanocomposites to determine its magnetic phase.

## Acknowledgments

The author (R M T Cumba) would like to thank DOST-ASTHRDP for the scholarship grant and the Division of Chemistry, Caraga State University for the use of instruments. CHED and DOST-PCIEERD is also acknowledged for providing research grant.

## References

[1] Lottermoser B 2007 *Mine Wastes: Characterization, Treatment, Environmental Impacts* (New York: Springer) pp 1-290
[2] Ledin M and Pedersen K 1996 *Earth Sci. Rev.* **41** 67-108
[3] Muhammad I, Ashiru S, Ibrahim I D, Salawu K, Muhammad D T and Muhammad N 2013 *Journal of Envi. Treatment Techniques* **1** 174-82
[4] Da Silva E F, Zhang C, Pinto L S S, Patinha C and Reis P 2004 *Appl. Geochem.* **19** 887–98
[5] Cobbina S J, Myilla M and Michael K 2013 *Int. J. Sci. Technol. Res.* **2** 221-7
[6] Smolders A J P, Lock R A C, Van der Velde G, Medina Hoyos R I and Roelofs J G M 2003 *Arch. Environ. Con. Tox.* **44** 314-23
[7] Marcus J J 1997 *Effects of Mining on the Environmental and American Environmental Controls on Mining* (California: Imperial College Press) pp 73-8

[8] Kurniawan T A, Chan G Y S, Lo W H and Babel S 2006 *Chem. Eng. J.* **118** 83-98
[9] Werkneh A and Rene E 2018 *Applications of Nanotechnology and Biotechnology for Sustainable Water and Wastewater Treatment* (Singapore: Springer) Chapter **19** pp 405-430
[10] Engates K and Shipley H 2010 *Environ. Sci. Pollut. R.* **18** 386-95
[11] Zhang G, Wurtzler E, He X, Nadagouda M, O'Shea K, El-Sheikh S, Ismail A, Wendell D and Dionysiou D 2015 *Appl. Catal. B- Environ.* **163** 591-8
[12] Lane D A 1991 *Int'l & Comp. L. Rev.* **14** 425-34
[13] Singh R, Barman P and Sharma D 2017 *J. Mater. Sci.-Mater. El.* **28** 5705-17
[14] Sabir S, Arshad M and Chaudhari S K 2014 *The Scientific World Journal* **8** 925494
[15] Shokouhimehr M, Piao Y, Kim J, Jang Y and Hyeon T 2007 *Angewandte Chemie* **46** 7039-43
[16] Dave P N and Chopda L V 2014 *J. Nanotech.* **14** 398-412
[17] Awwad A M and Salem N M 2012 *Nanosci. Nanotech.* **2** 208-13
[18] Amutha S and Sridhar S 2018 *J. Innovations in Pharmaceutical and Biological Sciences* **5** 22-6
[19] Namvar F, Rahman H S, Mohamad R, Baharara J, Mahdavi M, Amini E, Chartrand M S and Yeap S K 2014 *Int. J. Nanomed.* **9** 2479-88
[20] Makarov V V, Makarova S S, Love A J, Sinitsyna O V, Dudnik A O, Yaminsky I V, Taliansky M E and Kalinina N O 2014 *Acta Naturae* **6** 35-44
[21] Izadiyan Z, Shameli K, Miyake M, Hara H, Mohamad S E B, Kalantari K, Taib S H M and Rasouli E 2018 *Arab. J. Chem. (Preprint https://doi.org/10.1016/j.arabjc.2018.02.019 )*
[22] Makarov V, Makarova S, Love A, Sinitsyna O, Dudnik A, Yaminsky I, Taliansky M and Kalinina N 2014 *Langmuir* **30** 5982-8
[23] Salamat S, Younesi H and Bahramifar N 2017 *RSC Advances* **7** 19391-405
[24] Wang H, Yuan X, Wu Y, Zeng G, Tu W, Chong S, Deng Y, Chen F and Chew J 2017 *Applied Catalysis B: Environmental* **209** 543-53
[25] Kannan R R R, Stirk W A and Van Staden J 2013 *S. Afr. J. Bot.* **86** 1–4
[26] Chellapa M, Anjaneyulu U, Manivasagam G and Vajayalakashmi U 2015 *Int. J. Nanomedicine* **10** 31-41

# AUTHOR INDEX

# AUTHOR INDEX

**Institute of Physics**
Dirac House, Temple Back
Bristol BS1 6BE UK

ISSN: 1757-8981
ISBN 978-1-7138-0921-0